Western Philosophy Series

*We regret to announce that this is the last volume in the Western Philosophy Series that will benefit from the presence of Professor Stephen Korner and Professor H. J. McCloskey, both of whom died recently.

The Western Philosophy Series develops from the Nijhoff International Philosophy series originally published by Kluwer and previously entitled the Melbourne International Philosophy series. The new series continues to uphold the policy of presenting works in the tradition of exact, or 'scientific' philosophy based firmly in solid argument and strictest possible proof. The series continues a tradition of publishing titles concerned with prominent philosophers and includes collective editions of great logicians: the NIP series began with the 'Lesniewski Collection' (Vols 13, 24, 44, 54), and the 'Collected Works of Jan Lukasiewicz' are now in preparation.

Titles published in the Western Philosophy Series
(series volume numbers correspond to the original NIP series)

The Brentano Puzzle, edited by R. Poli, 1988 ISBN 1 84014 371 1 (vol 55)
Sociative Logics and their Applications: Essays by the Late Richard Sylvan, Edited by D. Hyde and G. Priest, 2000 ISBN 0 7546 1127 2 (vol 56)

SOCIATIVE LOGICS AND THEIR APPLICATIONS

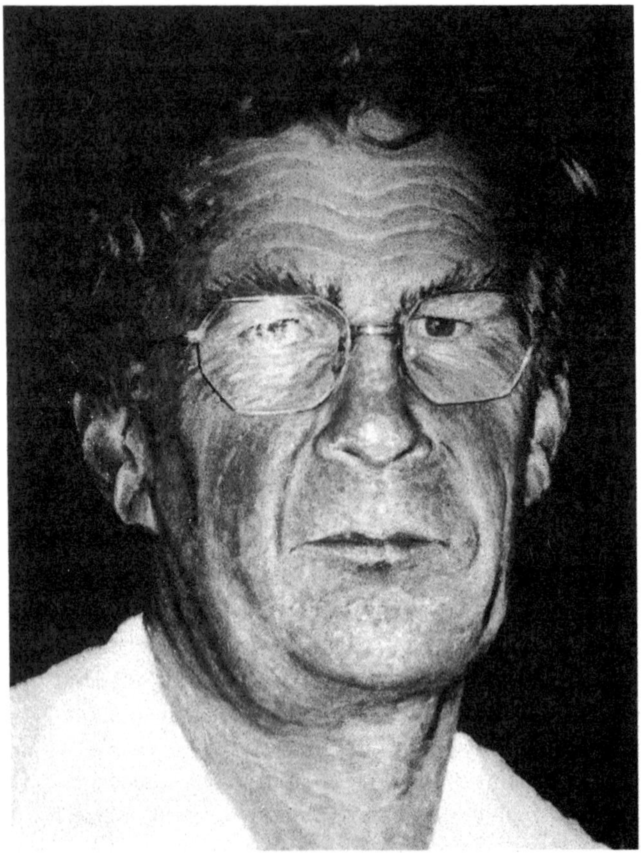

Richard Sylvan (né Routley)
1935-1996

(Portrait by David McClenaghan)

Sociative Logics and their Applications

Essays by the Late Richard Sylvan

Edited by

DOMINIC HYDE

GRAHAM PRIEST
University of Queensland, Australia

Ashgate

Aldershot • Burlington USA • Singapore • Sydney

Published by
Ashgate Publishing Ltd
Gower House
Croft Road
Aldershot
Hants GU11 3HR
England

Ashgate Publishing Company
131 Main Street
Burlington VT 05401-5600 USA

Ashgate website: http://www.ashgate.com

British Library Cataloguing in Publication Data
Sylvan, Richard
Sociative logics and their applications : essays. -
(Western philosophy series)
1.Logic 2.Relevance
I. Title II.Hyde, Dominic III.Priest, Graham
160

Library of Congress Control Number: 00-134826

ISBN 0 7546 1127 2

Printed in Great Britain by Antony Rowe Ltd, Chippenham

Contents

Acknowledgments

This book would not have been possible without the assistance of Richard Sylvan's wife, Louise Sylvan, and his literary executor, Professor Nick Griffin (McMaster University) who made available important manuscripts for inclusion in this book and gave freely of their time, supporting the project wherever possible. We are grateful to them for their help. Thanks are also due to the following friends and colleagues of Richard who co-authored material reproduced herein, provided assistance in the preparation of this book, or otherwise made time available for discussion: Ross Brady, Jack Copeland, Newton da Costa, Len Goddard, Chris Martin and Robert Nola.

Finally, our thanks go to Grace Dunn, Rachael Gibson, Arthur O'Neill and Koji Tanaka for their assistance in the preparation of the manuscript.

We acknowledge the kind permission granted by *Acta Analytica* to reproduce Chapters Fourteen and Sixteen, granted by *Logique et Analyse* to reproduce Chapter Seven, granted by *Studia Logica* to reproduce Chapter Twelve and Rodopi Publishers for permission to reproduce Chapter Ten. Sylvan may also have had material published unbeknownst to us and his literary executor; it may therefore be that we have inadvertently included material in this book which has previously been published elsewhere. We apologise if this has occurred.

This work was supported by a grant from the Australian Research Council. We are grateful for their financial assistance.

1 Editors' Introduction

On the June 16th, 1996, Richard Sylvan died of a sudden and unexpected heart attack. His death, at the relatively young age of 60, robbed Australasia of one of its greatest philosophers, arguably the most original that it has ever produced.

Richard was born Francis Richard Routley at Levin, New Zealand, on 13 December, 1935. He changed his name to Sylvan—much to the confusion of a number of people—when he remarried in 1983. After studying at the Victoria University of Wellington, he took a job at the University of Sydney, in Australia, where he lived for the rest of his life. He had several other jobs in Australia, including one at the University of New England, where, with Len Goddard, he trained a generation of Australian logicians. In 1971, he took up a position at the Australian National University, in Canberra, where he worked until his untimely death.

Richard was a most unusual person—by any standards. He would rise at dawn, write philosophy till lunch, then pursue his hobbies, like building houses or tending the forests that he owned, till the sun went down, when he would open a bottle of wine and relax. To those who did not know him well, he could appear blunt, aggressive, and he could certainly be a difficult person for university administrators to get on with. But those who got to know Richard well knew that he was a shy, caring, and warm person. For this he was loved by his many friends and students (not disjoint collections). One day, a biography will be written; but this is not it.

Richard made major contributions to many areas of philosophy: relevant and paraconsistent logic, Meinongianism and metaphysics, environmental ethics, to name but the most important. To others, these might seem a rather disparate collection, but for Richard they were all part of one seamless whole; solutions to (and the generation of) many philosophical problems presupposed particular solutions in areas others viewed as far afield. Many philosophical areas, he thought, suffer from the narrow scope of their authors' concerns.

Another of Richard's philosophical trademarks—in fact, perhaps the major one—was taking up grossly unpopular views and defending them doggedly. If relatively few people were ultimately persuaded by him, all had to admit that the views were not as silly and easily disposed of as they had hitherto thought.

Richard worked on many projects simultaneously. When he died, there were therefore several incomplete projects, including three or four books. One of these was a book he had provisionally entitled "Applications of Relevant Logic", known to him by its nickname, "Relapse". Most of Richard's technical work in logic had been driven, from its earliest times, by the philosophical need of a logical theory that would not make a hash of a number of important applications. *Relapse* was an acknowledgment that it was about time to set this all out in print. This book is an attempt to bring Richard's project to fruition.

When he died, there were, in his papers, many draft manuscripts intended as part of the project. At one time, we thought it might be possible to publish these in a suitably edited form, maybe filling in some of the clear gaps. But, it turned out, there were too few manuscripts, and in too preliminary a draft, to make this feasible. We would have had to write most of the book ourselves. And even had we covered the topics he had set out, we could not have said what he would have said—or in the way that he would have said it. It is a fortunate fact, then, that in many of the areas which were slated for the book, Richard had already written substantial essays, some co-authored with others. We have selected a number of these for reproduction here and have included bibliographic information at the beginning of each chapter.

Chapters Four, Five and Six have already appeared in preprint form. We hope that their full publication will result in their making a more lasting contribution to debate in the area. Chapters Seven, Ten, Twelve, Fourteen and Sixteen have previously been published however some appeared in rather inaccessible places and others were nonetheless chosen for reproduction because they illuminated key themes pursued elsewhere in the book. Chapters Eight, Eleven and Fifteen have not previously been published in any form. A number of further essays which we might have profitably included for their overall relevance to the project were, unfortunately, too large for inclusion in this already lengthy manuscript. Richard certainly did not intend to use these eleven manuscripts for the book; but no doubt many of them would have been rewritten and extended to provide chapters. We could do no better, then, than use these essays.

Together with the above material, we have included, as 'Fragments', such of Richard's draft manuscripts for the project that were finished enough to see the light of day. This includes a draft table of contents included in Chapter Two, so that readers can see Richard's vision of his project for themselves. Some of these fragments incorporate material from the eleven manuscripts used. We have chosen not to delete these repetitions as they are few in number, and also because they illuminate Sylvan's ongoing, integrated philosophical concerns.

The collected work thus represents material drafted between 1982 and 1996, some worked on just a few days before his death. As will be clear, the resulting book is not the one that Richard would have written, but we hope that, in the circumstances, it is close enough.

Editing the manuscripts threw up a number of problems. One of these concerned references, concerning which Richard was fairly causal. Even in the semi-published papers, he might refer to a book, when only a few pages of it were relevant; and, in many of his papers, works frequently cited were often referred to using the mnemonics that he favoured—for example "See RCR". Whilst retaining his use of mnemonics (see the Bibliography), we have done our best to make the referencing more user-friendly.

Another problem concerned logical symbolism. Not only did this sometimes vary from manuscript to manuscript, but Richard had a habit of proliferating new symbols to codify central concepts under analysis. We have taken the liberty of changing the symbolism sometimes to make the whole more coherent. We decided to leave unchanged, though, Richard's occasional (but non-arbitrary) use of U and P for the universal and particular quantifiers (more normally written as $\forall$ and $\exists$, respectively). For him, $\exists$ was too suggestive of the reading "there exists"—a reading that he sometimes rejected, preferring instead the existentially neutral "for some". (This 'preference' is explained and argued for in detail in JB.)

Finally, in accordance with the publisher's typesetting conventions, we have employed single-quotes for direct quotation and scare quotes, with double-quotes reserved to mark cases of phrases mentioned, not used.

As will be clear from the foregoing comments, the book will never win a prize for scholarship. That is not its aim, however. For Richard, it was the ideas that were important. Whether they were his or someone else's made little difference. He was always happy to give other people his ideas, and let them go and work them up into their own. So this is a book of ideas. Some of them may die a quick death; others will surely take on lives of their own. But at least they are out there now, in the public domain. We recommend them to you for your consideration.

Dominic Hyde
Graham Priest

2 Introductory Fragments

Included below are the extant introductory fragments of the proposed monograph.[1] *[Editors]*

2.1 Table of contents

Sylvan's intentions as to the contents of the proposed monograph at the time of his death are (at least partially) disclosed by the following Table of Contents, found amongst his papers. It was drafted in October 1995. [Editors]

APPLICATIONS OF PARARELEVANT LOGICS

---

[1] *Editorial comment has been kept to a minimum. However, where it is required within the text or footnotes it will appear as slanted text and signed by the editors.*

13. Building up that exclusive classical virtue, Rationality
14. On weight of argument
15. Universal theory achievement: a theory for every phenomenon

IV. Critical Addendum

16. Natural logic: another myth
17. Bourgeois logic from the Ivy Halls.

2.2 Introduction

The following incomplete draft of the Introduction of the proposed monograph gives an indication of the direction Sylvan intended it to take. It was drafted in July 1995. [Editors]

An organising theme of this book is that a certain small band of logics, namely pararelevant logics, offer solutions to many problems, puzzles and paradoxes in philosophy of science. While it is not maintained that the pararelevant logics provide a panacea, they do provide surprisingly many remedies. Most of those so far known are alluded to in the first part of this logical cookbook, and some are developed in detail in subsequent parts. Let us report upon what is included in these later parts, and what is excluded.

Pararelevant logics—logics within the overlap of relevant and paraconsistent logics—have a very extensive range of applications. While some sort of survey of the extent of presently known applications of these logics will be attempted in part I, only certain applications will be considered in detail in subsequent parts, primarily applications that would be regarded as falling within logics and methodology of science, or more generally within philosophy of science. That selection is made in terms of material that has not been published before in any accessible form at all (not even in those local publications that many academics, for various unbecoming reasons, resent). Indeed there is so much seemingly worthy but so far unpublished material on pararelevant applications, that it would be something of a scandal to exclude it, for instance by republishing previously published (even 'semi-published') work. Accordingly work that has appeared in some form elsewhere will be merely summarised, and not reproduced, even where useful improvements have been made subsequently.[2] What has been excluded however is material

[2] *In spite of these intentions, we have reproduced "semi-published" and*

where the burgeoning theory is not sufficiently well-developed so far. This includes most notably certain parts of non-deductive logic, such as those concerning induction. It is planned to treat these omitted areas some day in a sequel.

As a brash young man I set out from New Zealand for Princeton University with a project of trying to repair and renovate a derelict philosophy of science, resolving many of its evident problems, through an improved logical basis and corresponding theoretical setting. This project was hardly enthusiastically welcomed at Princeton, and I made comparatively little progress on it there (though I did make some small advances on the kind of logical basis needed). In the intervening thirty-five years, however, much has happened both in the advance of pararelevant logical theory and, more recently, in its applications. Now as a much older, perhaps still brash, person, I can bask a little, during the necessary continuing investigations, in the thought that the project can succeed and is beginning to come to fruition. It has taken so long not merely because I am slow (which I am, as well as working in several other areas) but also because it has taken a long time to reach an adequate logical basis (still not finally settled).

Many of the important applications of improved logical theory lie in philosophy of science. That area remains (like economics) a major refuge for positivism, a positivism attached to classical logic. It is not going too far to claim that crude empiricists and unreconstructed or only slightly reconstructed positivists are still enjoying field days in philosophy of science. These ideological luddites have resisted new logical technology, setting their fields up for technological invasion and take-over. (New science is one thing, and to be welcomed, new logic and logical methodology quite another, and to be avoided and rubbished.)

Presently we witness three broad styles of philosophy as it relates to logic, three styles that devolve to philosophy of science:

- that which proceeds with mainstream classical logic,
- that which eschews formal logical theory (because of the evident mess it makes) but

published work in what follows; moreover, work beyond the scope originally envisaged by Sylvan—i.e., beyond the philosophy of science. The fragmentary nature of the proposed monograph meant that much of it was unsuitable for publication and the bulk of Sylvan's writings in the area (published, "semi-published", or otherwise) are difficult to access. We have therefore made use of some of this work in an attempt to realise, as closely as possible, his intentions to argue for particular views in the proposed monograph. [Editors]

- which proceeds with informal logic, as an alternative, or
- which proceeds without logic at all (i.e., no logic as opposed to an informal logic option)

These alternatives tend to depend upon each other, defended by elimination of the others. As there is another excellent alternative, namely formal nonclassical logic (of some pararelevant sort or other), these cosy strategies break down. They are all, in any case, radically defective, as will become apparent.

Part I

Orientation

3 Orientational Fragments

As Part I of his proposed monograph, Sylvan intended to include general survey material. The extant draft fragments of these orientational chapters are reproduced below as individual chapter sections. Last worked on between March and July 1995—roughly a year before his death—they serve as an illuminating introduction to subsequent chapters in spite of their unfinished state. [Editors]

3.1 Kinds and features of relevant and paraconsistent logics

Relevant logics have intimate connections with paraconsistent logics, connections important for several applications. A *paraconsistent* logic is nothing but a logic that can support inconsistent but nontrivial theories, that is theories such that A and $\sim A$ both hold, for some A, but not every B obtains. Otherwise explained, a paraconsistent logic is one that does not trivialise inconsistent theories, because it does not admit the spread rule: $A, \sim A \,/\, B$. In more rigorous forms, paraconsistent logics can absorb logical and semantical paradoxes; it is these logics which admit dialectic applications. A central structural argument that mainstream logics are never paraconsistent exactly parallels a crucial historical argument for a negative paradox of implication—a paradox exhibiting blatant irrelevance of antecedent and consequent—a paradox that relevant logics must accordingly fault in one way or another. These parallels, shortly tabulated, expose well the close connections of relevant and paraconsistent logics. They also deliver helpful classifications of broadly relevant logics. A broadly relevant or *sociative* logic is one which extends conservatively a sentential part of the following sort: there are no theses of the form $A \rightarrow B$ (with $\rightarrow$ a distinguished implication or conditional) where A and B fail to share a variable (or literal). Thus, in particular, no logic containing such theorems as $p \,\&\, {\sim}p \rightarrow. \, q$, $p \rightarrow. \, q \vee {\sim}q$, ${\sim}(p \rightarrow p) \rightarrow. \, q$, $p \rightarrow. \, q \rightarrow q$, $p \rightarrow p \rightarrow. \, q \rightarrow p$, etc. is a sociative logic. So none of the present *mainstream* logics, classical and intuitionist logics, are sociative. Nor are main Brasilian paraconsistent logics such as the C-systems. Relevant logics proper are sorts of sociative logics. But there are as many *types* of sociative logics as there are genuine alternatives for defeating arguments to the negative paradox of implication $A \,\&\, {\sim}A \rightarrow B$. Since paraconsistent logics are bound to avoid the corresponding inferential form $A, \sim A \,/\, B$, there is evidently a close relation between broadly relevant and paraconsistent logics. Further, a typology for paraconsistent logics similar to that for sociative logics can accordingly be devised. The respective arguments and principles involved are now tabulated:

Table 3.1 Historic negative routes to logical disaster

To irrelevance and *implicational paradox*		*To explosiveness, i.e., non-paraconsistency*	
Principles involved	Implicational forms in argument	Derivational principles involved	Inferential form in argument
A. Simplification $A \,\&\, B \to A$ $A \,\&\, B \to B$	$A \,\&\, {\sim}A \to {\sim}A$	$A, B \vdash A$ $A, B \vdash B$	$A, {\sim}A \vdash {\sim}A$
B. Addition $A \to. A \vee B$ $B \to. A \vee B$	${\sim}A \to {\sim}A \vee B$	$A \vdash A \vee B$ $B \vdash A \vee B$	${\sim}A \vdash {\sim}A \vee B$
C. Rule Transitivity $D \to E$ $E \to F$ $\overline{D \to F}$	$A \,\&\, {\sim}A \to. {\sim}A \vee B$ $A \,\&\, {\sim}A \to A$ (by A.)	$D \vdash E$ $E \vdash F$ $\overline{D \vdash F}$	$A, {\sim}A \vdash {\sim}A \vee B$ $A, {\sim}A \vdash A$ (by A.)
D. Rule Composition $D \to E$ $D \to F$ $\overline{D \to E \,\&\, F}$	$A \& {\sim}A \to. A \& ({\sim}A \vee B)$	$C, D \vdash E$ $C, D \vdash F$ $\overline{C, D \vdash E \,\&\, F}$	$A, {\sim}A \vdash A \& ({\sim}A \vee B)$
E. Disjunctive Syllogism $A \,\&\, ({\sim}A \vee B) \to B$	$A \,\&\, ({\sim}A \vee B) \to B$	$A \,\&\, ({\sim}A \vee B) \vdash B$	$A \& ({\sim}A \vee B) \vdash B$
F.	$A \,\&\, {\sim}A \to B$ (from D and E by Transitivity)	Note: there are several variants upon such arguments (e.g. multiple on the right, plural consequent, versions).	$A, {\sim}A \vdash B$

The resulting types of sociative and paraconsistent logics are these:
*A. Non-simplifying, such as very typically, *connexive* logics.
*B. Non-additive, for example Parry logics, containment systems.
*C. Non-transitive and non-cut-free, for example relational logics.
*D. Non-compositive and non-adjunctive, for example confirmation logic, Jaśkowski logics.
*E. Non-dissembling, for example relevant logics proper, but also irrelevant 'positive-plus' systems such as minimal logic, C-systems, etc.
In brief, *relevant* logics are those sociative logics which reject Disjunctive Syllogism, $A \,\&\, ({\sim}A \vee B) \to B$, and associated forms such as Antilogism, $A \,\&\, B \to C \to. A \,\&\, {\sim}C \to {\sim}B$.

So far this classification conflates sociative and paraconsistent types. For instance, C and minimal systems, are (minimally) paraconsistent but not relevant; Parry logics are sociative but not paraconsistent. However the sorts are readily separated, and will be further isolated in what follows. Nor is the classification entirely exhaustive, for in the deployment of schemata, substitution, which could be questioned, has been implicitly presumed.[3]

As it turns out, *versions* of all these types distinguished can be simulated in the last type, relevant logics or modal elaborations upon them. Thus, for instance, versions of *B and *C can be obtained by adding a relational coupling, like this:

$$A \gg\!\!\rightarrow B \quad \text{iff} \quad A \rightarrow B \,\&\, ARB$$

for R an appropriate relation.
Versions of *D can be obtained similarly or through a strategy like this:

$$A\ J\!\!\rightarrow B \quad \text{iff} \quad \diamond A \rightarrow B$$

for $\diamond$ an appropriate modal functor.

As there is now a staggering array of logical systems to choose from, for various essential purposes, so, within that welcome plurality, there are *many* relevant logics, of different kinds. In addition to these relevant logics there are paraconsistent logics. In part the connection between them, though comparatively simple, is not well understood because paraconsistent logics have been confused with *dialethic* logics, logics that form only a very proper, though very controversial, subclass of paraconsistent logics.

Dialethic logics of interest are paraconsistent logics where there are inconsistent theses, i.e., both A and $\sim\!A$ are asserted for some statement A (utterly uninteresting dialethic logics are trivial). In the longer term, dialethic logics are going to prove extremely important in many roles, including for a general theory of computability and for computing science. What dialethic logic can accomplish in the way of paradox resolution and natural language representation, that nonparaconsistent logics *cannot*, is nicely

[3] My own initial investigations of paradox-free logics (in the late 1950s) focussed for a while upon *nonsubstitutional* logics. For example, from a strong consistency stance, the appeal that there is in conceding $p \,\&\, q \rightarrow p$ with p and q different does not extend to $p \,\&\, {\sim}p \rightarrow p$. Due separation of cases can be effected by, among other things, restricting uniform substitution.

A different classification of paraconsistent logics, based on extant forms, is offered in PL. It is also not exhaustive, omitting not merely nonsubstitutional systems but non-cut-free logics. Nonetheless it can be very usefully combined with an elaboration of the classification advanced in the text above.

illustrated through an argument van Fraassen advances for anti-realism—to the effect that there can be no complete correct description of the world, no such adequate presentation, or put differently, no adequate linguistic theory corresponding to the world.

> Suppose for a moment that there exists, or comes into being in the historical fullness of time, a language L so rich that every proposition is expressed by some sentence of L. Now draw a rectangle, and consider the proposition—call it *Pandora*—that no sentence of L ever written in the rectangle expresses a true proposition. Choose the sentence of L which expresses Pandora, and write it in the rectangle. Clearly it expresses a true proposition if and only if it does not. This thought experiment is one which, if you begin to execute it, leads you to the conclusion that you cannot execute it. It is certain that either the proposition Pandora, or the language L, does not exist. But if L existed, there would be nothing wrong with the described proposition. So the language L cannot exist. And therefore our natural language cannot be, and cannot become, like L (van Fraassen 1986, p.214).

Now, in defiance of van Fraassen, let us execute the experiment:

THE RECTANGLE

NO SENTENCE OF L (EVER) WRITTEN IN THE RECTANGLE EXPRESSES A TRUE PROPOSITION

The intended outcome is supposed to be:

Pandora is true iff *Pandora* is not true,

whence *Pandora* is true and *Pandora* is not true (given further rule Reductio and simplification). Thus execution, which is entirely feasible, produces *none* of van Fraassen's certainties. Rather what it shows is, firstly, that *Pandora* is an inconsistent but true proposition, and hence no doubt both unusual and anathema to mainstream logic: and further that L is a language capable of representing such unusual propositions. But so are natural languages, the resources of which we have not really exceeded. In short, then, in a dialethic setting van Fraassen's contentions (and further argument) fall to the ground.

This attractive illustration of the power of dialethic logic in removing alleged limitation and incompleteness (here as regards what can be expressed, and to discourse represention) is but a prelude to many more similar striking applications of dialethic logic. Very roughly, the *vast array of restrictive and limitative results, generated by clever contradiction-avoiding variations of self-referential paradoxes, is neutralized.* Thereby voided too are frequently acclaimed limits both to logic and to intelligence, limitations both to representation of the surrounding world and to discourse depicting it. More technically, thereby eliminated also are pressures for and main roles for order theory and type theories and their elaborations such as ramified type theories; thereby induced correspondingly is a welcome simplicity in "higher" order theories, in general theories of set, attributes and the like.

There need be *no serious tension* between relevant and paraconsistent logics. For satisfactory paraconsistent logics are relevant, and satisfactory relevant logics are paraconsistent. So we can work satisfactorily and comfortably in the overlap between the classes, in *pararelevant* logic. There are some little arguments to the unsatisfactoriness of systems outside the overlap. On the one side, relevant logics such as Ackermann's and Parry's which are not paraconsistent cannot carry inconsistent information without exploding, i.e., blowing out far too many claims. On the other side, the paraconsistent turn needs however to be a *relevant* one, else it will fall down badly in applications. There are arguments (to be sharpened considerably) to the effect that:

- Irrelevant paraconsistent logics cannot deliver requisite paradox removal or improved theories.

For irrelevant logics which are properly paraconsistent differ from relevant logics essentially by the addition of analytical junk: for instance, through principles like $A \rightarrow t$, $A \rightarrow A \,\&\, t$, where t is any tautology.[4] Naturally then positive paradoxes are inevitable, in many applications of such irrelevant paraconsistent logics. Further these paradoxical features make their way through into all applications that make use of parallel expansion principles such as $A \rightarrow. (A \,\&\, B) \vee (A \,\&\, {\sim}B)$ (for example to conditional preference, Bell's theorem). More generally still, these logics fail to adequately capture or

---

4 For details, see Urbas 1990. Urbas' main results (the arguments for which require some repair) are that

- *except* for conclusions which are theorems, strictly paraconsistent inferences from contradictory premisses conform to weak relevance; and
- *except* for conclusions which are theorems, all inferences in strictly non-explosive logics conform to weak relevance.

The further main argument is to the effect that the cost of the exception clauses is *far* too heavy. In particular, negation theory is severely damaged. But a good negation theory is much more important than weakening so as to carry analytic junk.

reflect incompleteness, partiality and the like, conspicuous features of much information and data that logic should be able to accommodate and access.

- Only a relevant band of paraconsistent logics can deliver the improved paradox-free theories sought.

For the logics must be paraconsistent to stop the spread of contradiction (both covered and not). Then they must be relevant by the previous argument.

Relevant logics (primarily those of the pararelevant overlap from here on) have *wide*-ranging applications primarily *because* of their paradox removing capabilities—many, many other paradoxes are generated by paradoxes of implication which relevant logics were designed to eradicate—and *because* of the capacity to handle incompleteness and inconsistency, incomplete or inconsistent data especially. Corollaries for philosophical theory and cognitive science can begin immediately.

The potential for applications was vastly expanded through the elaboration of semantics for relevant logics (from the late 1960s). The semantics can be seen as adding to possible world semantics for modal logic, two further types of worlds or situations:

- incomplete situations (where neither a statement nor its negation may hold), and
- inconsistent situation (where certain statements and their negations both may hold).

(The first type has since become familiar in North America through situation semantics and partial logics; symmetrising on negation gives the second type. Naturally the semantics did a little more than just adding these types; it also correctly introduced relations of more than two places for semantically assessing implicational-type connectives.)

Inconsistent situations enabled straightforward counterexamples to implicational paradoxes and related principles such as the controversial Disjunctive Syllogism (counters known since late medieval times, as it turned out). Consider then $A \,\&\, (\sim A \vee B) \rightarrow B$ (similarly the blatant paradox $A \,\&\, \sim A \rightarrow B$) and how it fares upon holding (a truth-value) preservation at a situation d which is inconsistent (in that conveniently, both A and $\sim A$ hold) but non-trivial, so not everything holds. Suppose in particular that B does not hold. Then A holds at d, $\sim A$ holds at d and so $\sim A \vee B$ also holds; but B does not. Thus the principle is counter-exampled. For $A \,\&\, (\sim A \vee B)$ holds but B does not. (Similarly $A \,\&\, \sim A \rightarrow B$ is counter-exampled, and along related lines the rule $A, \sim A \,/\, B$ gets upset.) Correspondingly syntactically, the main technical adjustment can be explained in terms of substitution principles. For the technical key to many improved applications is removal of modal substitution principles, which allow intersubstitutivity of strict equivalences (or of provable material equivalences).[5] In symbols, $A \leftrightarrows B, D(A) \,/\, D(B)$ (or $\vdash A \equiv B, D(A) \,/\, D(B)$. The relevant replacement conditions which require instead coentailment, i.e., $A \leftrightarrow B$, correspond semantically to situation widening. That is, there are situations which distinguish such strictly equivalent pairs as $A \vee {\sim}A$ and $B \vee {\sim}B$, A and $A \,\&\, (B \vee {\sim}B)$, A and $(A \,\&\, B) \vee (A \,\&\, {\sim}B)$, $A \,\&\, {\sim}A$ and $B \,\&\, {\sim}B$, A and $A \vee (B \,\&\, {\sim}B)$, A and $(A \vee B) \,\&\, (A \vee {\sim}B)$, and so on. Other crucial differences between modal and relevant logics can be explained as emerging from these substitution differences, notably (consequent) expansion and antecedent elimination. Thus the following principles, valid for strict implication, are invalidated in relevant logic:

Expansion-type: $A \rightarrow. (A \,\&\, B) \vee (A \,\&\, {\sim}B)$, $A \rightarrow. (A \,\&\, B) \vee (A \,\&\, C) \vee (A \,\&\, {\sim}C)$, etc. (obtained modally by replacing A by strict equivalents). Removal of these principles turns out to be important in relevantly removing puzzles in areas as diverse as decision theory and quantum mechanics.

Elimination-type: $(A \vee B) \,\&\, (A \vee {\sim}B) \rightarrow A$, $A \,\&\, ({\sim}A \vee B) \rightarrow B$, etc. Removal of these principles is essential for main paraconsistent purposes. Again there are far-reaching effects, extending as far as removing problems concerning determinism.

A simple illustration concerns *belief*, with functor Bel_a for agent a (but it could as well be *information*, for example that held by a). As beliefs simplify, $Bel_a(A \,\&\, B) \rightarrow Bel_a B$. So in particular, $Bel_a({\sim}B \,\&\, B) \rightarrow Bel_a B$. Thus if strict substitution conditions were to obtain, then as $A \,\&\, {\sim}A \leftrightarrows {\sim}B \,\&\, B$, $Bel_a(A \,\&\, {\sim}A) \rightarrow Bel_a B$, whence normally $Bel_a A \,\&\, Bel_a{\sim}A \rightarrow Bel_a B$, for any and every B. That is, *an* inconsistent belief trivializes an agent's beliefs. But such a result flies in the face of much evidence. Under relevant substitution conditions, such ridiculous results do not ensue. More elaborate examples, considering for instance *intention* and like functors from action theory, will be encountered subsequently.[6]

The remarkable pay-off from these easy preliminaries will come in the next chapters.

---

[5] Substitution is not merely a major method in logic; 'the process of inference always depends on the one same method of substitution' (Jevons). It is also critical in theoretical social sciences, where it is a major, though unexploited, key to improved theory. Needless to say, substitution does not hold everywhere in logic. It breaks down for more highly intensional functors, such as "a learns that", "is synonymous", etc: see DRL, p.329.

[6] *See Chapter Nine, section 1.2. [Editors]*

3.2 Wide-ranging philosophical applications of para-relevant logics

Many of the now standard problems of philosophical curricula are engendered by inappropriate logic, or by defective logic, by mainstream logic. These problems are variously dissolved, removed, or (at least) eased, by a switch to *a* relevant logic of some suitable sort. In one of their major roles, then, relevant logics are to serve as paradox and puzzle solvents. But resolving paradoxes is often but a prelude to grander philosophical enterprises.[7]

Relevant logics were primarily developed like early modal logics before them, in response to a widely felt need for entailment and implication connections, free of paradox, where relevant connections between antecedents and consequent components were decently retained, and where noxious paradoxes such as $A \mathbin{\&} {\sim}A \rightarrow B$ and ${\sim}(A \rightarrow A) \rightarrow. B \rightarrow B$ on the one side and $A \rightarrow. B \vee {\sim}B$ and $A \rightarrow A \rightarrow. B \rightarrow B$ on the other were voided. This they succeeded in doing, admirably (and to the satisfaction of increasingly many in the intellectual community), by contrast with modal logic which removed only certain paradoxes of material implication.[8] But it was soon realised that relevant logics could be applied to or adapted to a great many other regions of theoretical paradox and puzzlement than those of implication and entailment (i.e., necessary implication). It is these other regions that we aim to survey here, as the success story concerning entailment and implication has been told elsewhere.

For present organisational purposes, in showing what relevant logics can do, it is advantageous to group applications temporally, into *past, present,* and *future or projected*. (Later, when matters are clearer, better classifications will be possible.) *Past* comprises, say, up to middle 1980s;[9] *present* from then to now; *future* from hereon.

3.2.1 Past applications

Unremarkably, many of the applications, past and present, parallel applications, or intended applications, of modal logic, and upgrade these. Unremarkably, because modal logic was introduced by its prime mover, C.I. Lewis, to provide, through strict implications, a satisfactory explication of

7 For a different and fuller (if dated) introduction see UU.

8 With the exception of the despised, but philosophically more satisfactory, non-normal modal logics. For these non-normal logics also shed such paradoxes of strict implication as $A \rightarrow B \rightarrow B$ and ${\sim}(C \rightarrow C) \rightarrow D$, along with material paradoxes like $A \rightarrow. B \rightarrow A$ and ${\sim}C \rightarrow. C \rightarrow D$.

9 1987 if you think that was an important year, a *turning* year, as some Canadian environmentalists say.

entailment. Such an explication Lewis and many others insisted, classical Frege-Russell logic did not supply. Relevant logic tries to satisfy much the same market, but with a superior, paradox-resistant product. Some advantage can be taken then of modal logic applications and technology; but while the idea of applications transfers very nicely, much of the logical technology does not (because modal technology is heavily dependent upon tricks and shortcuts that paradoxical apparatus allows). Even so, with sufficient work, much of the enterprise of a long line of distinguished modal logic achievers can be replicated and much improved (as we may say in unflattering hindsight) in relevant logic: the work of Carnap, von Wright, Prior, Hintikka, Montague, Lemmon, and many others.

P(I) Enriched modal

Let us begin with applications of an enriched modal sort (such as those that build on Carnap's rich and voluminous work).

- *Content, information*, etc. Here all those bizarre modal results such as that analytic statements all have the same content, namely none, and that a single contradiction supplies total information, are peeled off. Inconsistent information too ceases to be an insuperable obstacle, or indeed an obstacle at all.[10]
- *Probability*, propositional *measure theories*, etc. So result non-classical theories of probability, and non-classical theories of and measures of coherence, desirability, volition, preference and so on, and therewith non-standard decision theory. Another application in this broad area is to triviality results concerning conditional probability and probability of conditionals, where triviality can be avoided relevantly.

In general, this area has not been given much publicity, and some of the results have not been published yet.

- *Belief, knowledge, and other epistemic functors*. Most work has been done on belief, where the main advantages over modal elaborations are removal of virtually all those limitations which render modal representation effectively inapplicable to more ordinary believers. For instance, requirements as to ideal believers, rationality, consistency, etc. The obstacle of inconsistent beliefs disappears.

Another application in this area is to perception, where again paradoxical modal features fall away.[11]

---

[10] Full details of these and several other relevant improvements upon modal theory can be found in UU.

[11] An example is the bizarre modal interderivability of the different acuity forms:

- *Deontic functors, legal logic*, etc. Relevant deontic logic affords a particularly successful application; it removes a sweep of paradoxes of deontic logics, and enables a straightforward and appealing technical treatment of moral dilemmas. But further it enables a reshaping of deontic theory, which disposes of the problematic business of *prima facie* obligations. Needless to say, in applications like those to *nihilism*, permissive ethics, possibilism and so forth, matters reach far beyond enriched modal extensions.

What is done for obligation theory is straightforwardly relocated for legal obligation, theory of contracts, and inconsistency in the law generally (for one development see IC, chapter 13).

- *Axiological functors*, such as preference, betterness, value etc. This run of theory, which has hitherto appeared rather uninteresting, is beginning to take on a new lease of life.
- *Tense theory, change and motion*. Relevant tense logic not only removes paradox, it enables the formulation of philosophical positions concerning time that have hitherto resisted exact formalisation. Further, it facilitates a dialectic treatment of problems of change and motion and of Zeno's paradoxes.[12] At this stage, of course, it is removed from the modal setting and spills over into the next section.

An amassing of all these interrelated applications suggests two (connected) grander applications: first, relevant formalisation of natural languages, or of deep relevant logical structure for languages, and second, universal semantics for languages. The second project has fared somewhat better than the first,

- *Universal semantics*. This by-product of relevant logic semantical analysis consisted in elaborating (vastly) the techniques that had worked in relevant logic settings to apply generally. Insofar as there are limits on results achieved they appear to be derived from syntax, not semantics.
- *Natural language formalisation*. The project differs from the mainstream project in proposing that deep structure is a relevant logic. The project, far from complete, persists, at a low level of activity.

P(II) Non-classical (inconsistency and incompleteness) issues

A different sort of grander application, than these grand language rehabilitation and formalisation programs is this: application of deep relevant logics to logical and semantical paradoxes and their limitative variants (these

---

$S_d(A \supset B) \rightarrow. S_dA \supset S_dB$ and $S_dA \,\&\, S_dB \rightarrow. S_d(A \,\&\, B)$ (for details, see PLI, p.288).

[12] For some detail see IC, chapters 11 and 12.

sort of grander applications are relevant to the previous ones, which have in any case to cope with paradoxical discourse in some way).

• *Logical* and *semantical paradoxes.* A main idea is the hitherto unthinkable one of simply letting the paradoxes stand. Given the logic is deeply relevant they cannot be spread around damagingly. In short the paradoxes are appropriately *contained* by the logic available: the one kind of 'solution' of logical paradoxes Russell did not develop at all in his early important classification of possible solutions. Apart from this dialethic offensive (which has mainly operated as a limitation-of-logical-apparatus approach anyway), the chief relevant approach (primarily pursued by Brady) has been one of limitation of apparatus.

• *Disarming limitative theorems.* Insofar as those theorems are smart classically-admissible variants upon paradox, they are not automatically transferable to non-classical settings. Very roughly, the less classical the setting the further the stock limitative theorems lose their force or fail. For instance, in a dialethic setting, it would be *expected* that a complete mathematics in inconsistent (because of logical paradoxes). Accordingly claims based on Gödel's first theorem about the limits of formalisation fall away;[13] these are but limits to classical-style formalisation. So far, however, but negligible work has been devoted to formalisation of complete dialethic mathematical systems, even a complete arithmetic (which would presumably include its own semantical theory).[14]

• *Consistency of arithmetic*, and more generally of classical mathematics. The key idea is that arithmetic (likewise, other parts of classical mathematics) is reformulated relevantly, where consistency is easily established. The task of recovering classical arithmetic (likewise other parts of classical mathematics) contracts to the much more accessible problem of proving rule γ admissible for relevant arithmetic. But the project as so far conceived has

---

[13] Thus, for instance: 'What Gödel showed was that any formal system of axioms and rules of procedure must contain some statements that can neither be proved nor disproved by means allowed with the system' (Smith 1990, p.22, repeating Penrose). No such result follows from what Gödel claimed to show. Still less do the anti-computational consequences that Penrose and others wish to draw, emerge: 'that the mind cannot be purely algorithmic, since it can decide matters (truth of undecidable statements) that cannot be decided algorithmically' (Smith 1990, pp.22-3). Similar nonsense is propagated by many other luminaries, for example Gardner, who offers up the standard mythology concerning Gödel's theorems (see Gardner 1983, pp.329-330).

[14] There are however some complete classical mathematical theories, due primarily to Tarski. A sustained dialethic investigation would not doubt look at methods applied in achieving such results.

failed: γ is not admissible for the relevant arithmetic (namely system $R^{\#}$) into which virtually all work went.

- *Foundations of mathematics* (see JB, p.900 ff; also the conclusion of DRL on a reinstated Hilbert program, etc.)
- *Incompleteness, vagueness.* These topics obtained little relevant elaboration. They are only now beginning to obtain that.[15] But a special case, so to say, did:
- *Relevant quantum theory* (JB, p.955 ff).

P(III) Relevance take-overs

Relevance has enjoyed a manifold and varied life, in philosophy and associated theoretical areas, independently of relevant logics. A survey of these roles and uses is given in BG, p.59 ff. I shall not attempt to repeat it here, though the survey can now be expanded and improved. Most roles are to *field delimitation* and (often as a special case) *paradox avoidance.*

Several of these uses of relevance are to issues we have already glanced at, to remove paradoxes induced by modal analyses and logics. But the appeal to relevance may be different; for example, a relevance screen or sieve on top of classical logic may be proposed, or different logics may be suggested. Present work suggests a synthesis of these different appeals, so far as they work, with relevant logics can often be effected.

P(IV) Relevant rehabilitation and semantical metamorphosis

Experience with relevant semantics and the ensuing universal semantics, along with the removal of many blockages and inhibitions through object-theory[16] (itself appearing to require relevant rehabilitation) led to the project of philosophical rehabilitation, rehabilitation of (philosophical) theories that had fallen on tough intellectual times. As well as object theory and nihilist positions,[17] early examples included dialectical logic (an application not all appreciated in Eastern Europe), extensional reduction, existential reduction, denotational theory, aboutness, verificationism, etc.

3.2.2 Present applications (c. 1985-1990)

Similar broad classifications to before will be attempted, with symbol '*N*' (for '*now*').

---

15 *On vagueness, in particular, see Chapter Sixteen. [Editors]*

16 *See* JB. *[Editors]*

17 *See Routley 1983. [Editors]*

N(I) Enriched modal

• Relevant *theory change and revision*. This is a relevant analogue of the AGM theory (as in Fuhrmann 1988), recently advanced in Restall and Slaney 1995.

• Assertion and associated *assertoric functors*. Included, as well as assertion itself, are affirmation, acknowledgment and assent, and subjunctivisations and commitmentisations of these (for example "d would be committed to acknowledging that").

• *Commitment*, its determinates, and associated relational functors. A basic functor is $\Gamma \mapsto_d B$, Γ commits d to B, where Γ is some class of statements (held).

Study of these functors leads out of the implications nexus to different developments, which can however still be accounted broadly modal analogues (because of the modal take-over of the philosophical logical world, which is still running ahead of relevant annexation).

• *Conditionals*. Relevant conditional theory stands to modal conditional theory (of Stalnaker, Åqvist, Lewis *et al.*) approximately as entailment theory stands to strict implication. The paradoxes and irrelevances of the modal theory are shed, and the exclusion of impossible conditions lifted. But semantically the relevant theory can proceed along very similar lines to the modal theory—once inconsistent and incomplete worlds again gain entry.

• *Conditional operations*, such as *conditional obligation*, *conditional probability* and *conditional preference*. Back of conditional theory comes a bunch of applications to conditional operations. These operations have been badly messed up by the modal theory, with the upshot that much modal epicycling has begun. Relevant relocation appears to remove most of the problems.

N(II) Variant relevant

• *Relevant containment logics*. The containment logics of Parry, analytic implication and bretheren, were shown to amount to relevant (variable-sharing) impositions upon modal logics (thanks to the labour of Fine, Dunn and Urquhart). These containment logics retained excessive modal logic garbage; their doppelganger modal analysis shows why.

New relevant containment logics seem to remove this modal pollution, while making tangential use of a doppelganger non-modal analysis; namely imposing constraints upon relevant implication (instead of strict implication). Such relevant containment logics have in their turn applications, two of which are worth mentioning immediately:

• *Frame problems* in AI.[18] To keep operations relevant, robots are programmed using relevant (containment) logic in place of irrelevant classical logic.[19]

• *Reasoning, reason* and *cause*. A main object is to track the logic of "A is a reason for [against] B". Investigations so far lead to the following sort of analysis:

A is reason for B iff A is (an implicational) condition for B and A enjoys priority over B,

e.g. for ponible reason:

$$A \mathrel{\supset\!\!\!\!\cdot}\, B \quad \text{iff} \quad A \to B \,\&\, A < B.$$

That is, explication makes this logic of reason a type of relevant containment logic.[20]

The logic of *cause*, investigated by analogy with causal implication, takes a similar containment form.[21]

• *Fallacy theory*. Several of the traditional fallacies are regarded as fallacies of relevance, so accommodation of some fallacies is to be expected. But in fact all can be accommodated under relevant theory. (See BG, p.50 ff).

• *Pragmatic theory*. Again applications are to be expected, simply by virtue of Grice's prime pragmatic directive: be relevant!

N(III) Non-classical issues

• *Dependence* on an argument, through relevance: *relevant predication, intrinsic properties* and *internal relations*. Much of what has been sketched above has been propositional (or expansive) in character, applying functors to propositional-type complexes. But the matter of *dependence*, as approached, is directly combinatorial and bears on predicate logic.

The essence of the main proposals, those of Dunn 1987 and 1990, is as follows:

---

18 Many applications of relevant logic in AI and cognitive science are evisaged, especially by those preparing programs or running out grant applications. Most of these applications amount to displacements of failed or bogged modal ones: knowledge representation, information and expert systems, data and belief bases, program verification, etc. The classic paper for relevant applications in computing science remains Belnap 1977; but it is getting old and is a very narrow exercise.

19 *See Chapter Seven for further elaboration. [Editors]*

20 *See Chapter Six for further elaboration. [Editors]*

21 *See Chapter Twelve for further elaboration. [Editors]*

- One-place: $(\rho x\ \varphi x)a =_{df}$ for all $x, (x = a \rightarrow \varphi x)$. This defines relevant predication of φ of a.
- Two-place: $(\rho xy\ \varphi xy)ab =_{df}$ for all $x, y, (x = a \rightarrow .\ y = b \rightarrow \varphi xy)$.

This defines relevant relationship of φ of a and b. These notions have deep metaphysical applications, particularly to intrinsic predication and internal relationships. According to Dunn 1990, the basic case of *intrinsic predication* is relevant predication as above; and the basic case of *internal relation* is that of relevant relation as above.

While some believe that these proposals for relevant predication and relevant relationship are astray (see for example BG, p.62), and that these notions should perhaps be directly defined in λ-calculus terms using relevant containment logic, nonetheless the accounts offered of intrinsic predication and internal relationship are fascinating and merit much further investigation.[22]

- *Confirmation* and surrounding functors. Although these resemble enriched modal applications, such applications depend upon going not merely beyond modal confines but beyond modal analogies to achieve sought outcomes.[23]
- *Freedom* and *determinism.*[24]

N(IV) Far-out philosophical applications

These are rather diverse and decidedly adventitious. They arise from glimpsing some feature of relevant theories (the logics, or their semantics, or of previous applications) that can be applied elsewhere fruitfully.

- *Coherence* theories and *truth* theories, especially a coherence theory of truth. What happens with the coherence theory is that semantical methods developed in relevant logic and universal semantics are adapted to coherence constructions. But the theories reach far beyond this. Analogous applications are expected for other truth theories (for example pragmatic theory, already investigated, though irrelevantly, by da Costa and associates), and for other coherence notions.[25]

---

22 Another possibility (dismissed by Dunn 1987) is to start from:

$\lambda x\varphi \leftrightarrow \varphi$, *where* φ has at least one free occurence of x,

as set with containment logic. Then with $(\lambda x\varphi)a$ we have relevant predication of property $\lambda x\varphi$ of subject a. It is important to establish that, in requisite containment logic, no equivalences of the form $\varphi \leftrightarrow$ combination (φ, Fx) hold. It should be observed that there are other reasons for moving to containment logic; there are further problems with Addition, apparent at the end of Dunn 1987.

23 *See Chapter Ten for further elaboration. [Editors]*

24 *See Chapter Fourteen for further elaboration. [Editors]*

25 *See Sylvan 1990 for further elaboration. [Editors]*

- Semantic metamorphoses and rehabilitation extended: *world representation*, and *idealism*.[26]
- Cosmological syntheses.[27]
- Pluralism.[28]
- Eutopian theory.[29]

3.2.3 Projected applications

Much of the projected growth has roots in the past or extends present growth. Running strongly into the future are applications as regards science, applications that are seriously incomplete or unsatisfactory. Thus a large region of the growth (always good for business!) is expected in philosophy and foundations of science, both general methodology and specific sciences.

F(I) Philosophy of science more generally
There are various groups of applications, which can be roughly classified as follows:

- Extensions and variations of *confirmation* and *evidence* to the following interrelated topics: *explanation, verification, verisimilitude* (and perhaps *corroboration*).

Some of these tasks have already been attempted. In particular, Mortensen tried to show that the stratagems that Miller and Tichy had independently deployed to show that Popper's apparently appealing and intuitive account of verisimilitude collapsed, were stratagems that were properly unavailable under relevant reformulation.[30] But Mortensen's attempt itself ran into serious trouble, of an analogous sort, leading Mortensen to conclude that a relevant rescue could not be effected. However this attempt was logically very confined; only standard relevance logics were really considered, although a crucial issue is readily seen to turn on the adequacy of the principle of Addition, $A \rightarrow . A \vee B$, as preserving *content*. Thus the investigation should have fallen within the reach of relevant containment logics. A related repair, removing the 'Tichy Miller paradox of verisimilitude' along what amount to relevant containment lines has been suggested independently in Austria (by Schurz and Weingartner).

---

26 *See* SM *and Sylvan 1990 for further elaboration. [Editors]*

27 *See* CS *for further elaboration. [Editors]*

28 *See* TM *for further elaboration. [Editors]*

29 *See Sylvan and Bennett 1990 for further elaboration. [Editors]*

30 Miller's transformations certainly do not stand relevantly.

- Investigation of *reasoning*, and therewith *rationality*, especially of *non-deductive* and *non-ponible* logics. Included in this area of investigation are: *induction*, relevant inductive logics, degrees of implication, weight of argument, and so on.
- A decent theory of *fallacies*. As well as successful reasoning of various sorts, it is desirable to develop an improved theory of failed or defective reasoning, in place of the utterly ramshackled development inherited from traditional logic.

Rationality concerns not only successful and unsuccessful non-deductive reasoning but also decision, another topic (it thereby links).

- Elaboration of *relevant decision* theory. In part this is a matter of extending relevant measure theory; in part it is a matter of tackling an array of paradoxes that infect classical decision theory. It is already clear that relevantisation can remove some of these paradoxes (for example that of Allais).

Since this investigation involves relevantisation of preference (and conditional preference) theory, it leads into foundations of economics. Non-classically this is largely unknown territory.

F(II) 'Foundations' for specific reaches of science

- *Foundations of economics* are but one set of foundations in need, not merely of scrutiny, but of upgrading and renovation, and perhaps of replacement.

Others include foundations of biology (intertwined with economics through maximisation theory, competition theory, etc.); but none more than those of physics. ("Foundations" is of course being used in a wide sense, compatible with the foundations being floating ones, like those of Arab river houses.)

- *Foundations of physics*. These include, above all, issues in quantum theory: inconsistencies, anomalies, Bell's theorem, etc. Proof of Bell's theorem, for instance, appears to depend, perhaps essentially, upon expansion principles (specifically $A \rightarrow. (A \,\&\, B) \vee (A \,\&\, {\sim}B)$), principles also intricated in renown quantum 'mysteries'. Other issues include the Kochen-Specker theorem.)

Other matters in physics, grandly conceived, include: global synthesis of quantum theory with relativity theory [!], and relevant cosmology.

3.2.4 Some additional scene setting

Part of the present exercise is a boring by-product of much more philosophically exciting nonclassical investigation of nonponible logics. There are many important connectives (>) for which modus ponens (in >) is in doubt or breaks down. Among the notions included are those:

- Evidential and case-making connectives such as plausibilification and probabilification: "A is evidence for B", "A confirms B", "A makes B plausible", "A probabilities B", "A makes B preferable", etc.
- Relevantly controlled connectives such as "A is relevantly related to B", "A makes B relevant".
- Similarity and indiscriminability connectives such as the functor "A is indistinguishable from B" of certain Sorites paradoxes.
- Default connectives, where a conditional connective which is normally satisfactory breaks-down because of trouble that emerges (for example inconsistency or paradox), identified as occurring earlier in the chain of argument.

It needs stressing that my survey is a somewhat idiosyncratic, regional one. What I know best, and so far like best, is Australian work. However relevant logic enterprise is now proceeding apace in many centres, and I lack good detailed information on what is happening in some of these places. In Moscow, for example, it is now seen clearly that classical thinking is *blinkered* thinking. It is apparently maintained that relevant logic represents a new logical ideology (a new paradigm, it is also claimed) which will supersede the present classical ideology (what Priest claimed for paraconsistent logic). But I lack further details and Russian supporting arguments.

This chapter, still little more than a set of notes in places, has been long in preparation, more than 6 years. It has too much of the air of an academic report (which I increasingly detest, especially when I have to do parts). Further, I have an Antipodean distaste for trumpet blowing. Moreover, what I don't care for, very much at all, is lists with lots of promises and little performance. Fortunately some low-level performance will follow in subsequent parts.

4 An Orientational Survey of Sociative Logics

This essay originally appeared in 1989 as chapter 2 of BG. *[Editors]*

4.1 Type of connection, and sorts of sociative and relevant logics

A *sociative* or connectional logic involves a definite connection in its implicational or inferential part, it is broadly relevant.[31] Trivial connections (such as being either the same or different, or as implying or not implying) are not definite; they provide no genuine linkage of components. Nor, though the claim does not go undisputed, do the implicational connections exhibited in the basic paradoxes of implication—$(A \,\&\, {\sim}A) \rightarrow B$ and $B \rightarrow (A \vee {\sim}A)$ in standard notation—afford definite connections. For A and B may have nothing whatsoever to do with one another, share no content. Later we will look hard at the implausible theme that the modal ecology of the universe is such, the modal web spun so tight, that impossible and necessary statements, whatever they concern, have definite connections with everything else. To begin however, we make the preanalytic assumption that implication requires connection. The main type of connection that has been emphasized recently is relevance, whence "relevance logic". But relevance does not constitute the only relation of definite connection, or the only one of historical importance—consider inclusion of content, sharing of terms, etc. Moreover, as is well known, relevance in general (not pinned down to specific determinates) is an elusive notion.

Because they include implicational and inferential paradoxes, the mainstream logics of the twentieth century, classical and intuitionistic, are not sociative, but *dissociative*.[32] By contrast the main relevance logics, E, R and T, are sociative because their implicational connective, $\rightarrow$, meets

---

[31] Much searching went into trying to find a satisfactory term to distinguish these logics. The eventually adopted term, *sociative*, derived through French from the Latin *sociare*: "to combine, unite, etc.", and *socius*, "companion". It now means "expressing or denoting association, conjunction, union": see OED. Occasionally, the now obsolete English noun, verb and participle, *sociate*, will also be deployed.

[32] These mainstream logics admit, however, through limitation of material detachment rules, of restricted sociative extensions. But in the process they are transformed into modus-ponens-modifying or -abandoning systems.

the requirement of weak relevance; that is, they have no theses of the form $A \rightarrow B$ where A and B fail to share a (sentential) parameter. But in their standard presentations, within only partly formalised methatheory (or epitheory), the rule structure or inferential component of these logics is not sociative (being classically expressible); for example, the rule, from B to infer $A \rightarrow A$, is a derived rule of R (though not a *normal* one).

The classification introduced depends of course on being able to distinguish connectives and meta-connectives as representing implication and inference. Typically this is given in advance or can be done. To circumscribe discussion let us concentrate, in the usual way, upon sentential logics which include, as well as truth-functional connectives &, $\vee$, $\sim$, an identifiable implicational or conditional connective, $\rightarrow$ say. (The usual formal-theoretic apparatus, of syntax and proof, is thus largely taken for granted.) Such a logic will be *sociative* (in $\rightarrow$) if it contains no thesis of the form $A \rightarrow B$ where A and B are not appropriately connected. That is, to display the principle involved for subsequent comparison purposes,

R. $A \rightarrow B$ is theorem only if $r(A, B)$, i.e., there is a definite connection between A and B.

There is more than one way to treat such a definite connection, r: for instance, to introduce it into the object language, as in relational logics, or to leave it in the metatheory, as typically with relevant logics. That issue (addressed later) is not of immediate moment. More important, definite connection can be guaranteed in a variety of ways, most obviously in these technological days syntactically, but by no means only in this way. There is, however, a long-persisting tendency, linked with the antagonism towards relations, for connections to vanish, if not entirely, at least into the *sharing* of factors.[33] Definite connection of statements, for instance, is reduced,

---

33 The explication of *relation* through sharing is not a peculiarity of philosophy; it runs right across the present reductive scientific disciplines. For example, in philosophy both identity and similarity (to which many other relations are collapsed in empiricism) are explained through sharing of properties; in physics forces are dealt with in terms of sharing of appropriate (sometimes specially concocted) particles; in chemistry bonds are accounted for through sharing; and so on.

As sharing itself is a relation, there is a really no prospect of eliminating relations in this fashion. But it does make sharing into a, or even *the, fundamental* relation (an idea highly appealing to those certain ideological persuasions). For there are evidently prospects for reducing the whole of relation theory to sharing, after the crude fashion that passes muster in classical logical theory. In fact if we can whittle sharing down to the determinate case of sharing elements, for example α shares element γ with β, then we can define membership $\gamma \in \alpha$ through such a connection as α shares γ with α, or an extensionalisation thereof, and

firstly to some sharing of content (substance, sense, etc.), and then, in conveniently circumscribed contexts, content is represented syntactically (so there is something hard and definite that logicians or computing machinery can go to work upon). In a third popular stage of reduction, the residual requirements reached are declared undesirable, unworkable or unnecessary, and thereupon junked. This latter stage of irrelevant degeneracy, often reached in these latter days, lacks any decent justification, as will become apparent; but the earlier stages of reductive explication are both important. The first stage issues in the rather traditional *content* principle:

CR. $A \rightarrow B$ is a theorem (or obtains, to put it in more traditional form) only if A and B share content (sense, use, or some such).

The sharing can take various different forms—inclusion, overlap, etc.—depending on what the presupposed account of *content* (or *use*) permits. Though content is at base a semantical notion, use is not; and in principle the further analyses of sharing can be accomplished in either sematic or pragmatic forms or reflected syntactically. All have been attempted, but syntactical reduction, with which we begin, has been strongly favoured.

The sentential situation, upon which we continue to focus, is crucial. Not only does it have to be included in any comprehensive story, but the rest can be seen as expanding it or can be contracted to it. For, in principle, quantification can be dealt with in an analogous fashion, for instance with predicate parameters generalising in a familiar way on sentential parameters (which serve as zero-place predicates). Or otherwise, further enlargements can be handled by the important strategy of *conservative* extension, that is, by the requirement that they do not induce new (irrelevant) results back on the sentential ground floor.

4.1.1 Syntactical connection, where statements share syntactical components

To render the connection nontrivial and stable (under elementary transformations) the components concerned must be more than punctuation, such as brackets. For suppose the formation rules of sentential logic were written so that they began with the clause that a sentential parameter enclosed in brackets is a wff. Then every wff of the logic would involve brackets, and A and B would always share syntax, much as all written English declarative sentences share full stops. Connection would be then trivially guaranteed—

---

then rely on the reduction of relations set theory is supposed to afford.

when there may be none of substance. Less artificially, any sentential logic where *variable* sharing holds at the first degree (i.e., the part of the logic where no nested $\rightarrow$'s occur), will exhibit a syntactical connection because higher degree wff will always share brackets. But sharing punctuation is a trivial connection; it affords no guarantee that A and B have anything to *do* with one another. Furthermore the notion would not be stable, because bracketing vanishes upon syntactical transformation, for instance upon re-expression in Polish notation. It is in part for these sorts of reasons that it is sometimes insisted that requisite connections must be semantical ones of some kind, of meaning or content. But in suitably restricted contexts, such connections may well be reflected in syntactical linkages, which *do* duly avoid trivially.

Such triviality problems are avoided in the most thoroughly investigated syntactical connection requirements, by restricting sharing to categorematic elements, to sentence variables at the sentential level. Thus the weak relevance principle:

WR. $A \rightarrow B$ is a theorem only if A and B are weakly relevant to one another, i.e., A and B have some variable in common.

Thus also the stronger proscriptive principle:

PR. $A \rightarrow B$ is a theorem only if all variables in B already occur in A.

The first reflects, in an approximate syntactical fashion, content overlap, the second, content inclusion.

Principles of these types enjoy a respectable history. The idea of weak relevance, rediscovered by Belnap in application to sentential systems, traces backs at least to the 15th century, and reappears periodically thereafter. To quote Gaulincx from the 17th century: 'in a logical argument some term recurs and is stated both in the antecedent and in the consequent'. It is an easy argument from the formation rules for traditional syllogism, that valid syllogistic arguments meet this requirement. Similarly, the proscriptive principle of Parry was anticipated by the Jesuits of Coimbra who defined a class of *argumentationes* where every variable in the conclusion is in the premisses but some variable in the premisses may not be in the conclusion (see further Thom 1981, p.28).

These syntactical relevance principles are far from trivial requirements; they are requirements which many systems fail. Evidently these principles, which are presented as but *necessary* conditions for theorem-status, do not exhaust the types of variable sharing principles of some technical interest:

a strong proscription principle, for instance, requires that A and B share all variables.[34] Another more jejune principle makes variable sharing sufficient (or almost sufficient) as well as necessary; and so on.[35] Each such relevance principle determines a corresponding class of systems; for example WR determines WR systems, weakly relevant sentential systems. The classes will include some strange systems, unless there are other controls. Some of them will have "implications" which fail to preserve values such as truth, necessity or meaning, or to meet other customary desiderata. Such strangeness only emphasizes however the severe inadequacy of the principles as *sufficient* conditions.

Leaving out syncategorematic expressions, such as connectives, from these relevance principles has an element of arbitrariness, technically convenient though it may be. Connectives are not without syntactically reflected meaning and content, which may afford connection. Nor is the omission altogether satisfactory, as the curious struggle to find grounds for excluding the Mingle principle $A \rightarrow .\ A \rightarrow A$ has revealed. Yet it is evident enough that where A is zero degree (i.e., contains no occurrences of $\rightarrow$), $A \rightarrow A$ contains an occurrence of connective $\rightarrow$ when A does not. The result is that where Mingle holds a nonimplication implies an implication, a result Ackermann, and following him Anderson and Belnap, certainly wanted to proscribe (see ENT, p.237 ff).

Obviously variable and connective sharing principles can be combined in a mix of ways, some of them are worth setting down. An ultra-weak relevance, or basic syntactic connection, is provided by the simple disjunction:

UR. $A \rightarrow B$ is a theorem only if A and B share a variable or a connective.

The I systems (studied by Routley, for example 1972, 1988) are sociative inasmuch as they meet this weak necessary condition. For these systems have the same first degree logic as relevance logics E and R, but diverge

---

34 The strong proscription principle can be satisfied in various weaker connexive logics. But in fact it is straightaway satisfied in the equivalential parts of Parry logics, i.e., the parts with connective set $\{\leftrightarrow, \&, \vee, \sim\}$ where implication is now construed as equivalence.

35 Copeland, to take an isolated example, has been interested in making relevance sufficient as well as necessary for implication—leaving out, or approximating, the further standard requirements on implication of truth or content preservation. But sufficiency would exonerate such paradoxes as $A \rightarrow .\ B \rightarrow A$ and $A \rightarrow .\ {\sim}A \rightarrow C$, but fail Detachment or Substitution, since these "relevant" principles, yield irrelevance immediately. Other less jejune forms of variable sharing have been considered by several other authors, for example Cleave 1974, Kielkopf 1977, Weingartner and Schurz 1986. Kielkopf 1977 presents a valuable, and duly critical, survey of several of these types of requirements (now quite fashionable in Europe).

from them at higher degrees in ways that violate WR. However the higher degree theorems involved in the divergence, such as $p \to p \to . \; q \to q$, always share connectives, largely by virtue of being higher degree.

There is a case for adopting UR as a *minimal* requirement for a sociative logic, and thus rendering UR systems *minimally sociative*. For it provides *a* basic connection of content syntactically-reflected. It also marks out an important watershed. Then not only are classical and intuitionistic logics and surrounding systems dissociative, not only are strict logics (with implication connective that of strict implication) and contemporary 'logics of conditionals' dissociative; so also are systems such as RM, i.e., relevance logic R together with Mingle, and their extensions (though sometimes accounted 'semi relevant', for example in ENT, p.375). For RM has, like 'logics of conditionals', theorems such as $p \,\&\, {\sim}p \;\to .\; q \;\to\; q$, where antecedent and consequent share neither variables nor connectives.[36]

It is one thing to guarantee some syntactical connection, which is all UR does at bottom; it is quite another to obtain intuitively *satisfactory* linkage of components. Basic connection not only fails to block higher degree irrelevance and suspect formulae such as $r \to s \to . \; p \to p$. It also fails on its own to exclude defective principles like Mingle which may induce no dissociativity in weaker logical settings than R. Worse, it fails to block the addition of bizarre principles such as $p \,\&\, {\sim}p \to . \; q \vee {\sim}q$. In order to be more exclusive connective combinations need to be scrutinized. Consider, for instance, the more discriminating mixed principle,

MR. $A \to B$ is a theorem only if A and B share a variable and all intensional connectives in B occur in A.

While MR duly excludes Mingle, the somewhat ad hoc restriction to intensional connectives avoides wiping out also such often-enough questioned themes as $A \to {\sim\sim}A$, $A \to . \; A \vee B$, etc. (given of course that $\sim$ and $\vee$ are not intensional).[37] But it is a crude and rather unsatisfactory way of getting at what is—independently—sought. For while it, of course,

---

[36] A proof sketch can begin from RM theorem ${\sim}(p \to p) \to . \; q \to q$, established in ENT, p.429. In R, and so in RM, $r \;\to\; s \;\to .\; {\sim}(r \,\&\, {\sim}s)$, whence contraposing $p \,\&\, {\sim}p \to . \; {\sim}(p \to p)$. Then use Transitivity. Note that this proof depends critically upon Commutation. In logics without forms of Commutation (i.e., Permutation), basic connection is restored (at least in negation weakened logics of this sort). For similar RM-style irrelevance in 'logics of conditionals', see recent surveys of dissociative 'conditional' logic, for example Chellas 1975, Nute 1984. For some of the belatedly burgeoning sociative theory, see DRL, and also Hunter 1983.

[37] Containment logics, of interest in Artificial Intelligence, conform to the requirement AR. $A \to B$ is a theorem only if all variables and connectives in B occur in A.

removes Mingle, it does not touch substitution instances of Mingle such as $A \to B \to .\ A \to B \to .\ A \to B$. (Nor would such a connective device discriminate this from what is often taken, in these warped export-oriented days, as even paradigmatically correct, namely $A \to A \to .\ A \to A \to .\ A \to A$).

More subtle connection principles must somehow combine degree layering of wff with variable sharing. This can be accomplished in various ways; but only some quite specific ways, answering to given relevant logics, have been investigated. One, established for relevance logics such as E, is Maximova-relevance:

MXR. $A \to B$ is a theorem only if some variable occurs as an antecedent part of both A and B or as a consequent part of both A and B (ENT, p.253).

But, on its own, owing to the cross-over fashion by which antecedent and consequent part are defined, the requirement fails to exclude weakly irrelevant formulae involving negation, such as the Minglish formula $\sim(p \to p) \to .\ q \to q$ (ENT, p.253). MX 'relevance' too has to be combined with other requirements of relevance, to exclude Minglish erosion for instance.

A much more demanding necessary condition for systemic relevance, not open to these sorts of problems, is depth relevance:

DR. $A \to B$ is a theorem only if A and B share a variable at the same depth, where what determines depth is numerical extent of intensional nesting, typically of occurrences of $\to$.

To be more exact about this important notion, let wff contain just connectives from the standard set $\{\&, \vee, \sim, \to\}$. *Depth of occurrence* of a sub wff of wff A is defined inductively as follows: The sub wff A of wff A is of depth 0 in A. If $\sim B$ is a sub wff occurrence of depth n in A then this occurrence of B is of depth n in A. If $B \,\&\, C$ (similarly $B \vee C$) is a sub wff occurrence of depth n in A then these occurrences of B and C are both of depth n in A. If $B \to C$ (similarly other 2-place intensional connections) is a sub wff occurrence of depth n in A then these occurrences of B and C are both of depth $n+1$ in A. Then A and B share a variable p at the same depth iff for some nonnegative integer n there is an occurrence of sub wff p in A at depth n and in B at depth n. Depth relevance of course implies weak relevance. *Deep systems*

Obviously there is a intermediate requirement of passing interest, viz.
NR. $A \to B$ is a theorem only if A and B share a variable and all connectives of B occur in A.

satisfy the depth relevance requirement, or (if they are super-propositional) conservatively extend systems which satisfy the depth relevance requirement.

Depth relevance enjoys, by contrast with some of the other rather arbitrary relevance conditions considered, significant theoretical linkages. A main result, adduced by Brady, is that the depth relevance condition affords a way, which is not *ad hoc*, through the logical and semantical paradoxes. Deep systems can nontrivially carry *unrestricted* comprehension principles and the like, while retaining extensionality principles in set theory (for full details see Brady 1984). The condition also turns out to circumscribe those relevant logics—the D systems—arrived at (as in RLR I) on independent adequacy grounds, as candidates for entailment and deducibility proper. Thus depth relevance exhibits features that make for an important distinction of systems.

Although depth relevance is a natural extension of weak relevance to combinations involving levels of intensionality, which does have its desired effect of ruling out Ackermann formulae such as Mingle and $A \rightarrow . A \rightarrow B \rightarrow B$, it runs into trouble with (now well established) relevance logics, and so will not appeal to exponents of these systems. For it also takes out much of what is distinctive about systems E and T, namely both *Exported Syllogistic* (ESyll) *forms*, $A \rightarrow B \rightarrow . B \rightarrow C \rightarrow . A \rightarrow C$ and $B \rightarrow C \rightarrow . A \rightarrow B \rightarrow . A \rightarrow C$, and *Contraction principles* like Absorption, $A \rightarrow (A \rightarrow B) \rightarrow . A \rightarrow B$, Importation, and Asscrtion, $A \,\&\, (A \rightarrow B) \rightarrow B$.[38] Dccp systems reject both classes of contested principles. The questioning of these contemporary principles, most of which enjoy little historical standing, had certainly begun when Lewis was puzzling over whether $S2$ or $S3$ came closer to capturing entailment (and wisely, given his limited irrelevant options, settled for $S2$). The critical investigation is taken much further in recent work on deep relevant theory, which much extends the typology of relevant systems in terms of the classes of principles they meet. In short, then, depth relevance removes practically all those parts of system E elsewhere argued to be in serious doubt for the intended explicandum, deducibility (see RLR I, chapter 3). But, because of its sheer exclusive power, such a requirement is inevitably controversial. Syntactical relevance requirements themselves require justification, nonsyntactical justification, for example in terms of what the requirements do do.

Syntactic connection certainly has its limitations. It depends (too) heavily on the syntactical forms of the systems investigated, and on form being

[38] There are already problems as regards the mere omission of Contraction in a logical setting as strong as system R. There Contraction is tantamount to Reductio, $A \rightarrow {\sim}A \rightarrow . {\sim}A$ —surely a mistaken equation of principles of different logical sense and force—and these in turn to the much less obnoxious LEM, $A \vee {\sim}A$!

appropriately displayed (and not condensed or hidden, definitionally or otherwise). While syntactic specificity is advantageous for proof purposes, too much may depend on a specific formulation of a system and not extend to equivalent systems (for example with different primitives); such defects afflict even the account offered of minimal sociativity. Despite what was said earlier, moreover, notions of syntactical connection do not always extend so easily or uniquely to more comprehensive logics, for instance to applied predicate logics. More important, syntactic connection is not the only type of connection, but, so to say, a surface representation (which syntactic forms may in fact hide) of other sorts of less superficial connection, such as connection of content or of argument. In particular, morphological linkages only indirectly, superficially and obliquely, represent what is crucial to implication and deducibility: preservation of (truth-)value and of content.

4.1.2 Semantic connection, through content or meaning

Much less work has been attempted on semantic and pragmatic connection of components of genuine implications than on syntactical linkages, despite repeated and regular traditional and twentieth century emphases on meaning or content coupling for conditionality, entailment and the like. But what has been attempted on the semantic front, primarily for relevant logics (but also to some extent for neighbours such as relational logics), can be extended, for what it is worth, to other sorts of sociative logics (those equipped with a tractable world semantics). Main attempts amount in fact to adaptations of what positivistic logicians (Carnap especially, but also Wittgenstein and others) tried less successfully to carry out on the bases of irrelevant logics (typically systems such as $S5$). These irrelevant explications of semantic notions such as sense, content, information, and so on, of course simply transferred the paradoxes of implication over to the semantical notions, and accordingly led to serious violations of the content principle, CR, and its analogues. While sociative explications can avoid such damaging outcomes (as that all necessary truths have the same content, usually and wrongly said to be none), the explications are sometimes not particularly helpful, in the way that syntactical principles can be, but partly parasitic on the notions of which a better understanding is sought.

The point applies especially to the first type of analysis, which characterises some notion of sense or content through implication itself, or a relation in its circle such as consequence. Typical of these *consequence* types is the account of the sense of A, sence(A), in terms of what A implies, for example $\lambda C(A \rightarrow C)$ or, under popular extensionalisation, $\{C : A \rightarrow C\}$.

(The neologism *sence* is used, because this explication doesn't quite exhaust sense in the way too often supposed.) *Every* logic with an implication can furnish such an account (one for each different implication). The accounts will be satisfactory, for instance in meeting a sence analogue of CR, to the extent that the logic already has a decent implication, and so is appropriately sociative.

A second superior type of analysis, which does make genuine use of semantics, ultimately encounters trouble also. Fairly representative of these *situational* types is the account of the content of A in terms of the situations, or worlds, A excludes. Most simply, $cont(A) = \lambda a$ (A fails to hold at a), i.e., extensionally $\{a : I(A, a) \neq 1\}$ where I is a situation relativised evaluation function, i.e., an interpretation function. Such accounts work very nicely for a *large* class of relevant logics. They incidentally vindicate, in a straighforward way, by contrast with modal logics, content principle CR (see UU). But, in a way, they work a little too well. They offer little or no discrimination between marketed relevant logics, and they extend to a wide range of other sociative logics, indeed to any—and that means extensive classes of virtually all types—that can be rigged out with suitable situational semantics. Yet, despite their lack of discriminatory power, such analyses have considerable point. They show, for example, that what has often been alleged cannot be done, can be done; that important traditional requirements upon implication as to (meaning) inclusion and sharing of content are sound, and can be made good, and so on (these points are elaborated in FD and RLR I).

Mistaken claims abound in the philosophical literature about what relevant logics cannot achieve in the way of truth-preservation semantics for implication, interrelations of these with content semantics, and so forth. The fact is that a truth-preservation analysis of relevant implication can now be straightforwardly provided, and straightforwardly integrated with an inclusion of content analysis, on the lines sketched and referred above. Of course, the analyses (better presented in terms of *holding* than *truth*) will make use of 'impossible situations' or some substitute for them or construction of them; but, as is now at last being realised, that is quite alright.

4.1.3 Use and argument connection

Prior cleverly claimed (as Bennett 1954 reported) that arguments to the basic paradoxes of strict implication simply show that necessary and impossible statements *are* connected to everything else. Of course the arguments do not show this (without a difficult further climb upward through the hierarchy of modal systems, using supports and ropes which give way in better systems, which duly separate implicationally distinct principles); for the paradoxes work in such weak modal systems as $S0.5$ and $S1$, which do not sustain the more sweeping modal ecology. A less ambitious orthodox retort, whose attempt to bag intuition will be disallowed, takes the following lines:- What, after all, could be better than linkage through intuitively valid and simple arguments such as those, known on and off from medieval times, for $(A \,\&\, {\sim}A) \rightarrow B$ and $B \rightarrow (A \vee {\sim}A)$? Because of these arguments (duly broken below) establishing appropriate nonparadoxical connection by way of argumentative linkage is an uphill effort. Circumscribed, or canonical, forms of argument—which however leave out arguments widely (if mistakenly) accepted nowadays as valid—have to be invoked. Whilst this can be done (and will straightforwardly emerge from the systematisation of sociative logics), the business of separating out the arguments which are good, from those which are not, involves appeal to other criteria, so leading beyond pure argument connection.

A more promising approach is that concerning *use* of argument components. Again there is a scattered tradition, but the approach has been pursued most diligently by Anderson and Belnap (with their strenuous work brought together in ENT. For there is a clear sense in which B is not used in the following sort of derivation leading to paradox:

A	$\Vdash A$	pristine axiomatic form
A, B	$\Vdash A$	weakening, or thinning
A	$\Vdash B \rightarrow A$	deduction principle
	$\Vdash A \rightarrow . B \rightarrow A.$	deduction principle

Given the way B is infiltrated, as an *extra* (which is merely conjoined and doesn't undercut other premisses), there is little excuse for then assigning B an implicational role as the deduction principle proceeds to do. In such simple cases, it is easy to see how to stop the rot; premisses not *used* in the argument cannot be assigned deductive (exportative) rules. One early problem is how to extend this simple and often obtained insight, even to expected full systems of sentential logic.

It soon becomes evident that the insight does not extend either easily or uniquely, even under some 'natural' controls. Indeed there are several ways to proceed, some of them fairly natural, especially once further connectives are introduced; and certainly with controls relaxed a plethora of systems can result (for example, to begin with, subscripted natural deduction systems or intensionalized Gentzen systems for hundreds of relevant logics). Worse, the insight does not lead, without much complication (for example through two types of coupling of premisses, or else through subscripting) in expected, *independently* motivated directions; and so far it has not yielded expected nice results for main systems (for example those that Gentzenisation often yields regarding proof structures, cut-elimination, and so on).

Use approaches, like argument approaches, do not live up to their initial promise. Though important, though a significant part of an integrated investigation, they cannot serve as a satisfactory starting point. For they do not stand on their own. Nor do they disclose motivation, or reflect semantical motivation, sufficiently well.

4.1.4 Pragmatic connection and relevance

Appeals to pragmatics, and injunctions to be relevant, can work in two very different ways. On the one side, they can be invoked to wave away demands for connection in logic, and for sociative logics. Everything is fine, or fairly fine, with mainstream syntax and semantics; connection is a pragmatic froth on top of the deeper structure. That foam can be satisfactorily skimmed off and taken care of along with other complicating froth of pragmatics. Such appeals to pragmatics are made not merely by defenders of the classical faith; but they also come from linguists who wish to use classical logic as deep structure, from computer scientistis who want to program their machines with a classical frame, from communication theorists, from cognitive scientists, and so on. To those who have glimpsed some relevant light on the scene, such appeals are unconvincing. They are unsatisfactory because they do not get things near enough right at a semantical level, or on more radical sociative thinking because they get things decidedly wrong (cf. BG, chapter 1). The paradoxes of implication are paradoxical because of semantical features, the semantics of implication especially, not because of some neglected nicety of speaker or hearer use not touching semantics. The pragmatic buffer-zone shielding mainstream logic does not render it impervious to sociative criticism. Yet, on the other side, much sociative argument relies upon pragmatic considerations. But from this side the pragmatic features are not semantically superficial; they reach down again to affect a properly-reflective syntactical analysis, whence requirements of syntactical connection.

Arguments to syntactical requirements of relevance are commonly of this sort. They begin with how discourse is used, how mathematicians talk and reason, above all from how individuals argue. They point out that these are not procedures which operate in irrelevant ways; if they were they would be open to legitimate criticism (as the famous paradox-applying mathematical author in ENT, pp.17-18 is wide open to editorial criticism).

Moreover, these constraints are not mere conversational conventions, they concern the context-independent meanings of key notions involved. Fallacies in argument, such as those involving relevance, cannot be set aside as pragmatically infelicitous, and immaterial to validity; fallacies bear directly on validity.

Such, in very broad outline to be filled out subsequently, are some of the reasons why requirements of connection—now pretty generally conceded at the pragmatic level, because hard to dismiss there—should be transmitted downwards to narrower semantical and syntactic levels. They by no means exhaust the arguments for connection at lower levels, as should now be becoming evident. Motivation for connection is commonly directly at the semantic level, already abstracted from rules for discourse users. And it is partly for this sort of reason that very little investigation has been made of specifically pragmatic criteria for connection. A different excuse, from the other side, is that it is very easy or convenient to neglect items shoved up into the pragmatic attic. In any event, there is little so far to consider in the way of accessible criteria for connection at the pragmatic level. What criteria are suggested drop down to ones like those looked at already; and the same applies to tests for relevance.

4.2 A working classification of sociative logics

The paradoxes of implication do have the virtue of delivering an exhaustive classification of sociative logics. For there are at least as many styles of sociative logics as there are distinctive ways of halting these arguments to dissociation, and of course no more. A little more precisely, an initial classification of sociative logics can be made in terms of the characteristic steps in derivations of the basic paradox arguments of strict implication at which these arguments are said to be broken.[39]

For an initial classification, the derivation of the more persuasive negative paradox, $A \,\&\, {\sim}A \rightarrow B$, affords a more revealing breakdown. The manifestly implausible expansion step, $A \rightarrow .\ A \,\&\, (B \vee {\sim}B)$, and its mate $A \rightarrow A \,\&\, B \vee .\ A \,\&\, {\sim}B$, of the standard positive paradox argument, is rather uniformly rejected by quite different types of sociative logics (though it is retained by certain nontransitive and relational logics).

---

[39] A similar approach was earlier adopted in RLR I, chapter 1, where however some forms of sociative logic are given only cursory discussion. The different emphasis in the material that follows will do something to compensate for what that one-sided approach neglated.

Apart from substitution—not to be neglected—one revealing derivation involves the following steps and principles:

Steps	Principles applied	Principle Names
$A \,\&\, {\sim}A \to {\sim}A$	$A \,\&\, B \to B$	Simplification (R)
${\sim}A \to. {\sim}A \vee B$	$A \to. A \vee B$	Addition (R)
$A \,\&\, {\sim}A \to. {\sim}A \vee B$	$\dfrac{A \to B, B \to C}{A \to C}$	(Rule) Transitivity
$A \,\&\, {\sim}A \to A$	$A \,\&\, B \to A$	Simplification (L)
$A \,\&\, {\sim}A \to. A \,\&\, ({\sim}A \vee B)$	$\dfrac{A \to B, A \to C}{A \to. B \,\&\, C}$	(Rule) Composition
$A \,\&\, ({\sim}A \vee B) \to B$	$A \,\&\, ({\sim}A \vee B) \to B$	Disj. Syllogism
$A \,\&\, {\sim}A \to B$	$\dfrac{A \to B, B \to C}{A \to C}$	(Rule) Transitivity

L abbreviates "on the left"; R "on the right". But the distinction will do little work; for, from antiquity onwards, communtativity and also associativity of connectives $\&$ and $\vee$ have been almost invariably, and correctly, taken for granted. The rule/thesis distinction is an entirely different matter, and most important.

Near enough, rejection of *any* of the principles involved yields a class of logics which halt the argument; and if the rejections are carried through in a systematic way, they yield classes of logics which exclude all such paradox arguments. Thus there is a class of non-Ψ logics for each principle label Ψ. That is, more colloquially, there are logics of all the following types, at least:- *nonsimplifying*, *nonadjunctive*, *nontransitive*, *noncomposing*, *nondissembling* (the latter where Ψ is Disjunctive Syllogism). Rejection of some principles makes room, in turn, for others. So logics of the classes discerned are by no means all subsystems of the strict systems which reject none of the principles involved. For example, nonsimplifying logics include as a fascinating subclass, connexive logics which underwrite principles like ${\sim}({\sim}A \to A)$; nondissembling logics make room for a decent investigation of properties of nonimplication, and of such principles as that nonimplications do not imply implications, for example the Anderson theme $A \nrightarrow B \nrightarrow. C \to D$. Indeed there are numerous interesting, often undreamt of, nonclassical principles for future logicians to play about with (the recently investigated $A \to B \to B \to A$ of Abelian logics is but a foretaste of this future: see PL, p.245 ff). The serious investigation of such radically nonclassical systems remains however in its infancy. A main focus of nonclassical (or nonmodal) logics has simply been on busting paradoxes, by logical means, rather than slumming it with them, and for this end rejections are what matter—*well*-motivated rejections if the idea is to obtain a good start.

Virtually every way of breaking the paradox argument by abondoning one of the principles cited has been attempted, except the rejection of historic communative features such as $A \& B \rightarrow B \& A$ connecting left and right forms, and the dropping of compositional principles such as $A \rightarrow B, A \rightarrow C \,/\, A \rightarrow . \; B \& C$. In effect, the latter too has been tried, though in a paraconsistent setting, with discussive logics, a leading feature of which is abandonment of Adjunction: $B, C \,/\, B \& C$. It is not difficult to show that (as based on normal modal logics), discussive logics also drop, what is closely allied, Rule Composition. The motivating idea is that what is guaranteed in separate discussions cannot necessarily be amalgamated (under the one guarantee) in a coherent discussion. But standard discussive logics, though paraconsistent, are not sociative, and indeed *retain* the negative paradox, $A \& {\sim}A \rightarrow B$, for discussive implication. While it appears no difficult logical feat to design sociative discussive logics (parasitic, on relevant rather than modal logics), and there is plausible motivation for attempting such a feat, the exercise has yet to be followed through properly. While all other paradox-smashing ways have been tried in recent times, mostly the attempts have been made with rather special forms of systems within the types distinguished. There is much more room for logical experiment within the sociative sphere. As a result of the specialisation, most of recent attempts have at least isolated or embryonic historical antecedents, some of them to be enlarged upon in Chapter Five. But (as with much earlier intellectual endeavour) even some of the attempts tried have not been pushed or explored very far, usually because they lack satisfactory motivation; and within the range of tried procedures there are some highly favoured moves. Thus, for example, while the idea is much favoured, especially by novices, of a logic which fails or qualifies uniform substitution (i.e., in effect limits applications of principles to cases), the idea has not been advanced very far, and often intended qualifications to the principle remain unarticulated. When duly clarified, the idea tends to fade away, or to disappear into other paradox resolutions.

Nonsubstitutional logics reject, or rather (if they are to get anywhere) qualify, the principle of uniform substitution upon sentential variables; not all substitutions are admissible. (So rejected thereby is the implicit generality interpretation of variables, the free algebra assumption, etc.) Thus equivalently these logics can be seen as prohibiting or restricting rewriting or relettering of principles, especially when the results identify components. The envisaged qualifications typically affect Simplification or Rule Transitivity, blocking their operation as regards inconsistent antecedents. For example, qualification of substitution (or rewriting) applied to Simplification, a quite popular move, stops derivation of $A \,\&\, {\sim}A \;\rightarrow\; A$ from $p \,\&\, q \;\rightarrow\; p$ (or from $A \,\&\, B \;\rightarrow\; A$). Similarly, restriction of Transitivity (i.e., Conjunctive Syllogism), $(p \rightarrow q) \,\&\, (q \rightarrow r) \rightarrow .\; p \rightarrow r$, blocks in particular substitution of $A \,\&\, {\sim}A$ upon p. Restriction of Rule Transitivity (normally presented in schematic form) stops analogous rewriting with an explicitly inconsistent antecedent.

Main motivation for both these sorts of restrictions characteristically derives from a severe unease with contradictions, from the idea that there are serious limitations upon where contradictions like $A \,\&\, {\sim}A$ can figure and what, if anything, they can imply. This disease terminates in the limit in the very difficult (anti-classical) position, not lacking a stream of philosophical adherents however, that contradictions entail nothing, are logically sterile. In part, this comes from overreaction to mainstream folly, with the net result a false all-or-nothing dichotomy: contradictions either entail everything or else nothing. In between these extremes lie more discriminating positions, like relevant ones, that contradictions entail some propositions (those that they do) but not others. But in part it derives from a traditional widely-held picture of negation, a neutralisation or cancellation picture diametrically opposed to the contemporary mainstream explosion picture (as explained in NC), according to which ${\sim}A$ cancels out A leaving nothing. But from nothing, it is always said, nothing follows. The picture is one thing, and easily grasped in outline; its formal development is quite another. Proposed substitutional restrictions, which are *one* way of trying to neutralise contradictions, have not been well-motivated, or closely linked to underlying negation pictures.

Substitutional restrictions tend to be tricky to manage formally, though not impossible as mixed experience with higher order predicate logic has shown. But as that experience has also come to show, substitutional restrictions can typically be removed by reaxiomatisation (for example restrictions on substitution rules become qualifications on comprehension schemes). So it is also with nonsubstitutional logics; upon clarification these transform into logics of other types.

What is proposed in nonsubstitutional logics is *qualification* of substitution, not its total elimation—which would stop logic in its tracks. But qualification amounts to substitution subject to provisos, i.e., to the restricted rule:

(i) $A \surd S^{p_1 \ldots p_n}_{B_1 \ldots B_n} A$ provided C

where $S^{p_1 \ldots p_n}_{B_1 \ldots B_n} A$ is the result of uniformly replacing variable p_1 by wff B_1, p_2 by $B_2, \ldots p_n$ by B_n throughout A. As is standard, the rule form $A_1, \ldots, A_n \surd B$ records that where $A_1, \ldots, A_n$ are theorems so also is B, whence B can be added to a proof sequence.[40] But the provisoed rule, in which C may depend on both A and $B_1, \ldots, B_n$, has a similar force to

(ii) $C, A \surd S^{p_1 \ldots p_n}_{B_1 \ldots B_n} A$

that is, an 'unrestricted' substitution rule with a further premiss. Of course to so introduce C into the object language *may* require additional symbolism. Furthermore, in many logical settings the additional premiss of (ii) can be shunted away, allowing restoration of the usual unadorned substitution rule, $A' \surd S^{p_1 \ldots p_n}_{B_1 \ldots B_n} A'$ (which then typically reduces to an iteration of single substitutions $A' \surd S^p_B A'$). This may be achieved by adding an appropriate form of C to *other* rules as a further premiss. Alternatively, in certain cases, the result may be achieved by conditionalising the axiom schemes to which the qualifications are supposed to apply. Then the proviso of (i) is distributed across the rule forms, yielding $C \rightarrow A \surd S^p_B (C \rightarrow A)$, which is simply a case of the usual rule. Looking at the main example where restriction of substitution is tempting in this setting should help to clarify the more abstract argument. A common objective (perhaps with a history stretching back to before the twelfth century) is to ensure that Simplification, $C \,\&\, D \rightarrow C$, only operates provided the addition of D is compatible with C. Instead of curtailing substitution everywhere, it is enough to restrict applications to $p \,\&\, q \rightarrow p$, where however the intended effect can be obtained, in the form: $C \circ D, p \,\&\, q \rightarrow p \surd S^{pq}_{CD}\, p \,\&\, q \rightarrow p$, where $\circ$ is a standard compatibility (or consistency) symbol. In some settings a conditional form may prove satisfactory, for example $p \circ q >. \, p \,\&\, q \rightarrow p$, for some conditional $>$, whereupon qualified substitution can be replaced by usual substitution. For many pruposes (but not some finitary ones) it is preferable to move to schematic formulation (which gives some of the effect of substitution across rule signs), for instance to $A \circ B \surd A \,\&\, B \rightarrow A$. In these sorts of ways, through formal precisification, nonsubstitutional logics can be transformed into logics with unqualified substitution.

40 In settings with appropriate apparatus available $A_1, \ldots, A_n \surd B$ gets defined as $\vdash A_1, \ldots, \vdash A_n / \vdash B$. The intended contrast between / and $\surd$ is of course this: whereas / may operate hypothetically, $\surd$ does not. Most of the book follows, however, the usual practice of papering over this important contrast, assigning symbol / to cover both types of rules.

Nonsubstitutional logics designed to halt the 'hard' paradox argument by qualifying substitution upon Simplification give way to nonsimplifying logics and very typically to connexive logics, while those qualifying substitution upon Transitivity give way to nontransitive logics. As to the first, the grounds for qualifying substitution ensure that such characteristic principles as $\sim(A \rightarrow \sim A)$ hold. For if A were to imply $\sim A$ it would have $\sim A$ as part of its content, and so would amount to the null content $A \& \sim A$ which does not imply $\sim A$; so A cannot imply $\sim A$. (Such an argument does not pass uncontested however; there are problems, for example, with such reductio methods in connexive settings.)

Nonsimplifying logics include not only connexive logics, but a variety of other sorts of systems as well. They include all sociative logics which are thoroughly traditional in adhering to the rule applied in indirect reduction of syllogism, now known as Antilogism: $A \& B \rightarrow C \,/\, A \& \sim C \rightarrow \sim B$ (cf. BG, chapter 1). For given Simplification, $A \& \sim A \rightarrow \sim B$ is but a single step. So, not surprisingly, many of the systems implicit in ancient logical theory, by no means all of them connexive, are nonsimplifying (cf. Chapter Five). Nonsimplifying logics also include all those which find something wrong with going anywhere logically from or within an explicit contradiction, such as $A \& \sim A$—indeed this is a main motivating idea behind such logics. They thus include logical elaborations of positions as seemingly remote as those of orthodox medieval *obligationes* theory, which prohibited posit of such assumptions, and those of Wittgenstein, Körner and others, which called for logical procedures to halt (and backtrack) at such contradictions (see for example PL on Wittgenstein). A common basis for most of those sorts of positions derives, once again, from the traditional cancellation picture of negation, that $\sim A$ in some fashion cancels or erases A, leaving no A-content, whence $A \& \sim A$ can neither include nor imply A, and similarly $A \& \sim A$ cannot involve or sustain $\sim A$. Of course, such a cancellation picture (elaborated in NC) leads back to connexivism; for A cannot include what cancels it, its negation $\sim A$.

Among the oldest ways of trying to resolve the paradoxes are such connexive ways, which supply some of the oldest sociative logics (as will emerge in Chapter Five). *Connexive logics* are distinguished by two connected features: their restriction of Simplification, and their espousal of nonclassical connexive principles such as $\sim(\sim A \rightarrow A)$ and $\sim(A \rightarrow \sim A)$. It is easy to see that such principles trivialise mainstream logics and dialethize (i.e., render inconsistent) various sociative logics. A typical argument (already known in medieval times) to such a result need use only some of the most venerable of logical principles: Contraposition, Double Negation and Transitivity, along with Simplification. From Simplification, $A \& {\sim}A \rightarrow A$ and also, as in the paradox argument, $A \& {\sim}A \rightarrow {\sim}A$, whence by Contraposition (and Double Negation) $A \rightarrow {\sim}(A \& {\sim}A)$. So by Transitivity, $A \& {\sim}A \rightarrow {\sim}(A \& {\sim}A)$, contradicting Abaelard's connexive principle ${\sim}(C \rightarrow {\sim}C)$ for C of the form $A \& {\sim}A$. Consistency generally—with the principle of Noncontradiction a pre-eminent example—was a major requirement of the dominant traditional paradigm; something had to give: Simplification. Simplification of redundant components has, and had, no traditional standing. (For example, the number of premisses of an argument, as well as their genuine use, was an important consideration in both Peripatetic and Stoic logics.) The way was clear for an obvious traditional exit from implicational paradoxes worked out to some extent in the twelfth century, then partly abandoned under strict occupation of central medieval logical territory—namely, qualification of Simplification.

Whereas Simplification and its effects have been under suspicion since ancient times, Transitivity, the other likely casualty of rectified nonsubstitutional approaches, enjoyed unquestioned standing in the ancient classical world. The leading nontransitive idea is that the trouble in the paradox argument comes not from the immediate steps, but with chaining them together by Transitivity. The trouble, or alleged trouble, with transitive chaining is regularly traced to epistemic sources, especially by those mainstream proponents who regard the 'trouble' not as genuine logical trouble, but as trouble for certain users and uses of the arguments. Such questions about Transitivity, though with medieval antecedents (for example in Strode and in Sermonete), are largely modern issues, flowing from the epistemologization of logic and of thought generally (the broad source of intuitionisms and idealisms). As typically doubts about Transitivity arise from epistemological springs, so normally epistemological currents motivate *nontransitive logics*. Thus, for instance, it is claimed that a 'good implication' enables natural (epistemological) transit from the antecedent to the consequent, not transit simply (as with material implication) on the strength of knowing the falsity of the antecedent or knowing the truth of the consequent; but a 'good' implication, so epistemologically characterised, does not conform generally to transitivity. With Frege and others came, however, a strong reaction to the modern psychologization of logic. Objections to classical logic generally were seen by Russell and most logical empiricists, quite erroneously, as drawn from epistemological sources (the way they were disposed to view modality also), and were accordingly dismissed—from *within* the broader epistemologization setting—as part of the damaging epistemologization of logic. With notable exceptions such as intuitionism, wrongly so dismissed. Still, one of the regular and tedious reactions from tougher promoters of classical logic to objections to the 'paradoxes of implication', remains of course that these objections rest on a confusion: a logical/epistemological confusion.[41] Such an argument from dichotomous confusion, worthless against most sociative targets, is too blunt an instrument to cope with the finer features of nontransitivism. While there are, naturally, perfectly good logical routes, by deducibility, along long classical chains, which epistemically ideal creatures would grasp without hesitation, ordinary mortals might well not grasp these chains as a whole or from end to end. But the chain effect is achieved by Transitivity. Accordingly, iterated uses of Transitivity fail epistemologically (the theme obtains reinforcement from Sorites paradoxes, where a chain effect multiplies indiscernable or barely discernible differences into conspicuous differences). While obviousness undoubtedly gives out under long chaining—hence one source of the considerable capacity of deductive arguments to yield new knowledge (as opposed to new logical content)—it is hardly convincing to claim that a five-step argument from an explicit contradiction to any proposition whatsoever is so long that evidence has been irretrievably lost along the way. While it is true that the implication (or inference) from $A \,\&\, {\sim}A$ to B is not *immediate*, except in provocative resystematisations, explication of immediate implication (or of one- or three- or five-step inference) is not the objective. Implication just is the closure of immediate implication under such operations as chaining.

There has been a series of more recent attempts to repair such an epistemic grounding for nontransitive logics; amusingly, main earlier approaches as elaborated by Lewy in fact ended up underwriting first degree relevant logics, i.e., non-dissembling logics, rather than nontransitive ones (see RLR I on Lewy). These earlier approaches started once again from some account of an epistemically acceptable implication $A \to B$, as being one where an epistemic agent could not come to know $A \to B$, recognise its truth or whatever, simply through coming to know B or through coming to know ${\sim}A$. As it happens, by a little variation, such an account (already doctored to

---

[41] The distinction is heavily deployed, for instance, by Iseminger against relevance logic. Of course that isn't the only supposedly dichotomous and heavy distinction that is regularly deployed in accommodating or in attacking critics; the object language/metalanguage distinction, subsequently concocted (through again with isolated historical anticipation), was also worked very hard. Any established paradigm gets surrounded by a battery of distinctions, which are regularly deployed by those faithful to the paradigm.

avoid the assumed unknowability of contradictions) can be doctored to yield different sociative outcomes, for example parts of relational logics.

Though relational logics tend to fall among nontransitive systems, as that is where the main forms recently advanced belong, strictly they yield a cross-classification of the typology offered. For how such logics break the paradox arguments, *if* they do, turns on the condition the relation involved satisfies. If the relation is transitive as well as reflexive and symmetric, then a relational logic based on a modal (or classical) logic collapses, by the paradox arguments, back into modal logic. Something else has to give, for example symmetry, in which case resulting relational logics may well resemble containment logics, i.e., nonadditive forms.

The nonadditive idea is that the trouble with the paradox arguments sets in with the tacking on of B which may be no part of the content of A; with the addition of an arbitrary, perhaps irrelevant B (as with $A \rightarrow. A \vee B$), and with the widely repudiated $A \rightarrow. A \,\&\, B \vee. A \,\&\, {\sim}B$). The main forms of *nonadditive logics* so far investigated are systems in the vicinity of Parry's systems of analytic implication. It is now known that these systems, directly tied to modal logics, form a rather special class of nonadditive systems.[42] The motivation for these systems is also tied to what would have been counted, in the bad old days, as epistemological considerations, and would perhaps now rank as semantic or pragmatic matters. Whether the motivation, in terms of containment of content and concepts, set out first by Parry for analytic implication, ranks as epistemological or not, it does not withstand careful examination if intended for a theory of deducibility, implication or conditionality (see RLR I, p.96 ff).

Even so, a theory of containment type is important for other purposes, as will start to appear, both traditional logical purposes and new purposes. That is not to suggest that 'analytic implication' is a theory with ancient historical roots. While some of its concerns, such as content, are old, it is a modern theory; its rejections are modern. For as well as jettisoning Rule Antilogism, it rejects Contraposition, an implication principle largely unquestioned until modern times. Of course a nonadditive logic, different from analytic implication, can retain Contraposition; but then either it will be a nonsimplifying logic also, and so already spoken for (connexive logics typically reject or qualify Addition), or it will break now standard connections between disjunction and conjunction through negation, such as ${\sim}({\sim}A \vee {\sim}B) \leftrightarrow . A \,\&\, B$ and the like ('extended De Morgan' laws). There are

42 *A somewhat more general class of containment logics is introduced in Chapter Seven. [Editors]*

many precedents, reaching back apparently to Stoic times, for abandoning the latter linkage. Apart from a period in later medieval logic, it is only really in this century that disjunction has regularly been treated truth functionally along with conjunction. But while an intensional construal of disjunction makes for interesting complications and defences (for example that in ENT of Disjunctive Syllogism as involving an intensional linkage), it does not materially affect the generality of the discussion, or treatment of paradox removal. For the whole previous argument and typology can simply be repeated with a new connective, $\cup$ say, defined $A \cup B =_{df} \sim(\sim A \,\&\, \sim B)$. At most there will be requirements on negation (now in doubt under intuitionism and its duals) if $\cup$-disjunction principles are to be derived from conjunctive mates.

Nondissembling logics are not entirely a contemporary innovation; Disjunctive Syllogism (DSyll) does not enjoy the uncontested history the (broad) classical paradigm mythologically ascribes to it. At many times in the history of logic, it was at the very least under a cloud (as Chapter Five begins to reveal). There is a significant contemporary literature questioning or rejecting DSyll. It is questioned, for example, in Jeffreys 1942, for essentially the right reasons, the same as those advanced in Cologne and elsewhere in the fifteenth century. Namely, if both A and $\sim A$ are assumed, then A cannot also be used to knock out $\sim A$ in $A \vee B$ to arrive at B. Nor, in twentieth century systematisations of non-syllogistic logic is the rejection of DSyll something recently concocted by Ackermann and seized upon by Anderson and Belnap and others. DSyll was, rather accidentally omitted from an early relevant logic formulated by Lewis, and it was rather more deliberately omitted from the system of Halldén 1951. Halldén's observations, which elaborate upon Duncan-Jones, hit important sociative nails on the head. But Halldén did not follow through on his formal innovations, which in fact adopted a combination strategy, of rejecting both Adjunction and DSyll principles, specifically Antilogism.[43] Thus his very weak system, $S0$, meets

[43] In early contemporary days there was considerable confusion, well exemplified by Duncan-Jones 1935, as to how much of classical (or modal) logic had to be 'given away' to avoid paradox. (Thus too the approach, which should have placed more emphasis on working upwards from correctness, was defective.) Such oppositional muddlement, characteristic of Anglo-American philosophy when Moore was influential, of course made the crude mainstream cause so much the easier. Duncan-Jones requires a meaning connection for entailment but says nothing explicit about relevance; where p entails q then 'q arises out of the meaning of p'. While in principle a modal account of 'entailment' is not so excluded, Duncan-Jones is rightly and Moorishly certain that q does not arise out of the meaning of $p \,\&\, \sim p$. He is also certain that Antilogism must go, but quite unclear as to where to stop the modal rot, the 'independent arguments' of Lewis. In fact he attempts obliteration bombing of the

both relevant and containment requirements. It is thus a relevant connectional logic, a type of much interest (further investigated in Chapter Seven), but representing overkill from a paradox-removal objective. As it happens, DSyll is not rejected just by relevant logics; in its main form it is also contested by both minimal and intuitionist systems, in a way that interestingly parallels relevant rejection (cf. Chapter Thirteen, section 2), and it is jettisoned in main discussive systems (cf. OP, p.50).

Standard relevant logics represent a maximal-style of nondissembling logics—maximal in the inexact sense that all first degree principles other than those in the immediate vicinity of DSyll are retained. A more exact characterisation of relevant logics, as conservatively extending Distributive Lattice Logic (system DLL), comes to much the same (the characterisation is elaborated in RLR I, p.153 ff). Plainly, then, there are nondissembling logics which are not standard relevant logics. But, apart from the significant class of relevant logics, these other types all fall under classes already considered, as for instance relevant connectional logics fall into the nonadditive class.

The classification of sociative logics elaborated is primarily orientational, not critical. Requisite critical work can be found elsewhere (see especially again RLR I). On some of the types of sociative logic discerned, little or no investigation has been carried out. This is true not only for types which are admittedly not of great interest other than for limited technical purposes (such as those dropping commutativity of conjunction), but also for types of long-standing historical interest, such as connexive logics.

---

logical landscape, doing *all* the following things: exhibits strong connexivist leanings with the claim that $p \& \sim p$ does not entail $\sim(p \& \sim p)$, aligns himself with nonadditive practice in forcing out Contraposition along with Antilogism—practice he reneges upon at the very end by conceding that it may be possible to get cases of Contraposition back again—and leaves room for a nondissembling approach, by not endorsing Disjunctive Syllogism. More clarity about entailment, its 'analysis' (or 'nonanalysis') and properties, had to await a different philosophical tradition.

5 A Preliminary Western History of Sociative Logics

This essay originally appeared in 1989 as chapter 4 of BG. *[Editors]*

Contemporary logical investigations enjoy the advantage of vastly improved logical technology as compared with all earlier terrestrial times. Yet, by comparison with earlier periods of high logical activity, the twentieth century is anomalous in its heavy mainline concentration upon classical logic, and, as a result, appears stodgy and unadventurous. For the deadening effect of the wide educational imposition of a narrow and intellectually disastrous dominant logical paradigm, classical logical theory, has (again) destroyed much logical expression and adventure. Rival logics have become very much a minority and esoteric activity, not even incidental to the serious affairs of life; no longer do even the city crows converse over logical issues. The full flourishing of sociative logics, in their rich variety, has yet to occur.

So far as we know, there have been three main periods in the long history of Western logic when the central issues of logic, as to what makes an argument valid, when deducibility obtains, and whether these connections can be captured in true or necessary conditionals, have been vigorously discussed. The periods are these: around the third century BC when Stoic logic flourished, in the medieval period, especially the twelfth century AD, and in the present century.

The logical investigations carried out in these three significant periods are, thus far, substantially independent. The Stoic enterprise of the third century exerted little or no influence on medieval thought, and indeed details (such as they are in Sextus Empiricus) were not available until after the seminal work of Abaelard's school and rival schools had already been accomplished. Of the main contemporary strands of sociative logic, only connexive and nontransitive logics have clear historical representation, and even there main investigations have proceeded substantially independently of historical inputs. When history has been appealed to in support of relevance logics, for example, it has been rather peripheral and, too often, historically dubious.

Though the main historical settings for sociative logic presupposed a heavy consistency assumption (in particular throughout medieval times), and though paraconsistent logics tend to be missing in expected areas of application (such as treatment of semantical paradoxes), nonetheless the history of sociative logics is deeply interwoven with that of paraconsistent logics. One major reason for the intertwining is of course that a crucial issue for sociative logics is what—by contrast with strict and classical spread and collapse—happens with impossible premisses and assumptions. How is loss of connections to be avoided there? A special section of the theory of obligationes (or suppositional reasoning and commitment) was devoted to this issue in the Middle Ages; and a similar division of research, plainly parasitic on classical logic however, can be seen in contemporary North American research (such as that of Rescher and Brandom and of Woods and Walton). These pretty unsatisfactory ways of shunting off, and sidetracking, significant logical problems fortunately by no means exhaust feasible lines of approach, as the rich history of paraconsistent logics helps disclose. That history has already been documented, admittedly also in a very preliminary fashion, elsewhere (especially in PL and its extract OP, which should be read in conjunction with this material—conversely, this material enriches PL, which is scanty on several topics of relevant interest). The entertaining story, which overlaps and complements the history of sociative logics, will not be repeated, but elements of it will be drawn upon where appropriate (and readers who seek a fuller picture to begin upon their own investigations should consult that story, as well as, of course, but cautiously, standard texts).

Influences in historical investigations run, so to say, two ways. Not only can historical work influence later work, though almost invariably in a highly selective way; thus, for example, the influence of Aristotle's work and Stoic logic on the contemporary path of connexive logic (for example in Angell 1962 and in McCall 1967). Later work can also have a heavy impact on the interpretation and construction of earlier work, which is often read as anticipating the moderns or is boxed in by later conceptions of the alternatives. The mid-twentieth century interpretation of Stoic logic affords a striking example. Both Mates 1961 and Bocheński 1961 attribute to Chrysippus a theory of strict implication—despite his explicit requirement of connection—seemingly because that was the only box they had available into which to fit him! And Mates' tale that Chrysippus is committed to 'strict implication' (p.4) is simply reiterated in a sequence of subsequent writers (for example Bocheński 1961, De Rijk on Abaelard, and Gould 1967). Now that a much wider range of logics is recognised (than was apparently available to Mates), rather more justice can be done to original logicians like Chrysippus. But, of course, while we still lack any satisfactory picture of the scope of logics, and while many quite basic logical notions remain without adequate investigation, the distinct possibility of pushing ancients, and especially original thinkers, into preconceived boxes which they do not fit, remains.[44] Stoic logicians are by no means the only casualties from classical times. Aristotle himself, subsequently the towering figure in Western logic (also 'the author of darkness' according to followers of Ramus), may also have been a casuality, as will soon appear.

The misrepresentation of prominent figures in the history of logic is one aspect of mythologic; another is the setting up of a myth supporting dominant logic, confirming its supremacy and all-encompassing character, and illustrating its historic progress. The history of medieval logic, especially, has suffered from this sort of Whiggish treatment. Yet another aspect of mythological distortion in favour of the conventional wisdom consists in the elevation of thinkers, perhaps much more minor in their times, who appear to anticipate elements of the present dominant framework, and to neglect or demote those pursuing rival themes. Naturally imperialism in logic has much in common with imperialism elsewhere; mythologic, an approved progressive (Whiggish) history of logic, is but a part of the pattern of domination. Control of or influence upon the subject and its presentation through appointments, patronage, journals, publications, grants and support, are other features, as old as the history of logic. While these devices are not always successful, particularly in the longer term, they have impacted on the lives of many significant logicians, especially alternative thinkers such

44 The positions of original researchers are mispackaged not only because of lack of boxes, but because of the presuppositions researchers and historians brought and continue to bring with them. For instance, the recent American history of the long debate on universals has been *seriously* distorted by the ontological assumptions that the Americans typically bring with them. The result is that significant positions get eliminated from consideration; for instance, those according to which universals are perfectly good objects of their own sort, not names, not mental constructs, not something else, which furthermore do *not* exist. Each generation needs to rewrite history not merely because of new informational bases (which may represent contractions on former bases), but to get past the presuppositions of predecessors (only of course to impose some of their own).

It should be evident that the present minority sketch of some logical history—which *is* highly selective, with a view to beinging out some of the neglected sociative tradition—has presuppositions of its own. To some extent, usual Whiggish rankings are reversed. For example, a Pseudo-Scotus (or Buridan or another), who tries to make the paradoxes of implication an inescapable feature of the consequence relation, is not here seen as making a significant advance on Abaelard who is committed to some sort of paradox-free connexive structure; Pseudo-Scotus so far from representing historical progress takes us step backwards. Not only pre-moderns get this sort of vulgar (Whiggish) treatment and distortion; more recent examples are Bolzano and MacColl, Peano and Cantor.

as Abaelard and MacColl and dubious ones such as Ockham; those affected include, appropriately enough, Aristotle (for details, consult what information we have on their lives).

5.1 The initial classical period: Aristotle and early Peripatetic logic

The Western history of sociative logics undoubtedly reaches back well before Aristotle. It is sociologically implausible to suppose that workaday logic sprang up, full-clad like the goddess Athena, in a single philosopher. More substantially, we have solid indirect evidence that many of the pre-Socratics and Sophists were bound to adhere to nonstandard logical theories, in order to support their strikingly unorthodox philosophical positions, positions which were, in several interesting cases, apparently paraconsistent. For details see PL, p.5 ff). Given the tight control Greek thinkers liked to keep upon assumptions and arguments, given that they did not appear to approve or tolerate regular suppression of assumptions in argument practice, it is likely that the implicit logics underlying many of their very distinctive positions were not merely paraconsistent, but sociative. The positions involved include not only those of several Sophists, but also of pre-Socratics like Heracleitus and of later schools such as the Megarians. But such claims as to sociative, or other, underpinnings seem bound to remain speculative; thanks partly to celebrated barbaric ancestors, there is insufficient literature surviving to enable much worthwhile logical reconstruction.

The significant surviving literature for the recorded history of sociative logics, begins, like much else, with Aristotle. Nothing as crude as classical logic (even in 'natural deduction' disguise) is to be found in Aristotle's work. Rather there is a formally elegant and tightly controlled theory of syllogism, which meets demanding standards of relevance between premisses and conclusions; and there are some proposals for extending that theory, for example in intensional directions, the case of modal syllogisms being worked out to some considerable extent. But there is also, coupled with this syllogistic theory and reduction of figures, the rudiments of a theory of argumentation, of good and fallacious argument, several elements of which were very important for the subsequent direction of sociative logics.

Included among these elements are Aristotle's account of progressive reasoning and argument, his theory of topics, and his theory of fallacies. Another, which helped shape recorded logical work on implication, especially that of Boethius in the fourth century and the sustained investigation of twelfth century, is Aristotle's apparent commitment, certainly taken for granted in the twelfth century, to connexive principles, that is to certain nonstandard principles of conditional argumentation. To this limited extent, Aristotle is the originator of connexive logic.

While Aristotle offered no systematic treatment of conditional statements, despite heavy use of them in his logical theory, and no specific propositional logic, he was responsible for the directions taken in early medieval discussion. Most important was the connexive principle of the *Prior Analytics* (II4 57b43): '...it is impossible that the same thing should be necessitated by the being and by the not-being of the same thing. I mean, for example, that it is impossible that B should necessarily be great since A is white and that B should necessarily be great since A is not white'. In Boethius's formulation, the principle became, 'it is impossible that the same thing is by necessity from something when it both is and is not'. Demodalised, statementalised, and put in implicational formulation, this is

AR1. $\sim((A \rightarrow B) \,\&\, (\sim A \rightarrow B))$

Aristotle argued for this principle, apparently in a quite general way,[45] effectively as follows:- Suppose on the contrary, $(A \rightarrow B) \,\&\, (\sim A \rightarrow B)$. Then as $A \rightarrow B$, $\sim B \rightarrow \sim A$, by contraposition. Since also $\sim A \rightarrow B$, by transitivity, $\sim B \rightarrow B$, something Aristotle took as absolutely impossible. AR1 then follows by contraposition, from the simpler (equivalent) principle

AR2. $\sim(\sim A \rightarrow A)$

These difficult principles, too quickly set aside in contemporary logic, bedevilled early medieval investigations.

---

[45] Though the argument proceeds through an example, the principle involved is stated in a quite general way. Certainly the principles and its derivation were taken to be unrestricted in Boethius and in medieval authors, both those who adopted it and those who later criticised it. But see Montgomery and Routley 1968 on the ways of evading generality here. Thus, while Aristotle's work permits of, and mostly encourages, a connexivist interpretation, it is not entirely decisive.

The argument presented for AR1 also makes some defensible assumptions about Aristotle's notion of negation. Indeed the issue of negation is critical to the question (see NC). By and large, Aristotle's discussion of negation notions supports a cancellation story and connexive themes. For instance, 'Affirmation is the statement of something of something. Negation, the statement of something [as subtracted] *from* something' (28b, 17a15-6). According to Martin 1991, in Aristotle's discussion of repugnance, an external negation is used to cancel the negation of a consequent and the result is claimed to be equivalent to an affirmative hypothetical; i.e., effectively $\sim(A \rightarrow B)$ is equated with $(A \rightarrow \sim B)$ as in hyperconnexivism.

Aristotle's theory of syllogistic and its surrounds involved, implicitly, a propositional logic of some sort, which was applied, for example in deriving some arguments and figures from others. While it seems unlikely that any such logic can be precisely determined, especially as many inessential principles are optional extras (which might have been selected by Aristotle, *if* he had considered the matter), certain minimal systems can presumably be tentatively figured out, and some bounds imposed. A minimal logic in $\{\rightarrow, \&, \sim\}$ would no doubt include, in some form, the principles tabulated in table 5.1.

The 'certainly applied' forms yield, ideally after a little organisation, a minimal logic—of a *sociative* sort. Moreover, this minimal logic represents, while connexivity stands, some sort of connexive logic. The point applies not merely to what it includes, for example also $\sim(A \rightarrow \sim A)$, but what is apparently omitted and *not* endorsed in Aristotle, for example cases of $A \rightarrow A \& A$, of $A \& B \rightarrow A$, and no doubt of $A \rightarrow A$. Thus the minimal logic differs in *two* important respects from contemporary connexive logic. Firstly, it appears to be progressive, or noncircular, excluding mere repetition—a *Peripatetic* feature. Secondly, Aristotle's implicit propositional logic appears to be significantly weaker and more rule-oriented than better-known connexive logics, in particular than those which have been shown to be adequate for the quantificational representation of the full theory of syllogistic (independently by McCall 1967 and by Angell 1962). However, substantially weaker logics than those deployed would serve for such a contemporary reductionistic representation; and weaker logics again would suffice if, more authentically, syllogistic forms are taken as primitive and not quantificationally characterised.

There are several reasons, apart from connexivity, then, for inferring that Aristotelian propositional logic differs not only from strong connexive logics, but also differs markedly from mainstream logics. Firstly, Aristotle's presentation includes none of the higher degree principles, listed as unlikely forms. Indeed such forms scarely enter the logical record, except by occasional accident, until contemporary times. The point is enough to defeat recent formal reconstruction of syllogistic theory. Secondly, there is Aristotle's much quoted account of what [conclusive] reasoning consists in, presented at the beginning of the *Topics* (100a 25): 'Now reasoning is an argument in which, certain things being laid down [perhaps as assumptions], something other than these necessarily comes about through them'. The account, repeated in essentials elsewhere, appears to make reasoning, the main type or argument upon which Aristotle concentrates, irreflexive. The

Table 5.1 Distilling Aristotle's implicit statemental logic

Principle	*Certainly Applied Forms*	*Likely Accepted Forms*	*Unlikely Forms*
Transitivity [b,c]	$\dfrac{A \to B,\ B \to C}{A \to C}$	$(A \to B)\&(B \to C)$ $\to. A \to C$	$A \to B \to. B \to C$ $\to. A \to C$ $B \to C \to. A \to B$ $\to. A \to C$
Contraposition [a]	$\dfrac{A \to B}{\sim B \to \sim A}$ and variants	$A \to B$ $\to. \sim B \to \sim A$ and variants	
Connexivity [a]	$\sim(\sim A \to A)$	$(A \to \sim B)$ $\to. \sim(A \to B)$ and variants	
Non-Contradiction [a]	$\sim(A \& \sim A)$		
Modus Ponens [b]	$A,\ A \to B \ /\ B$	$A \& (A \to B) \to B$	$A \to. (A \to B) \to B$
Reductio Forms (Modus Tollens [a])	$\dfrac{A \to B,\ \sim B}{\sim A}$ (from Contrapos.)	$(A \to B)\&(A \to \sim B)$ $\to. \sim A$ $\dfrac{A \to B,\ A \to \sim B}{\sim A}$ (from Contrapos. and Transitivity)	$A \to B \to. A \to \sim B$ $\to. \sim A$
Double Negation [c]	$A \to \sim\sim A$ $\sim\sim A \to A$		
Antilogism [c] and "Disjunctive" Syllogism	$\dfrac{A \& B \to C \quad A \& \sim C}{\sim B}$	$\dfrac{A \& B \to C}{A \& \sim C \to \sim B}$ $A \& \sim(A \& \sim B) \to B$	$A \& B \to C$ $\to. A \& \sim C \to. \sim B$
Adjunction	$A,\ B \ /\ A \& B$		$A \to. B \to A \& B$
&-Replacement	$\dfrac{A \& B \to C,\ D \to A}{D \& B \to C}$ $\dfrac{A \& B \to C,\ D \to B}{A \& D \to C}$	$(A \& B \to C)\&(D \to A)$ $\to. D \& B \to C$	$A \& B \to C \to. D \to A$ $\to. D \& B \to C$; etc.
Praeclarum	$\dfrac{A \to B,\ C \to D}{A \& C \to B \& D}$	$(A \to B) \& (C \to D)$ $\to. A \& C \to B \& D$	
Antecedent Commutation [c]	$\dfrac{A \& B \to C}{B \& A \to C}$	$A \& B \to C$ $\to. B \& A \to C$	

KEY

a Explicitly formulated principles (cf. Bocheński 1961, p.97, who also cites the modal principle $A \to B \ /\ \diamond A \to \diamond B$).

b Fairly general formulations are stated (cf. Bocheński 1961, p.89).

c Consciously presupposed rules, according to Bocheński 1961 (see p.77, p.78). (It has been suggested that some principles Aristotle used regularly, such as Antilogism, apply only to syllogistic reasoning and theory. But in Aristotle's terms that strictly makes no qualification, since syllogistic, as distinct from term logic, embraced the whole of logical reasoning.)

striking result is that Identity, $A \to A$, fails. (Note that this implicational failure does not imply the breakdown of the traditional law of thought, $A = A$, Identity proper.) There is indeed considerable independent evidence that Aristotle is concerned with *progressive* reasoning, which would certainly reject $A \to A$. The fallacy of 'begging the question', which he is concerned to emphasize as involving *invalidity*, says as much. So does his stress on reasoning as leading to new knowledge. In fact the fallacy delivers more, leading to the rejection of degenerate syllogisms with an 'identical' premiss (such as when A belongs to B and B belongs to B, then A belongs to B) as begging the question (65a 10). Thus Simplification in the form $C \,\&\, D \to C$ is also rejected, at least sometimes, as would be now expected on connexivist principles. Perhaps Rule Simplification, $C \,\&\, D \,/\, C,\ D$ is retained—it would simplify the theory of inference (while upsetting ancient requirements on the number of premisses)—perhaps not.

Consider, furthermore, the immediate historical development of Aristotle's account of reasoning, that by the Peripatetics. They almost certainly construed reasoning as progressive, and rejected such conditionals as $A \to A$ (for example "If it is day, it is day": see below). The comparative complexity of a logic of progressive reasoning—a matter still by no means worked out properly now more than 2000 years later[46]—does something to account for the fact that the Peripatetics did not, so far as we know, formulate a propositional logic which can easily be compared with Stoic propositional logics.[47]

Nothing in Aristotle's theory commits him to the tenets of contemporary mainstream logic, such as the classical logic Łukasiewicz and many others of note would foist upon him in reconstructing and axiomatizing or natural-deductionizing syllogistic.[48] Not only is Aristotle's connexivism incompatible with such logics, and the room he makes for future contingents

[46] For some of the difficulties see Martin 1983, and Thom 1981, p.243 ff.

[47] There was more to it than that. There was also a much discussed rivalry between Peripatetics and Stoics; on this intriguing feud, touched upon below, see for example Frede 1974.

[48] Thus a rich, and not easily absorbed, new satellite city of literature upon Aristotle's syllogistic, mostly bent upon fitting it within mainstream logical theory, is by-passed—conveniently for energy—and time-expenditure. Both the axiomatic and natural-deduction approaches, now often presented as exhaustively competing to explicate and formalise Aristotle's theory of the 'assertoric' syllogism, would foist upon Aristotle principles and procedures there is little or no evidence he accepted, or would have or should have accepted. Aristotle's underlying or operational logic, or 'propositional logic', certainly was not that which Łukasiewicz ascribes to him, classical two-valued logic or alternatively, if future contingents are duly allowed for, a three-valued logic. Nor was it any many-valued logic neatly contained within classical logic. Nor was it such a logic reformulated in a natural deduction

apparently incompatible with principles of classical logic; but more, the whole tenor of Aristotle theory of reasoning and argument, from initial definitions on, is inimical to the highly suppressive features of mainstream logics (see further RLR I, p.141 ff). The step-by-step progressive elaboration of necessary truths typical of elementary geometry, was a leading model for conclusive reasoning in classical Greece (cf. Kneale and Kneale 1962, p.3 ff). It would not have been at all readily conceded by Aristotle, or by most other Greek logicians, that these demonstrations could be truncated, since any necessary truth yielded any other. Nor did Aristotle grant that a contradiction leads anywhere; to the contrary he argued, in his famous defence of Non-Contradiction and of consistency themes, rather that contradictions led nowhere, nullifying themselves (see further Łukasiewicz 1971 and OP, p.28). In short, it seems clear that the paradoxes of implication would have been roundly rejected by Aristotle. Nor does his decidedly relevant theory commit him to them.

---

guise. All those include not only implicational paradoxes and irrelevance, and associated bad news such as the deduction theorem (an integral part of any 'natural deduction' approach, and a straightforward outcome of logics favoured on axiomatic approaches); but they also include a variety of circular statements derivable in the run of syllogistic systems, such as "All A's are A's" and "If all A's and B's and all B's are B's then all A's are B's". As Corcoran, one of the leading proponents of the ill-named 'natural deduction' approach, admits, statements such as those that he represents as "Axx" obtained mention nowhere in Aristotle's writings. (But Corcoran, apparently oblivious of progressive reasoning, uses *this* to suggest that 'Aristotle had no idea of logical axioms' or 'logical truth', Corcoran 1973, p.216!)

As to rule or conditional formulations of syllogistic, evidently Aristotle and the Peripatetics deployed both (certainly if the conjunctor "therefore" is taken as signalling a rule). But the rule formulations are not those of 'natural deduction' in a contemporary sense; nor are the conditional formulations suggested strictly axiomatic in a modern sense. No doubt it is feasible to develop a theory of syllogistic, much like Aristotle's which furthermore does not presuppose propositional logic (cf. the propositional-free formulation of algebras in RLR I, p.117). Such a system may or may not be axiomatic; it could, for instance, be 'genetic', in the way mathematics was before the fuller rise of the axiomatic method last century.

The mixed strategy of Thom 1981, which aims to combine and duly restrict the satisfactory elements in the main approaches, both rules and axioms, no doubt leads to a reconstruction of non-intensional syllogistic significantly closer to Aristotle's theory of non-modal syllogisms than the work he builds upon. The reconstruction properly renders the theory both progressive and relevant. But the attractive reconstruction in the first part of his text is not sustained in the later part of the text, where Thom's own propositional theory is elaborated. There (in fine accord with a proclaimed dogmatic procedure, Thom 1981, p.11), connexivism is out, progressivism is out, identity is in, etc. Thom loses touch with Aristotle and Peripatetic traditions, but surfaces bearing the implicational part of relevant system $R\text{-}W$—no great gift, but at least weakly relevant. More authentic is the approximation to Aristotle's implicit propositional logic which Slater 1979 proposes. But Slater's approach remains unduly algebraic and reductionistic, perhaps collapsing altogether. The problem of more satisfactory reconstructions remain open.

On the contrary, the notion of connection, as required in demonstrative and hypothetical reasoning, surfaces at several points in his work, particularly in this theory of fallacies and sophistical refutations. The fallacy of non-cause as cause in both its original and subsequent forms affords a striking example. According to Hamblin 1970,

> *Sophistical Refutations* refers quite literally to the Fallacy as the one 'about the non-cause'..., but makes it clear that a logical interpretation is intended by, later in the book, referring to it merely as 'insertion of irrelevant matter'. Irrelevant matter can be inserted in an ordinary argument without prejudice, but it is methodologically dangerous to permit it in a *reductio*. However, some later writers objected generally to insertion of irrelevant matter: cf. the Stoic Fallacy of Superfluous Premiss (p.79, incorporating a fn).

The logical bearing of the fallacy is almost immediate if the wide sense of "cause" is recalled, which encompasses reason. Reexpressed in terms of reason, the fallacy is, according to the initial reference, that of 'stating as reason what is not the reason' (166b 27). As Aristotle goes on to explain, insertion of what is not a reason (a 'false cause', 'irrelevant matter') in an argument *ad impossible* is fallacious. It is not just 'methodologically dangerous', but unsound. Hamblin's treatment here, as elsewhere (for example with begging the question), reveals his classical prejudices. For the type of argument Aristotle is rejecting is classically sound. Let B be false cause, inserted in a reductio from A, which is assured, to C, which proves impossible. Then the type of argument can be presented: $A \rightarrow C \,/\, A \,\&\, B \rightarrow C$; insertion of irrelevance (which *is* connexively invalid). But now C is impossible, and A holds by assumption; therefore, as $A \,\&\, {\sim}C \rightarrow {\sim}B$ by antilogising, ${\sim}B$. (Of course there are classic ways out of this little hole, for example A can no longer be maintained, since it implies C without B.)

Contemporary selection procedures, for instance in the face of logical paradox, illustrate the original 'non-reason as reason' fallacy quite neatly. These involve picking out one factor from a whole group of factors, as cause of the paradox, though it is perhaps irrelevant or not the reason as variants on the paradox show (thus for example Frege's 'way out' of the Russell paradox). What the fallacy of non-cause as cause quickly and subsequently became, as opposed to what it originally was, so Hamblin further explains is that of inflating a sequential (material) association into a cause or reason; it is the mistake of assuming that 'because B happens *after* A, it happens *because of* A' (*Rhetoric*, 1401b 30). It is thus very similar to the fallacious inference from $A \supset B$ (or its quantified, formal analogue) to $A \rightarrow B$. Like most natural language connectives, *because of* requires connection beyond material 'association'.

The sort of progressive connexive reasoning that Aristotle's underlying propositional practice appears to lead to, has yet to obtain proper investigation.[49] The relevant theory is neither connexive logic, as normally presented, because that includes redundancies such as $A \rightarrow A$ (and perhaps $A \& A \rightarrow A$ and $A \rightarrow A \& A$, etc.), nor straight progressive reasoning, since that need not be connexive. Rather, it is a *fusion* of these two separate sociative theories, both of which were to persist after Aristotle. The fusion itself also persisted and flourished, so it appears, firstly in the logical work of Theophrastus who seems to have elaborated and made more explicit Aristotle's implicit propositional, predicate and class logic, secondly in the theory of the long-maintained Peripatetic school, and, to jump further ahead, in variant forms in the later logical enterprises of Boethius and Abaelard.

5.2 The richer classical period: the great debate on implication

The ancient development of sociative logics really blossomed, in Alexandria, with a major debate about conditionals, a debate which was continued through the main era of Stoic logic. Progressive reasoning of some sort appears to be one of the four positions cited in the famous debate on conditionals, a position now commonly, and appropriately, ascribed to the Peripatetics. 'The Stoic controversy over implication was by no means restricted to the Philonian [material implication] and Diodorean [tensed formal implication] views. In a very interesting and important passage Sextus states and illustrates four distinct definitions which were discussed by the Stoics' (Mates 1961, p.47). It is the third and fourth positions that matter for sociative logic:

> [3] And those who introduce 'connection' or 'coherence' say that a conditional holds whenever the denial of its consequent is incompatible with its antecedent; so that, according to them, the above-mentioned conditionals do not hold, but the following is true: "If it is day, then it is day." [4] And those who judge by 'suggestion' declare that a conditional is true if its consequent is in effect included in its antecedent. According to these, "If it is day, then it is day," and every

[49] Some small beginnings only have been made on the task. As well as the items referred to in fn 46, see Martin and Meyer 1983 and RLR I, chapter 11.

> repeated conditional will probably be false, for it is impossible for a thing itself to be included in itself.[50]

The full passage gives the further information that the third position rejects mere material juxtapositions of truths, such as "if A then B" when both A and B are (contingently) true, and also rejects paradoxical features of Diodorean formal implications, namely conditionals which hold by virtue of always false antecedents or always true consequents. Regettably, the passage does not decisively inform us whether the third position, usually attributed to Chrysippus, also rejected, in its requirement of *connection*, other paradoxes of strict implication. Mates assumes not, and proceeds, on the basis of but scant evidence, to identify the third, genuinely Stoic position, with strict implication.[51]

But Diodorean implication constitutes a kind of *strict* implication. 'For, according to Diodorus, whatever is true for all time is necessarily true; thus, any conditional which would satisfy his requirements for truth would also satisfy his requirements for necessary truth' (Mates 1961, p.47). Putting Diodorean components together, the Diodorean account is simply: "if A then B" holds iff it is impossible that both A and not B, i.e., $\Box{\sim}(A \,\&\, {\sim}B)$ for Diodorean necessity $\Box$, i.e., *a* strict implication. Mates, however, goes to some trouble to evade what is thus obvious, and to set aside the prevailing view that Diodorean implication is 'the ancient counterpart of strict implication' (pp.49-51): the distorting assumption that there is only one strict implication is made throughout. He also tries to offset his own opinion that Diodorus's opinion would be that Diodorean implication is strict implication, by counterbalancing it with the claim that 'it is doubtful that any modern logician would recognise it as such' (p.47). Unfortunately for Mates, modern tense logic was in the making; it was not long before several competent modern logicians, Prior and his former students, were treating Diodorean implication as an interesting kind of strict implication.[52]

---

50 The original much-quoted passage comes from Sextus Empiricus 1960, Bk II, pp.110-112. Much of what we know about the on-going conditionals debate is conveniently collected in the Appendices of Mates' important *Stoic Logic*, i.e., Mates 1961.

51 Bocheński likewise suggests, wrongly, that 'connexive' implication, the implication of the third position, is an ancient form of *strict* implication (Bocheński 1961, p.119). But he offers no supporting evidence or argument; and it looks as if he is prepared to contrue 'strict' implication so widely that it includes connectional implications of a non-Lewis type. Gould (for example 1967, p.153), De Rijk, and several others also repeat Mates' misconstrual of the third position.

52 For details see, for example Prior 1957. Gould is, like Mates, confronted by the gratuitous problem of how to distinguish Chrysippus from Diodorus, given that both are

It seems decidedly unlikely that the great debate on conditionals was over *kinds* of strict implication, and much more likely that significantly different types of implication were under consideration. Most of the evidence points that way. Mates' case for treating 'the third type of implication' as 'the ancient version of strict implication' (p.49) is exceedingly weak. At bottom the case for a strict conditional consists in imposing a narrow *modal* interpretation on the crucial term translated as "incompatible" or "inconsistent".

> Judging from the position of this type in the list, which obviously was intended to proceed from weakest to strongest [note: The examples make it obvious], we are led to suppose that 'incompatible' is used in its ordinary sense, according to which incompatible propositions cannot both be true, i.e., their conjunction is logically false. The example bears out this interpretation (Mates 1961, p.48).

Even granting the inferred ordering of strength (which is not obvious and is in doubt, especially as to the fourth position), it is more than a little obscure how this can lead to an allegedly ordinary sense of "incompatible", a modal one! Perhaps it is taken to suggest that there is no evidence for "incompatible" being construed in an unusual way, that everything said is compatible with a strict interpretation. While everything said, including the example given, is so compatible, it is also compatible with more interesting rival constructions, such as a connexive interpretation (proposed by McCall) and a relevant interpretation (suggested in RLR I, p.83). Nor does the ordinary sense of "incompatible" independently underwrite Mates' modal reduction, of compatibility to conjoint possibility (i.e., of $A \circ B$ to $\Diamond(A \,\&\, B)$). According to standard English dictionaries, "incompatible" means "inconsistent with something else; incapable of subsisting with something else; incongruous" (thus *Concise English Dictionary*; similarly OED). Not only is incompatibility regularly of less than logical (or analytic) strength; more important, since the modal $\Diamond$ can also be construed as of less than logical strength, incompatibility (like incoherence) is thus ordinarily *relational* ('*with* something else'), in a way that the modal reduction, to a one-place functor $\Diamond$, removes. Let us accordingly represent incompatibility by the relational functor σ, and correspondingly compatibility by the more

committed to strict implication. He fails the test. For not only is what he says (on the last page of Gould 1967) implausible, insofar as not obscure; furthermore his modelling is simply defective. Note too, Cicero's argument, supposed to bolster the strict case, where incompatibility is replaced by impossibility of a conjunction (step (4) to (5) in Gould 1967, p.158) does not require anything like a strict account; for it only uses $A \rightarrow B \dashv\!\exists \sim\!(A \,\&\, {\sim}B)$ and $\sim\!(A \circ B) \dashv\!\exists \sim\!(A \,\&\, B)$.

familiar fusion connection $\circ$ (introduced by Lewis for consistency, but now mainly treated in a fashion that breaks free of his modal reduction). Incompatibility is uncontroversially analysed as the negation of compatibility, i.e., σ spills out as $\sim\circ$.

The third, Chrysippean position thus furnishes the connection

CD. $A \rightarrow B$ iff $\sim(A \circ \sim B)$,

(given that denial is represented as negation). Then the example, which Mates claims bears out his interpretation, falls out of *any* interpretation, such as main sociative ones, in which propositions are never compatible with their own negations. Moreover, Mates' assumptions that the ordinary sense of "A is incompatible with B" is "It is logically false that both A and B", $\sim\Diamond(A \,\&\, B)$ in standard symbols, is tantamount to the assumption that implication is, ordinarily, strict implication, and just as controversial (see further RLR I, p.361 ff). For $A \circ B$ iff $\sim(A \rightarrow \sim B)$ from CD, i.e., iff $\sim\sim\Diamond(A \,\&\, \sim\sim B)$ by the infiltrated strict equation, i.e., (by further traditional negation transformations) $\Diamond(A \,\&\, B)$. In sum, $A \circ B$ iff $\Diamond(A \,\&\, B)$. Mates' reasoning begs the interpretational question at issue.

The evidence against a strict interpretation of the third position stacks up much better than the slight evidence is its favour. In the first place, the Stoics, like many subsequent schools, considered that natural laws supplied true conditionals. Such conditionals do *not* admit of a strict interpretation (in the sense of Lewis and Mates)—though they may admit of a (still unsatisfactory) modal analysis when the modality $\Diamond$ is not interpreted logically, but rather naturally, for example as "it is scientifically possible that", "as far as natural laws go, it is possible that". Evidence that the Stoics exceeded strict logical bounds, both in their treatment of conditionals, and in their accounts of incompatibility, is assembled in several authors (for example Long 1974, p.144, Gould 1967). More detail than Long offers on the incompatibility typically being *empirical* is provided in Gould 1967 (p.154). Most of the sample Stoic arguments still extant involve contingent or lawlike connections, for example dark with night, milk with conception, etc., not logical ties. The incompatibility of "Fabius was born at the rising of the dogstar" and "Fabius will die at sea", for instance is not certainly logical. Gould 1967 calls it empirical, explains why it is contingent (since relying on an inductively established principle, but then ties it with natural connections and laws (pp.159-60).

In the second place, there is the fundamental requirement of connection, specifically imposed, which strict implication and all modal accounts wrongly abandon (except under low Priorean redefinition). Thirdly, interlinked, there is the crucial matter of invalid arguments. Two important classes of the four classes of invalid arguments comprise components which are strictly valid, i.e., valid according to the canons of modalism. These classes are important not only in countering strict interpretations and impositions, but also both in indicating the relevant character of Stoic logic and in delimiting its type. They include the following:-

> 1. Incoherent arguments ... are arguments which are invalid because there is no logical connection of the premisses with one another or with the conclusion.
>
> If it is day, then it is light
> Wheat is being sold in the market.
> Therefore, Dion is walking
>
> 2. Redundant arguments ... contain a premiss which is not necessary for drawing the conclusion.
>
> If it is day, it is light.
> It is day.
> Dion is walking.
> Therefore, it is light.
>
> If it is day, it is light.
> It is day.
> Virtue is beneficial
> Therefore, it is light.
>
> By all usual tests [sic!] these would be perfectly valid arguments, though inelegant. Perhaps Sextus made a mistake here, or perhaps he was following an inferior handbook (Mates 1961, pp.82-3).

In a similar caustic vein, Mates remarks that 'the principle, if any, which was used by the Stoics in their classification of invalid arguments is hard to detect' (p.82). Is it? Isn't the principle but a tight requirement of relevance, for use? Redundant or unnecessary or disconnected premisses are not needed, and need not be used, in reaching the conclusion; accordingly they are, under such a use criterion, irrelevant. (Nor are such extras always merely an inelegance, as mainstream positions would have; sometimes, as in connexive logics, they can result in triviality: see Montgomery and Routley 1968). Interestingly, Thom 1981 uses such a lack of variable sharing, as in 1 above, to characterise incoherence, a strong form of irrelevance, and specifically links it with one of the Stoic kinds of invalidity, incoherence, which he takes as a sort of relevance violation. But such a presentation in terms of relevance (which the Stoics do not explicitly mention) can be avoided. What matters is that arguments *without* connection of premisses with their conclusion, or even with one another, as in some forms of Simplification (for example cases of $p \,\&\, q \;\therefore\; p$ where q has nothing to do with p) or *with* redundancy of premisses, as in other forms of Simplification (for example $p \,\&\, p \;\therefore\; p$ and $p \,\&\, q \;\therefore\; p$ where q follows from p or vice versa), are *invalid*. As they are strictly valid, Stoic argument does not correspond to strict implication. Most important, arguments without connection of premisses with their conclusion are invalid (most unfortunately no examples decisive against strict theory are given, though the principle is clearly stated). Thus Stoic logic is rendered appropriately sociative.

Much more congenial to a sociative story than the presently entrenched strict interpretation of the third position is a connexive interpretation. Such a very different interpretation of the third position has been pushed by McCall, who (in several publications) contends that Chrysippean logic is connexive, and even that 'connexive logic represents an attempt to formalise the species of implication recommended by Chrysippus' (ENT, p.435). One countervailing fact is that many researchers arrived, rather independently, at distinctively connexive principles with little or no input from Stoic logic, in particular, the contemporary 'founder' of connexive logic, Nelson. Another problem for McCall's proposal is that the (excessively) strong systems of connexive logic, which he presents in his full proposals, are substantially irrelevant, and thus include Stoically invalid arguments. Moreover, if Chrysippus did espouse connexive logic, that information, long lost, has only recently been disclosed.

How does McCall, who starts out correctly with the Chrysippean definition CD, manage to ascribe connexive principles to Chrysippus? He pulls off this feat by importing, from nowhere, 'the plausible thesis that if A implies B, A is compatible with B' (McCall in ENT, p.435), i.e.:

M^c. $\quad A \rightarrow B \rightarrow . \, A \circ B,$

a thesis which at least imposes an interesting constraint on the problematic connection: namely, compatibility. From M^c it is an immediate step, applying CD (for example reformulated as a definition of $\circ$), to the connexive principle:

BT. $\quad A \rightarrow B \rightarrow . \sim\!(A \rightarrow \sim\! B)$

a principle found in Boethius and Abaelard, and called by McCall *Boethius's thesis* (though the more difficult converse, $\sim\!(A \rightarrow \sim\! B) \rightarrow . \; A \rightarrow B$, is what is really distinctive of Boethius's theory). What immediately yields an implausible principle is itself rather implausible; and indeed the proposition that M^c is plausible would not win much contemporary support (from those who understand it). M^c is itself a connexive principle. It yields at once, given identity, that *every* statement is self-compatible, *even* contradictions; and in any case, it tells us that whatever a contradiction implies is compatible with it! *It* stands in need of more independent support than an allegation of plausibility.

More relevant here is: what is the evidence that Chrysippus adhered to this 'plausible' principle? McCall adduces none, and there appears to be none. Rejection of the principle is compatible with all the information we have on the third position or on Stoic logic (as a relevant modelling will show). Furthermore, if the fourth, Peripatetic position was, as already hinted, a connexive position, then it is unlikely that the third position was also (though again the contrast made between positions could just have been between progressive and inclusive connexivism).

Nor is it simply that there is little or no evidence that Chrysippus and other third positioners would have accepted M^c or the distinctive nonclassical theses, BT and $\sim(A \rightarrow \sim A)$, of connexive logics. It seems that such ancients as the Megarians were committed to non-connexive implicational principles like $A \& B \rightarrow A$ (Simp) and $A \rightarrow . A \vee B$ (Add), which quickly lead to the rejection of M^c. As was known in the thirteenth century, $A \& B \rightarrow . A \vee B$, whence using normal negation principles again, $A \& \sim A \rightarrow \sim(A \& \sim A)$. Therefore, assuming M^c, $(A \& \sim A) \circ \sim(A \& \sim A)$, i.e., where C is $A \& \sim A$, $C \circ \sim C$, whence, $\sim(C \rightarrow C)$. But the Megarian position, like the third position, and by contrast with the fourth did accept Identity. Accordingly, McCall's assumption M^c has to be given up, on Megarian as on contemporary preconceptions. But it is unlikely that such a point of difference between Megarians and Stoics would have passed unnoticed. The type of connection Stoics like Chrysippus apparently insisted upon is not that of connexive logic so McCall-construed. What sort of connection was it?

An overlooked problem—overlooked because of the removal of *connection* with the strict equation—is this: *how* does Chrysippus's logic avoid the paradoxes and retain connection? It is a problem because the Stoics accepted, on the face of it, many of the principles involved in the derivation of the difficult negative paradox. Yet there is no indication that Stoic logicians confronted, or were aware of the problem, though the derivation was hardly beyond their considerable logical ingenuity. Why not? However it is approached, the matter calls for some extra-historical construction; various possibilities are open. What seems the most likely reconstruction will be developed here; other possibilities will merely be indicated. It needs remarking that Stoic logical systems, in which demonstrations were carried out, comprised 'indemonstrables', corresponding to axiom schemata and yielding rules, and 'themata', corresponding to further inference rules. It is not known exactly what these themata were; but, until some better grasp is gained upon what they were, there is no certainly about what can—and, more important here, what cannot—be proved in logical frameworks such as that of Chrysippus. As a result, assessment of the Stoic claim of the 'completeness' of their logic for instance, remains a difficult exercise.

The standard derivation of the paradoxical inference $A, \sim A \,/\, B$, which circuits through disjunction, is in fact easily averted in the Stoic scheme of things. Certainly *a* rule form of Disjunctive Syllogism was adopted in the fifth Stoic 'indemonstrable' (i.e., axiomatic first principle), namely:

5 *Either the first or the second; but not the second; therefore the first.*

But much turns on how "or" is construed. The evidence indicates that disjunction was *not* intended truth-functionally.

> A disjunctive proposition..., i.e., a proposition such as we express by the use of "or", was said to involve a complete opposition, ... of its disjunction ... . The expression we have translated by "complete opposition" must surely be understood to mean incompatibility, i.e., more than mere de facto separation (Kneale and Kneale 1962, pp.160-1; see also pp.162-3).

Moreover, Galen stated that the disjunctive statement "Either it is day or it is night" is equivalent to "If it is not night it is day" (p.162).
So construed, the fifth indemonstrable amounts rather to a variant on Modus Ponens, i.e., the first indemonstrable. Under such an intensional construal of disjunction, Addition (for example $A \therefore A \vee B$ and its mates) fails, even strictly. For A does not at all guarantee that not-A implies B or that not-A is incompatible with any arbitrary statement. Thus the standard demonstration of $A, \sim A \therefore B$ breakes down, as with containment logics (Parry logics and others) and certain relational logics, on Addition.

Another wave at once forms to overwhelm connectional interpretations of Chrysippean logic. For the paradox argument can be rerun without disjunction, intensional or other. The third indemonstrable:

3 *Not both the first and second; but the first; therefore not the second.*

an orthodox equivalent of a Disjunctive Syllogism rule, does not on the face of it admit of similar intensional reconstrual. For, by contrast with disjunction, the Stoics characterised a conjunctive statement as one which is true if both components are true, and false if one component is false, i.e., essentially truth-functionally (cf. Kneale and Kneale 1962, p.160, Mates 1961). It is important to observe that such an account does not make conjunction two-valued; the recipe is compatible with, and indeed adopted for, four-valued interpretations of relevant logic (under the American-plan for the semantics; RLR I, p.319). And it would equally fit with Stoic truth-theory, which appears to have been three-valued, allowing for neither true not false statements. Nor, even more critical, does the account of conjunction guarantee all the strict principles concerning conjunction. These depend also on the semantical accounts given of implication and inference. (For example, the account of conjunction does not guarantee, $A \rightarrow . B \rightarrow A \& B$, and quite rightly.) Still one fairly certain result has to be confronted: given known Chrysippean principles, not all strict, not even all normal, conjunction principles can be retained without sacrifice of connection. The paradox argument to this conclusion is given in implicational form, but it has an inferential analogue, which should be not lost sight of, because it is also damaging unless faulted.

The implicational form of argument to loss of connection runs as follows:

1.	$\sim A \rightarrow \sim(A \& B)$	by Contraposing Simp, $A \& B \rightarrow A$
2.	$A \& \sim A \rightarrow A \& \sim(A \& B)$	by Composition, Factoring, or Praeclarum, using 1
3.	$A \& \sim(A \& B) \rightarrow \sim B$	by the third indemonstrable (for implication)
4.	$A \& \sim A \rightarrow \sim B$	by Transitivity from 2 and 3.

The argument, together with what we think we know about what the Stoics accepted, leaves little room from manoeuvre. A form of Contraposition was the second indemonstrable, a principle that appears unquestioned throughout antiquity. Transitivity, although not an indemonstrable, was regularly applied and appears to have been a Stoic thema (cf. the third principle and accompanying discussion in Kneale and Kneale 1962, p.164). By elimination (Stoically admissible) then, the trouble lies either with composition principles (&-introduction group) or simplification principles (&-elimination group). While it is not entirely clear which must go, the latter seems the decidedly likely candidate.

There are several reasons for supposing that it is &-elimination (or Simplification) that fails, beginning with a theoretical consideration, that the building-up character of composition, as distinct from decomposition, can hardly lead to trouble through suppression of information. A more direct consideration is that the Stoics appear to have accepted composition principles (see Kneale and Kneale 1962, p.169). But there is considerable evidence indicating that they did not accept simplifying principles without some qualification. Firstly, there are matters concerning number of premisses (generally required by Chrysippus) to be amalgamated to one, through $A \& B \,/\, A,\ B$, and would admit redundancy by allowing $A \& B$ where A sufficed as premiss. At least in inference, the Stoics took inclusion of redundant or otiose premisses to be fallacious. Indeed the Stoic theory of invalidity appears to be decisive against Simplification, as we have already seen. Secondly, there is the matter of content, a notion important in Stoic logic. Implication involved connection (and probably *inclusion*) of content, but in premisses like $A \mathbin{\&} {\sim}A$, ${\sim}A$ undercuts the content of A, leaving no content to be transmitted in simplifying implications like $A \mathbin{\&} {\sim}A \rightarrow A$. But those are Peripatetic ideas; it is not certain that Stoics like Chrysippus would have accepted all of them. One Aristotelian principle the Stoics did accept is however almost decisive against Simplification. That is the principle of Antilogism, the first of the Stoic themata, assumed by Aristotle in his indirect reduction of syllogisms: 'If two propositions entail a third, then either of those two together with the negation of the third entails the negation of the remaining one' (Kneale and Kneale 1962, p.169, quoting Apuleius), i.e., in intended inferential form:

RAntil. $A \& B \rightarrow C \,/\, A \mathbin{\&} {\sim}C \rightarrow {\sim}B$

But given Simplification it is an immediate inference to disconnection, It is an inference regularly rediscovered in recent times (for example Nelson, Duncan Jones), which could scarcely have escaped the astute Stoics, who constructed some much more elaborate propositional proofs (see especially the Kneale and Kneale's 1962 reconstruction of Stoic theorems, pp.171-2). The argument simply takes C as A. Then, supposing $A \& B \rightarrow A$, $A \mathbin{\&} {\sim}A \rightarrow {\sim}B$, and also $A \mathbin{\&} {\sim}A \rightarrow B$ (putting ${\sim}B$ for B and applying double negation, i.e., 'super negative' principles as they were called by the Stoics, who would have "1", "2 ", "3" for "A', "B", "C" respectively). Whence disconnection. So by Contraposition, or by Reductio (of which the Stoics much approved), Simplification sometimes fails.

As the elementary *watershed* argument from RAntil and Simp shows, these are two strikingly different routes the quest for logical connection can take, the traditional Nonsimplifying direction, imposed historically by Aristotle's authoritative imposition of RAntil, and the very modern

Nonantilogising direction. An interesting cross-classification of sociative logics takes the following shape:

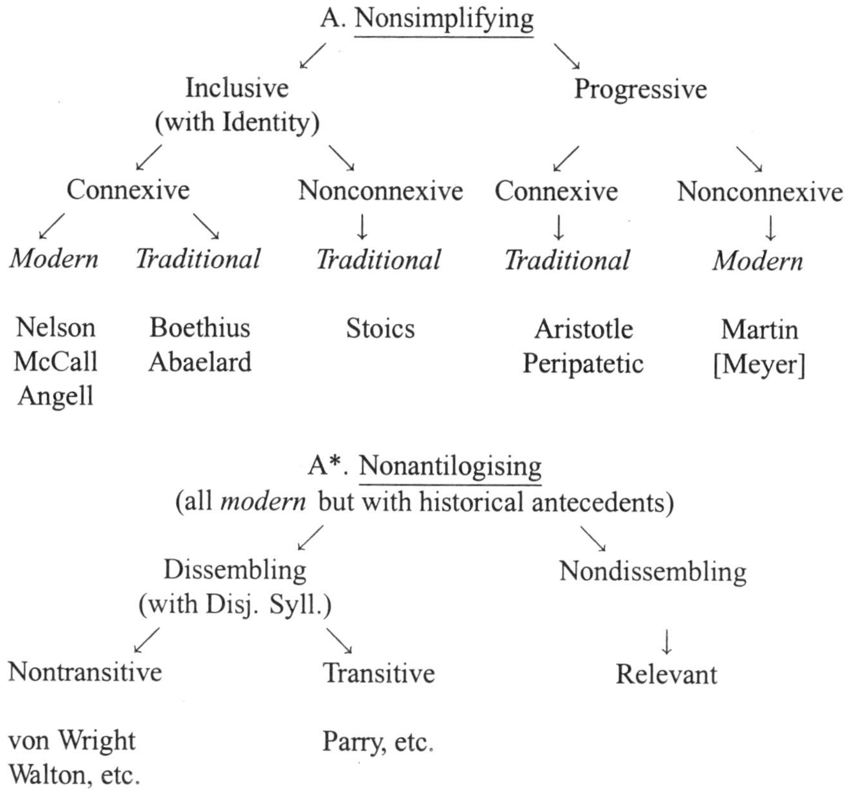

Figure 5.1 A binary cross-classification of sociative logics

Although virtually all of Chrysippus's extensive works have been lost, the general shape of his propositional logic has been roughly patched together from reports in subsequent, often ill-disposed or hostile, commentators (for conjectured details of this logic, see Kneale and Kneale 1962 and work cited therein, and also Corcoran 1974). It is commonly assumed, and argued, that Chrysippus's logic represents both the third position and mainstream Stoic logic. These assumptions are not however essential in what follows, which mainly concerns what is ascribed to 'the Stoics', as distinct from other Schools. Traditional Stoic propositional logic appears to admit of the following sort of formalization:- As well as initial propositional parameters, named as "the first", "the second", "the third" and so on, it included the (apparently independent) connectives: $\rightarrow$ (conditional), & (conjunction), $\cup$ (exclusive disjunction) and $\sim$ (negation), in modern representation. The conditional $\rightarrow$ is definable in terms of a further primitive, $\circ$, of compatibility and negation, thus: $A \rightarrow B =_{df} \sim(A \circ \sim B)$. Inversely, $\circ$ is definable through $\rightarrow$. (It is somethimes suggested that $\cup$ can also be defined in terms of $\circ$, but in the logic it appears to have played a significant independent role.) Standard formation rules for well-formed formulae (wff) are adopted. But the formation rules should strictly extend beyond those for well-formed formulae, to include rules. For there were special conditions upon rules, restricting basic rules to two premisses and one conclusion, which are not matched in contemporary free-wheeling procedures that permit much suppression of premisses. The two-premiss formation rule of Chrysippus seems to be as follows: where A, B and C are wff, then A, B / C is an (object) rule (with "/" read "therefore".) The best preserved part of Stoic logic is a list of object rules, the 'indemonstrable moods'.

The postulates of the logic can be divided into three groups:-

I. *Indemonstrata*, or primitive *object rules*:

I1. $A \rightarrow B, A \,/\, B$ I2. $A \rightarrow B, \sim B \,/\, \sim A$
I3. $\sim(A \,\&\, B), A \,/\, \sim B$ I4. $A \cup B, A \,/\, \sim B$
I5. $A \cup B, \sim B \,/\, A$

These schemes, traditionally stated with "therefore" as conjunctor, are taken to supply both (normal) rules and theorem schemes, for example to consider the third, the theorem scheme, $\sim(A \,\&\, B) \,\&\, A \rightarrow. \sim B$. The connection is given through a relevantly correct principle of conditionalisation, namely, in many premiss form,

$$A_1, ..., A_n \,/\, B \quad \text{iff} \quad A_1 \,\&\, ... \,\&\, A_n \rightarrow B$$

or similar, for example with a meta-rule notation replacing "iff" (for the connection, see Mates 1961, p.75; Kneale and Kneale 1962, p.159). 'According to the Stoics any argument is valid if the conjunction of its premisses forms the antecedent and its consequent forms the conclusion of a true conditional' (Frede 1974, p.4, who cites many references). Now one half of this rule is straightforwardly derivable (metatheoretically), using equipment Stoic logic should have supplied given its oft cited claim to 'completeness'. For suppose $A \,\&\, B \rightarrow C$, to take the appropriate doubleton case. Then:

$A, B \,/\, A \,\&\, B$	by Adjunction
$A \,\&\, B \to C, A \,\&\, B \,/\, C$	by I1 (i.e., Modus Ponens)
$A, B \,/\, C$	by a Transitivity (or Cut) rule

Such a Transitivity rule was supplied (as a thema); and such a (two premiss) Adjunction rule is derivable from an Antilogism thema and I3, which yields $B, {\sim\sim}A \,/\, {\sim\sim}(A \,\&\, B)$, given Double Negation principles it is known the Stoics endorsed (for example Kneale and Kneale 1962, p.169). Such little arguments also help reveal what else Stoic logic, as complete, no doubt included. Then too the conditionalisation principle can be reduced, in the two premiss case, to the relevant principle:

CD. $$\frac{A, B \,/\, C}{A \,\&\, B \to C}$$

This rule is, it should be noted, very different from the mainstream rule $\frac{A, B \,/\, C}{A \,/\, B \to C}$ which is relevantly invalid, and in Chrysippus even ill-formed.

II. *Themata*, or *meta-rules* (sometimes described as those of 'exposition' or 'analysis', etc.):

Thema 1. 'If some third is deduced from two, one of the two together with the opposite of the conclusion yields the opposite of the other' (Bocheński 1961, p.127 citing Galen). Thema 1 thus comprises these two rules (given, in view of the separation of I4 and I5, ordering of premisses is respected):

$$\frac{A, B \,/\, C}{A, {\sim}C \,/\, {\sim}B} \qquad\qquad \frac{A, B \,/\, C}{{\sim}C, B \,/\, {\sim}A}$$

Thema 1 yields, from the first indemonstrable I1, the Stoic theorem (of Counterexample): ${\sim}B, A \,/\, {\sim}(A \to B)$.

This simple example illustrates the grand Stoic thesis that 'any valid argument whatever is made up out of elementary, indemonstrable syllogisms' (i.e., arguments). Such reduction always involves, however, further rules: themata for central syllogisms, (unformulated) suppression rules for enthymemes or methodological conclusive arguments, and substitutivity of equivalence principles for 'hyposyllogistic' arguments (see Frede 1974, p.5). In fact, as the simple example indicates, it would simplify presentation of the Stoic syllogistic system to introduce the supplementary thema:

S1. $$\frac{A, B \,/\, C}{B, A \,/\, C}$$

Then half of Thema 1 and of Thema 3 could be simply derived. But Stoic logic no doubt contained enough resources (somewhere) to derive S1, as a methatheorem. (It would take a little work, since S1 permits some neat derivations. For instance, from Adjunction by S1, $B, A \,/\, A \,\&\, B$, whence by CD, $B \,/\, A \to A \,\&\, B$.)
Thema 3. 'If some third is deduced from two and one (or the two) can be deduced syllogistically from others, the third is yielded by the rest and those others' (Bocheński 1961, p.128, Kneale and Kneale 1962, p.169). The principle *appears* to yield such themata as $\frac{A,B \,/\, C \quad E,F \,/\, A}{E,F,B \,/\, C}$, so violating a strict two premiss requirment. There are two escape routes: relax the two premiss requirement for a multiple premiss requirement (perhaps with $n \geq 2$); or assign further work to conjunction, &. The former, easier, course is needed if there is to be much prospect of proving implications like those of Associativity (and Distribution). The latter, tighter course would result in such themata as $\frac{A,B \,/\, C \quad E,F \,/\, A}{E\&F,B \,/\, C}$, i.e., effectively rule CD is applied together with a prefixing principle, $\frac{D \to G \quad G,B \,/\, C}{D,B \,/\, C}$. Details of the second and fourth Stoic themata are nowhere recorded, but apparently they comprised composition and cut rules rather like the third, Thema 3.

Taking advantage of CD, it is tempting then to adjust, or brazenly reformulate the thematic structure as follows: S1 together with the antilogism principle $\frac{A,B \,/\, C}{A,{\sim}C \,/\, {\sim}B}$ and the following composition principles:

$$\frac{D \to G \quad G,B \,/\, C}{D,B \,/\, C} \ \text{(Prefixing)}$$

$$\frac{C \to E \quad G,B \,/\, C}{G,B \,/\, E} \ \text{(Suffixing)}$$

Note that the other Prefixing rule:

$$\frac{D \to G \quad B,G \,/\, C}{B,D \,/\, C}$$

is derivable using S1 (and assumed iterability of rules). Also Suffixing would be derivable, were Contraposition available (as it no doubt should be, given ancient perceptions of negation), along with Double Negation elimination, thus:

$$\frac{C \to E \quad G,B \,/\, C}{{\sim}E \to {\sim}C \quad G,B \,/\, C}$$

$$\frac{\sim E \rightarrow \sim C \quad \sim C, B \,/\, \sim G}{\sim E, B \,/\, \sim G}$$

$$\frac{\sim\sim G, B \,/\, \sim\sim E}{G, B \,/\, E}$$

But of course derivation of Double Negation elimination itself presupposes Suffixing in one case, or *else* an analogous replacement rule.

So far the systematic details available for Stoic logic—exception for CD—provide *only* for a pure rule system (as Corcoran 1974, p.179 has observed). There are initial rules, under I, and rules for deriving more rules, under II. Whence only rules result. But it also is known that the Stoics asserted several propositional principles, as well. They espoused not only those statements derived by rule CD, but such principles as Identity, Excluded Middle, Double Negation, and so forth. To capture these latter, a third class, of axiomata, is required.

III. *Axiomata*, or *truisms* (from assorted sources):

$A \rightarrow A$	$\sim\sim A \rightarrow A$
$A \cup \sim A$	$A \rightarrow \sim\sim A$
$(A \rightarrow B) \,\&\, (B \rightarrow C) \rightarrow. \, A \rightarrow C$	$A \rightarrow B \rightarrow. \sim B \rightarrow \sim A$
$(A \rightarrow \sim B) \,\&\, (A \rightarrow B) \rightarrow. \sim A$	

No doubt the list, not particularly well organised, could be expanded. But there would be even greater danger of exceeding the meagre historical basis (which has already been stretched, here and there). Nonetheless the general, and impressive, shape of the propositional system is evident. It is organised and determinate enough for several features of the logic, not hitherto made much of, to be remarked.

It is strikingly conspicuous that the propositional system so elaborated is not strict implication—however it is varied here or there in historically justifiable or plausible ways. It is commendably weaker than strict implication systems, in particular as regards paradoxes. It also involves an unusual disjunction, not present in Lewis modal logics, which do not validate rule I4. In fact, Stoic logic is a sublogic of several different sorts of sociative logics, and can perhaps be analysed as a certain intersection of those logics (as a doppelgänger logic in the fashion of Chapter Six). For it can be mapped into sublogics of relevant logics and deeper connexive logics. An immediate corollary is: *mainstream Stoic logic is relevant.*

Consider the following connective mapping of Stoic logic into relevant logic (formulated at least in terms of connectives, $\rightarrow$, $\circ$, $\sim$, $\leftrightarrow$): $\rightarrow$ maps to

$\rightarrow$, i.e., to itself; & maps to $\circ$, i.e., to fusion; $\sim$ maps to $\sim$, i.e., to itself; $\cup$ maps to antiequivalence, i.e., $A \cup B$ transforms to $\sim A \leftrightarrow B$ (i.e., to strong fission). Finally transform $A, B \,/\, C$ to $A \circ B \rightarrow C$ (i.e., in effect $A \rightarrow . B \rightarrow C$). It is a straightforward matter, mainly of inspection, to show that all the transformations of the principles of Stoic logic hold in (many) relevant affixing logics, for example for convenient definiteness, in system R. Only the transformations of & and $\cup$ are even distorting, and that for &, into intensional conjunction, is both familiar, and has some basis in Stoic theory. 'Chrysippus, with reference to the ... conditional, "If anyone is born under the Dog Star, then he will not be drowned in the sea," recommends that it be expressed as a negated conjunction, "Not both: someone is born under the Dog Star and he will be drowned in the sea".'[53] If the conditional is intensional and not material, as all the examples given in Cicero's text suggest, then the conjunction involved *is* an intensional conjunction. Only the transformation for Stoic disjunction, not heavily controlled by rules and regulations, is a trifle unusual, mainly because again of I4; otherwise $\cup$ could less deviously be represented as fission, intensional disjunction (on intensional conjunction and disjunction, see RLR I, chapter 5). But the biconditional rendition has a solid basis in Stoic logical theory (as will soon appear).

Now suppose Stoic logic voided decent relevance, to the extent that there was some Stoic thesis $A \rightarrow B$ where A and B failed to share a propositional parameter, or some proper Stoic rule $C, D \,/\, E$ where C and D failed to share a parameter with E. Then $\vdash_R t(A) \rightarrow t(B)$ or $\vdash_R t(C) \circ t(D) \rightarrow t(E)$, $t(F)$ the relevant transformation of F, where in the first case $t(A)$ and $t(B)$ fail to share a parameter and in the second case $t(C)$ and $t(D)$ fail to share a parameter with $t(E)$. But both cases are impossible as system R is weakly relevant. So, Stoic logic does not violate minimum decent relevance. Since, furthermore, Stoic logic, by contrast with Peripatetic and connexive logic, admits such relevant embedding, it is not straightforwardly connexive.

Stoic logic was neither a single nor a static affair. Though it perhaps peaked in the extensive systematic work of Chrysippus, it was continued, varied and augmented, long after Chrysippus. Two of the interesting developments were the enlargement of the acclaimedly complete set of indemonstrables (doubtless suitably complete only for the connectives they captured), and the further intensional interpretations offered of connectives

[53] The story is told by Cicero. The term omitted from the quote, drawn from Mates 1961, p.55, is the bracketed "(material)". Contemporary, and dubious, assumptions are again being interposed.

such as conjunction and disjunction. Conjunction had been explained along truth-functional lines by the Stoics. That explanation applied of course only to actual situations (to the base worlds of the semantics); it did not determine the interaction of conjunction with intensional connectives, such as implication, where evaluation looks to other situations as well. But however conjunction behaved, disjunction was different. Disjunction did not inherit (De-Morgan-wise) the qualified truth-functionality conjunctions sometimes enjoyed.

Though the possibility of giving a truth-functional reading to disjunction has been realised since ancient time, for most of the long recorded history of logic, disjunction has not been construed merely truth-functionally, but intensionally in one way or another. This fact, shortly to be illustrated, as a significant bearing on recent appeal to the traditionality of Disjunctive Syllogism (Syll). These appeals get revealed as fallacious. For what is to be defended is tf DSyll, i.e., DSyll with (inclusive) disjunction construed truth-functionally; but what tradition, from the Stoics on, more or less uniformly supports is *not* tf DSyll, but nf DSyll, i.e., an exclusive form with disjunction commonly intensionally construed. A corollary is that Chrysippus's famous and enlightened example of The Dog (laboriously evaded in ENT, p.296 ff) is no evident support for the tf DSyll at all.

Disjunction was usually regarded by the Stoics as at least exclusive. The fourth indemonstrable (A or $B, A \therefore \sim B$) was regularly put up along with the fifth, which is said to underlie The Dog's reasoning: thus not only earlier Stoics, but later writers in the tradition such as Cicero. But, further, disjunction was regularly taken as hypothetical, and normally as not truth-functional. However, there appears to have been a controversy over disjunction corresponding to that over implication, 'but unfortunately we do not know the details of this argument' (Mates 1961, p.51). It is known that 'there were some among the Stoics who did not regard a disjunction as true unless the components were incompatible', which Mates again tendentiously glosses in modal style, 'i.e., unless the components could not both be true' (p.52). The phrase 'some among the Stoics' is not just roundabout, but misleading also; for the mainstream Stoic position appears to have required not merely such incompatibility but rather stronger connecting conditions (as the passages from Gellius, Galen and others which Mates 1961 quotes make plain; pp.52-5). What emerges (from such later work) is that Stoic exclusive disjunction amounts to an antiequivalence: A or B is tantamount to, iff not A then B, i.e., $\sim A \leftrightarrow B$, where connective '$\leftrightarrow$' is a non-truth-functional biconditional, often rendered as in English as a conditional (cf. the discussion in Mates 1961, p.56). Under this construal, the fifth indemonstrable becomes

$\sim A \leftrightarrow B, \sim B \therefore A$, about which there is no issue, at least no issue similar to that about tf DSyll. The Dog can surely apply biconditional Modus Ponens. But The Dog would not have reasoned in accord with the truth-functional addition principle, A so A or B. Few before the thirteenth century did.

It is only in contemporary times that a truth-functional rendition of disjunction has become entirely dominant, a rendition that does not however reflect at all well ordinary language uses of disjunction (as Strawson and others have explained). Even in Boole, one of the multiple progenitors of contemporary extensionally-ruinous logical theory, disjunction does not appear in extensional form (or implication really at all). Only in the later degenerating days of medieval logic does truth-functional disjunction come to have a recognisable role—a period to which contemporary apologists now look to draw (a pretty limited and insipid) inspiration and some historical underwriting for tf DSyll.

To gain an impression of the other unregimented unreconstructed side of the long story of disjunction, consider for instance Whately 1868 on disjunction. Whately refreshingly proceeds (p.70) to divide disjunctions into two classes:

(i) those where there is a natural connection, where a disjunction states an alternative and will not be true *unless* one of the members of it is true' and ('on the other hand')

(ii) those where there is 'no such natural *connexion* together as to warrant their being proposed as an alternative' as "either Britain is an island or a triangle is a square". Such a proposition would rather be called nugatory or absurd, than false...'[!].

Then he appears to set type ii aside. 'Such propositions are often colloquially uttered in a kind of jest' (p.70), but not to be taken seriously logically. So much for the negative paradox argument depending on DSyll. Contradictions would only yield—what is still too much—all 'naturally connected' (or relevant) propositions. Where there is a connection, i.e., type i obtains, that licences an inference. 'If, therefore, one of these categorials be denied (i.e., granted to be false), you may infer the remaining one, or (if several) *some one* of the remaining ones, is true' (p.70). Thus, 'it is evident that a disjunctive Syllogism may easily be reduced to a *conditional*, by taking as an Antecedent the *contradictory* of one or more of the members; for example, if it is not spring or summer, it is either autumn or winter' (p.71). That is, disjunction is once again intensional; DSyll amounts once again to Modus Ponens for a suitable connecting conditional. Whately even advances within hailing distance of the mainstream Stoic position, in observing that disjunction is also mostly *exclusive*, so justifying in appropriate situations such inferences as: A or B, A; therefore not-B.

Not only was disjunction generally construed intensionally; conjunction sometimes was also. Herein lies a straightforward explanation of the problem of the two additional undemonstrated schemes that later Stoics such as Cicero added to the original five of Chrysippus. The additional schemes are these:

6.	Not both this and that. But this. Therefore, not that.	7.	Not both this and that. But not this. Therefore, that.

Cicero's list has presented a considerable problem, because it looks as if his sixth scheme is but a variable-amended version of the third scheme, while the seventh is patently invalid. But if instead Cicero has adjoined schemes for a further connective, a different (intensional) "and" signalled by use of different variables, then there is no problem of redundancy. Moreover, for certain choices of conjunction connective, the puzzle of invalidity of scheme 7 disappears; both schemes 6 and 7 can be validated.[54] Since there are few constraints on the connective involved but for 6 and 7 and some lax notion of 'conjunctiveness', many choices are possible (including extensional ones). One obvious choice, with a fair pedigree, connects Ciceronean conjunction, symbolised $\cap$, with Stoic disjunction $\cup$, in an entirely standard way, i.e., $A \cap B$ iff $\sim(\sim A \cup \sim B)$. Thus, in the stronger setting of system R, $A \cap B$ is tantamount to $\sim(A \leftrightarrow \sim B)$, i.e., to a nonequivalence of one component with the negation of the other. Then, with such a Ciceronean conjunction both 6 and 7 follow, in effect, just by negation transformations from equivalential Modus Ponens. Or more simply, 6 and 7 reduce definitionally to schemes 5 and 4 respectively. This reduction suggests another straightforward choice, which obtains interesting historical confirmation:- Define a 'conjunction' $\underline{m}$ thus: $A \underline{m} B$ iff $\sim(A \cup B)$, i.e., unscrambled as before, $\sim(\sim A \leftrightarrow B)$. Then schemes 6 and 7 formulated with $\underline{m}$ reduce at once to schemes 4 and 5 respectively, and so are valid for similar reasons. In weak relevant logic

[54] The resolution of the problem proposed shares significant common ground with the earlier resolution of Frede 1974, who reconstrues the connective used in schemes 6 and 7 as subdisjunction (which amounts truth-functionally in the two argument case to $\sim(A \,\&\, B)$). But taking conjunction as suitably intentional avoids the remoteness of subdisjunction and the implausibility, forced by a classical framework, of representing conjunction as a special sort of *dis*junction. The weakness otherwise of Frede's resolution, and the serious defects of shuffles like that of the Kneales, has been explained by Stump.

settings these are *equivalent* choices, furthermore, since $\sim(A \leftrightarrow \sim B) \leftrightarrow \sim(\sim A \leftrightarrow B)$, i.e., $A \underline{m} B \leftrightarrow . A \cap B$.

Revealing support for a two-conjunction resolution of the problem derives from Boethius's treatise *In Ciceronis Topica*, which however gives a striking slant to the matter: *both* conjunctions are intensional (as you no doubt already guessed). The Boethian conjunction of schemes 6 and 7 is the negation of the disjunction of schemes 4 and 5.

> The sixth and seventh modes are derived from the disjunctive proposition of the fourth and fifth modes by adjoining a negation, withdrawing the disjunction from the propositions which were inserted previously in the disjunctive proposition, in the following way. "It is not the case that it is day and that is night"; this was formerly a disjunction of this sort: "Either it is day or it is night" (Boethius 1988, p.141).[55]

In short, $\sim(A \underline{m} B)$ iff $A \cup B$. It is a corollary, moreover, of the further data Boethius provides that the conjunction concerned is not truth-functional. For, on the one hand, 'the stipulations he places on propositions which can serve as the conjunctions in the last two modes show that the conjuncts in his view meet the conditions ...: one and only one of those conjuncts can and must be true' (Stump 1987, p.27). Thus the truth conditions for Boethian conjunction are those for exclusive disjunction, i.e., for nonequivalence; but according to Boethius's definitional account it is the negation of that, i.e., that for equivalence. It cannot be both. But, in any case (whether Stump is right or not about Boethian conjunction having such, or similar, conjunction properties), Boethius would not have accepted disconnected truth-functional complexes as supplying a conjunction or disjunction.

Naturally, the reduction of schemes rather diminishes their undemonstrated character (and destroys their indemonstrability), as does Boethius's reconstruction of the third scheme, which introduces distinctively hyper-connexive elements. For Boethius not only explains the conjunction of scheme 3 in terms of implication in a now familar intensional way, i.e., $A \circ B$ is tied to $\sim(A \rightarrow \sim B)$; but as well he imposes his strong connexive principle $\sim(A \rightarrow \sim B) \leftrightarrow . A \rightarrow B$, with the (disconcerting) result that intensional conjunction gets equated with a conditional, $A \circ B$ with $A \rightarrow B$. That is, $\sim(A \rightarrow \sim B)$ is equivalent not only to $A \rightarrow B$ but

[55] The quotations are Stump's translations of Boethius's *In Ciceronius Topica*. This whole section on Boethius on Cicero is *heavily* dependant upon Stump 1987; nonetheless it does not come down quite where she does.

also to $\sim(A \circ \sim B)$, i.e., $\sim$ (A and $\sim B$) (to put it the way Stump does, p.24; but note her suggested modelling is not satisfactory). Then the third scheme follows by negation principles from the first scheme. The connexive interposition also explains Boethius's otherwise puzzling statements about conditionals showing connection better than equivalent conjunctions do. For instance, Boethius *re*formulates Cicero's example of the third scheme, which exemplifies the standard major premiss form, with the major premiss, 'It is not the case that if the silver was bequeathed, the coin was not bequeathed'. As he remarks:

> Cicero himself, however, formulated the proposition in this way: "It is not the case that the silver was bequeathed and the coin not bequeathed"; but one added the casual conjunction "if" in order to show the genus of such a proposition. For an incompatible arises from a connected proposition with the addition of a negation. But no conjunction can show a connected proposition as well as "if" can, although a copulative conjunction might produce the same proposition since things that are connected are also understood to be conjoined ... (Boethius 1988, p.147).

Similar equations were envisaged by Alexander of Aphrodisias, who wondered whether the third indemonstrable was not the same as the first. He appears to have been toying with what amounts to the equation of $\sim(A$ and $B)$ with $A \rightarrow \sim B$, an equation an intensional conjunction ($\circ$ of R for instance) could immediately supply.

Stoic logical theory, despite a marked reductionistic cast, included perceptive remarks about many connectives beyond standard contemporary sets, not only intensional conjunctions and disjunctions, but also for example on the behaviour of reasoning connectives like *since* and *because* (see for example Diogenes Laertius 1958, pp.180 ff). Furthermore, the apparently restrictive canonical theory of hypothetical syllogism, examined above, was extended by a theory of enthymematic argument and as well by a theory of 'hyposyllogistic' argument. Nor were Stoic logics confined (any more than other historic sociative logics) to the propositional level. They included also the rudiments of a *neutral* theory of quantification, apparently supplied through a combination of propositional logic with indefinite descriptors (for some discussion see Hay 1969, p.152 ff). Certainly some such logical treatment of universal and particular judgements is required to justify the widely reported Stoic boast to be able to treat all valid arguments, including those of Aristotelian syllogistic. While the latter more modest claim may have some basis—since the confined forms of the 'assertoric syllogism' are amenable to sociative analysis—the larger claim now looks mere bravado; it is a bit like boasting that quantificational logic is adequate for such purposes. It would be tempting to set the important matter of descriptors and quantifiers aside, to pretend that they make no difference to the range of Stoic sociative logics and their rivals. But such a convenient fiction simply is not true to the inconvenient facts still coming into view. Descriptors can work in surprising, and nonconservative, ways in sociative logics. Indeed the effects of descriptors in intensional logics are not yet much investigated or particularly well understood. There are further chapters on Stoic logics, unfortunately premissed again on very scanty data, to be written, or, where written, rewritten.

In this respect, what holds for Stoic logic, and the elaboration of the third position, applies also to its traditional rival, Peripatetic logic, to the retrieval of relevant details of Peripatetic logic and reassembly of fragments of the fourth position on implication. The fourth position involved an inclusiveness or proper containment picture of conditionality. Under it the consequent is *implicitly* contained in the antecedent: 'judging by "suggestion"' is assessing by implicit containment. Evidently, proper containment was required, and mere repetition excluded; the analytic unravelling of what was implicitly contained in the antecedent had to yield different information. Accordingly the position was almost certainly coupled to the progressive reasoning which the Peripatetics opposed to Stoic hypothetical argument. If the Peripatetic coupling stands, it is the fourth position, not the third position, that there is a solid case for accounting as connexive. But the coupling is circumstantially based only. And the connexivism involved is progressive connexivism, deriving from Aristotle's theory of noncircular reasoning, not the reflexive connexivism, involving duplication, which McCall would attribute to the Stoics.

In any event, the fourth position was not a Stoic position. For the Stoics apparently accepted Identity (cf. Mates 1961, p.49), though they rejected—what is different—irrelevant premisses in arguments. Although the Stoics typically required (at least) two premisses in an argument, they allowed that arguments with redundant premisses are valid, for instance $A \rightarrow A, A \,/\, A$ and $A \rightarrow . \, B \rightarrow A, A, B \,/\, A$. But the Peripatetics did not; they rejected such redundant arguments. Like Aristotle, the heavyweight source of the loosely-affiliated school, they insisted that the conclusion of a correct argument be different from any of the premisses.

The fourth position represents an early stage in the long and far from finished business of explaining inference and implication (or varieties thereof) in terms of containment or potential containment or proper potential containment. Its contrast with the third position indicates already that there are significantly different ways in which containment can be elaborated. It is tempting to speculate that the fourth position evolved as some variation upon the extended term logic and embryonic relational logic of Theophrastus (as pieced together by Bocheński), which was in turn an elaboration of 'Aristotle's hints about "syllogisms from hypotheses"' and his underlying theory of progressive reasoning, i.e., syllogisms (the quotation is from Bocheński 1961, p.104). It is also tempting to conjecture that the fourth position was not well developed technically because, in part, of the sheer difficulty of the business, even with contemporary logical technology (to reiterate an earlier point). Given the very little we presently know about the fourth position, many sociative explications, all much embroidering the position, are feasible (at least three different relevant explications are available, viz., Chapter Seven; RLR I, chapter 11.2; and Martin and Meyer 1983). It is a corollary of those explications that the fourth position does, like the other positions, have a non-vacuous representation of sorts in terms of worlds (*pace* Mates 1961, p.49); for its approximate explications all do.

Given the different and apparently conflicting objectives underlying the third and fourth positions as tentatively expanded, and the different ways of effecting these objectives logically, it is hardly surprising that they contributed to a major dispute over general logical framework that broke out and persisted between Stoics and Peripatetics. So viewed, the long-running dispute, often portrayed as one over priorities, emerges not primarily or at all as one over priority (not a matter of such consuming importance to the ancients), but as one over logical fundamentals, implication especially, and the proper development of these basics, over the correctness of apparently rival fundamentals. procedures, and methodologies (see further Frede 1974). For instance, the Stoics insisted upon narrow canonical forms, such as the indemonstrables, and reduction by certain prescribed rules, such as the themata and substitution principles, to these circumscribed forms. By contrast, the Peripatetics tended to focus primarily or exclusively on the orthodox categorial syllogism, with the first figure as canonical; but, even where they *were* more generous in initial forms, they still only admitted argument forms which did show something, which yielded further information, which progressed.

The contrasting features pertaining to argument and conditionality get reflected in different demands imposed upon the *use* of premisses or antecedents. Whereas the Stoics operated a use condition in avoiding utter irrelevance, the Peripatetics adopted a use condition to exclude redundancy, to ensure that each premiss was separately used and pulling its weight. These contrasting use requirements, hardly much developed, were intimately tied to what it was supposed argument is about or doing. According to later Peripatetics like Alexander of Aphrodisias the role of argument is 'to show or establish something', something different from what is given or assumed. Whereas categorial syllogisms 'show or establish something', the Stoic 'hypothetical syllogisms, taken by themselves, do not show anything' (Frede 1974, pp.25-6, reporting Alexander). Genuine argument has to be progressive and not beg any questions.[56]

> ... Alexander's attitude towards noncategorical inferences and especially Stoic syllogisms, seems to be determined by two assumptions:
> (1) Only inferences of a form such that an inference of that form can show or establish something can be called syllogisms. To show or establish something is supposed to entail: what is supposed to be shown is not used as a premiss; what is supposed to be shown is not presupposed by any of the premisses in such a way that one has to accept or know the conclusion in order to accept or know the premiss; one can fail to know what is supposed to be shown even if one knows the premisses.
> (2) Hypothetical premisses, at least in hypothetical syllogisms, are not treated as assumptions about facts but as assumptions about the way one can argue which are to be exploited in the course of the argument in which they are made explicit. It is on the basis of these assumptions that Alexander may think that only Aristotelian categorical syllogisms are really syllogisms, though he is quite willing to admit that there are many other forms of valid, and even logically valid, inference (Frede 1974, summing up, p.29).

What amounts to a distinction between facts and hypotheticals, which are like argument warrants, appears, in one form or another, in much Peripatetic

[56] Mills's notoriety-grabbing theme that all argument involves *petitio* lurks in some of the background here. Galen recognised such a problem with truth-functional arguments like the Stoic third indemonstrable, where knowing the conjunctive premiss truth-functionally involves knowing the conclusion; he proposed avoiding the difficulty by intensionalising! Hence too part of the motivation for later amendment of the Stoic 'indemonstrables'.

theorizing. It appears for instance, in the proposition that hypotheticals must be supported by topical rules. These distinctions between facts and hypotheticals, along with special roles assigned to certain rules, topical rules or inference tickets and certain major premisses, broadly anticipate elements of Ryle's study in Ryle 1950, technically explicated, in one interesting way, in Anderson and Belnap's ticket entailment (ENT, p.41 ff). The explication, through labelling or subscripting methods which track use of different types of statements, admits of interesting adaption and variation, from the implicationally strong ticket system T, to weaker and different theories of topical implication, irredundant argument, and unrepetitive conditionality.

Evidently then there are various contemporary ways in which significant features of main ancient theories of consequence can be further explicated. For one thing, the use-tracking methods can be applied at different points, for instance proof-theoretically or semantically (the latter would however be decidedly remote from ancient practices, *except* as adapted to containment models, whereas elements of subproof methods do appear in Peripatetic work). For the tricky Peripatetic requirements of separate use, not only would every hypothesis in a proof scheme, even if repeated, require a new (ticket) label, and the differentiation affected in T between major premisses and minor ones (or 'facts') copied, but the rule of repetition would be sharply curtailed—so failing the 'archetypal implication', $A \rightarrow A$, not a 'foundation' for any sort of genuine Peripatetic reasoning. Indeed subscripting constraints are so severe, that the resulting reconstructed Peripatetic logic though not a subsystem of system T (because of different negation rules appropriate for RAntil), is bound to be sociative.

From the end of the fruitful period of Stoic and Peripatetic logic until the early Middle Ages there was little original Western work in sociative logic, so far as we presently know, except for that of Boethius. The long period may not have been a really creative one for logic of any sort, as Bocheński explains (1961, p.134 ff). But as Bocheński also observes it is 'a period into which hardly any research has been done'. What has been done since 1961, for instance by Ebbesen appears to confirm Bocheński's main bleak conclusions (see for example the discussion of Aristotelian scholasticism, pp.64-5). But Boethius is certainly one late exception: Apuleius, who influenced Boethius, may be an earlier exception (for some orthodox background see Sullivan 1967): and there may well be other exceptions. For it seems sociologically unlikely that logical investigation sank for so long into an intellectual void, eventually to spring again from the void, especially when (unoriginal) work was undoubted proceeding in some centres.

Boethius, himself influenced by both Peripatetic and Stoic factions, served unwittingly as a major cultural transmitter, relaying ancient logic to the Middle Ages. He certainly exerted a significant influence upon twelfth century logical enterprise, largely but by no means only though his transmission of Aristotelian thought by way of detailed commentaries; for he also sketched out the rudiments a remarkable logical theory. By contrast with the Stoics, there is no well organised, clearly articulated framework of logic in Boethius, except for borrowed work. Like Wittgenstein, much of Boethius' output is substantially disorganised; like Wittgenstein too, it has proved very suggestive to subsequent researchers, while looking naive or simplistic to those trapped in mainstream thought. From Boethius' work likewise, an interesting collection of proposals can be pulled together, and a highly nonclassical protologic outlined. Features of especial sociative interest to be drawn from Boethius include these: the accounts offered of connectives, the association (or affective connection) view of conditionality, and the strong connexive principles. There are other elements which may also turn out to be relevant as sociative investigation proceeds, for instance aspects of his historically influential but dense theory of topics.[57] As well there are matters we now tend to take for granted that may have originated with Boethius, like the distinction between logical and factual implication. Finally, of paraconsistent importance, there is the licence given for reasoning from impossibility. Reasoning from the impossible, which anticipates the medieval *positio impossibilis*, takes place when there is agreement to suppose that something which could never in fact occur holds. Boethius then describes how we may start from a recognised impossibility aiming not so much to derive an explicit contradiction but rather to explore the logical structure and features of the impossibility (for further details and some discussing of an elaborate original example see Martin 1999). Unfortunately for the larger paraconsistent history Boethius does not describe the logical procedures admitted and excluded.

Boethius is often looked down upon, by contemporary scholars of lesser stature, as a patient but rather dull scholar who simply laboriously assembled

[57] The theory of topics covers several matters which ought to be of more contemporary interest, but have largely slid off the logical agenda. In Boethius, one is that of relevant enthymematic argument; another (following Cicero) concerns methods of argument discovery. A main method of argument discovery in fact amounts to ascent up a proof tree; for what are surveyed at each node are the sources that would lead from above, by recognised rules, to the sort of proposition under scrutiny at the given node. A similar general procedure can be (noneffectively) applied in filling out enthymemes. For other roles of topical argument, see Martin 1991 and Woods and Walton 1982.

ideas of his predecessors, and was historically lucky (seen from *their* angle) to be in the right place at the right time. Boethius's astonishing elaboration of connexive logic, dismissed of course as a gross mistake, should have been enough to overturn such patronising approaches. Boethius presented as patterns of valid reasoning not only connexive principles such as $A \rightarrow B \rightarrow. \sim(A \rightarrow \sim B)$, and consequences of them such as $A \rightarrow (B \rightarrow C), B \rightarrow \sim C \,/\, \sim A$ derived using Contraposition, but strikingly he strengthened the first to a biconditional, i.e., $A \rightarrow B \leftrightarrow. \sim(A \rightarrow \sim B)$, a hyperconnexive principle (already encountered) which he systematically applied. The logical theory thus indicated contains not only usual connexive principles, but as well stronger, not to say strange, nonclassical principles, most notably the connexive converse, $\sim(A \rightarrow \sim B) \rightarrow. A \rightarrow B$. These further principles of *hyper*connexive logic, occasionally reappearing in subsequent work, have so far substantially resisted contemporary attempts at explication and elucidation. They are technically interesting at least for that reason. But the principles of hyperconnexivism, straightaway plausible for an 'equivalential' implication, also appear in other logical settings, old and new. For one thing, hyperconnexivism, whatever its (considerable) difficulties, allows for an entirely naive representation of syllogistic in propositional-quantifier form. Just as "every A (item) is (a) B (item)" transforms directly into the form $(Ux)(Ax \rightarrow Bx)$, so "some A (item) is (a) B (item)" transforms to $(Px)(Ax \rightarrow Bx)$. But by virtue of hyperconnexivism, this is tantamount to a *conjunctive* form $(Px)(Ax \circ Bx)$.[58] Moreover, all the logical connections of the square of opposition are preserved without any hassle—including, obviously, inference from *every* to *some*, a traditionally valid principle knocked over in the modern mainstream rush.

Apart from this 'naive' representation of syllogistic, a familiar castigated classroom temptation, there are two other places where hyperconnexivism is beginning to make a contemporary restricted appearance: in theories of subjunctive conditionals, and in the embryonic theory of *invited* inference. Boethius's biconditional can be reexpressed in the form, $(A \rightarrow B) \cup (A \rightarrow \sim B)$, which is a strong version of the principle of conditional excluded middle (CEM), which identifiable classes of conditionals are said to satisfy, at least in material form (including the Philonian conditional). But if $A \rightarrow B \equiv \sim(A \rightarrow \sim B)$, for example, holds in a certain logics of conditionals (for example Stalnaker systems), then it holds as a matter

---

[58] *As mentioned in Chapter One, the existentially-neutral quantifiers U and P replace the more usual (and, in Sylvan's view, existentially-loaded) $\forall$ and $\exists$ respectively. [Editors]*

of logic; so the corresponding strict form should hold. But such a strict form is, it may be argued, yet another modal approximation to a proper implicational form, namely Boethius's biconditional. At first sight, the theory of invited inferences is far removed from that of certain everyday subjunctive conditionals. The idea there is that some statements, conditionals especially, *invite* inferences. For instance, the statement "If you mow the lawn I'll give you \$5" invites the inference "If you don't mow the lawn I won't give you \$5", "Dogs that eat Opla are healthy", invites the inference "Dogs that don't eat Opla aren't healthy", *and conversely*. Plainly, a theory beginning with a very weak logic of conditionals and closing under invited inference, would be a hyperconnexive theory ascribing equivalential features to its resulting conditional. Thus the theory would have salient features in common with Boethius's.[59]

Several characteristic connexive features merge in Boethius, along with presentation of strong connexive principles and projection of implication towards equivalence; namely, a containment picture of implication and conditionality, which is straightforwardly combined with an account (like the main Stoic position) of implication in terms of incompatibility: and a clear cancellation picture of negation (negation consists of removal, withdrawal, or the like). Both these intuitive modelling features were transmitted to the early Middle Ages, and clearly received. Furthermore, whatever the fuzzy shape of Boethius's logic, it can be conjectured with some confidence that it was sociative in intention as well as fact. For one reason, there were tight requirements of connection imposed not merely upon conditionals, but apparently upon all connectives. The demands often said to be made in everyday natural discourse upon connections of components of connectives, are to some extent matched in Boethius. For another, the propositional part of Boethius's logic appears to be embeddable in a fragment of equivalential-negation logic (connectives $\circ$, $\cup$, $\cap$ are definable). But even in this classical theory, of which Boethius's is only a quite proper part, significant restrictions upon parameter occurrence are met (see Prior 1962, pp.306-7), which already suffice to remove some noxious paradoxes of implication ($\approx$ equivalence).

It is the modesty of explicit conditions upon connectives that makes Boethius's hyperconnexivism easy (by contrast with recent connexivism) to allocate to a sociative box. Seemingly, *none* of the standard propositional connectives get assigned their contemporary regimented roles in Boethius,

---

59 What is indicated in the text is the easy part of the story, which was related to us by Miguel de Castro (who referred back on invited inference to Geiss and Zwicky 1971); the hard part is making it work.

whose theory takes decidedly more account of grammar and natural language than its recent successor. Even the connective *and* is not truth-functional and is subject to restriction in well-formation rules. Indeed Boethius appears committed to the proposition that no standard connectives are simply truth-functional (thus anticipating, but out-distancing, Strawson and van Dijk). Such a claim would not have astounded the ancients: for conjunction, perhaps along with negation, was the only connective regularly assigned to a truth-functional basket. Dispose of conjunction, as Boethius appeared to do, and the main case for intensionality is done. Disjunction, for example, was always coupled with conditionality, and (as noted) but few before the Parvipontani in the late twelfth century would have accepted the addition principles, $A \rightarrow . A \vee B$ and $B \rightarrow . A \vee B$, that a truth-functional story (among others) characteristically underwrites. Conditionality itself was generally given a non-truth-functional rendition, though Philo's radicalism was well-known. Boethius himself, though he noted and perhaps incorporated a modal account, appears to have reached a rigorous association linkage.

According to Boethius, if A then B (for example "if it's day, it's light") does not assert that both are [the case] but rather that if one is, then the other follows, that both come together in a certain understanding. Here and elsewhere what Boethius says suggests a double-banger analysis of implication (of a type considered in Chapter Six) which runs as follows: $A \rightarrow B$ iff $A \dashv B \,\&\, A \sim B$, where $A \sim B$ symbolises the association element: A is associated in understanding with B. Certainly Boethius makes *some sort* of strict requirement a necessary condition for the truth of a conditional, and he repeats the association condition (though sometimes as agreement in nature, rather than sense or understanding). The same double conditions are repeated in Boethius's explanation of *affective connection*, which appears to correspond to the (semantical) conditions for conditional statements. The double conditions for affective connections are again the strict conditions, the impossibility of one component without the other, and the association condition, the inevitable conjunction in the understanding of the components (cf. Martin 1991).[60] Given the equivalential cast of the auxiliary connective

---

[60] Martin tries to extract a containment account of conditionals from Boethius's obfusicating prose, an account which would reveal a natural historical source for Abaelard's containment theory. At one stage Boethius perhaps appears to extend the association condition to an implicit containment interpretation, when he remarks that it may not occur that something is understood as antecedent unless in the same what is consequent is understood. But as Martin honestly observes, this remark is ambigious, and the larger context appears to resolve the remark decisively against a containment-in-understanding point, that understanding the anecedent involves understanding the consequent, in favour of a contrast theory point, that

$\sim$, the analysis already has several of the right features for connexivism. But while the analysis eases the problem of validating connexive principles, it does not assist so well in vindicating hyperconnexivism. For all the impressive power of recent logical technology, a coherent synthesis of Boethius's sociative ideas on argument and conditionality remains to be achieved, if it can be.

5.3 The early medieval period: Abaelard, his rivals, and successors

Neither in Alexandrian times, nor during the medieval period when debates about the fundamentals of logics were intensive, did classical-style logic attain the dominance it has now achieved. Theories of entailment and conditionals were certainly an important issue in the twelfth century. There was a major debate, centred around Abaelard (as we now see it), over connexive logic. There was also an emerging theory of argumentation from false and impossible assumptions, the new supplementary theory of obligationes.[61]

There are also other little remarked features of medieval investigations worth emphasizing. The logical principles deployed were almost invariably of low degree (mostly at first degree, occasionally at second). Conditionals were seldom nested; pure consequence statements never were. Coupled with this, there was no emphasis on strength of system, but a heavy stress on adequacy to apparent data. Thus, by contrast with contemporary system building, there was no pressure to formulate, for example, a logic of pure conditionals. Logical principles investigated mostly contained other connectives, such as "and" and "not", and very often syllogistic material. (Not made were contemporary artificial separations of sentential and quantificational principles, divisions into syntactical, proof-theoretic and

'antecedent' cannot be understood independently of 'consequent' and vice versa.

Elsewhere Boethius gives what is, at first sight, a different account of 'the' conditional: 'The sense of the conditional proposition is this: that it would be the one thing just so long as it were the other...', an account suggestively similar to modern truth-preservation-over-situations analyses. There is no conflict of accounts. For it is now realised, in some places at any rate, that such a situational conjunction or overlap analysis can be seamlessly joined with other analyses, such as association or containment ones.

[61] Much of this section follows Chris Martin's detailed discussion in EA (unpublished parts of which he also kindly made available). Martin suggests that investigations of the sort he has helped initiate will result in a new appreciation of the vacillating fortunes and features of logic, at least up to the end of the Middle Ages. We suspect he was right. Much of what is offered here, supporting the same bold theme, is heavily parasitic on his pioneering work, especially the summary of the twelfth century debate on conditionals.

semantical theory, and so on.) The upshot was that what logical systems did emerge, or rather can subsequently be reconstructed, were very weak, at least by contemporary overpowered standards; indeed they were generally of lower power than that at which most recent theory functions well. Firstly, then, it is unlikely that the pure implicational theory would contain many principles beyond $A \rightarrow A$; that is fine, as there are interesting logics, such as those in the vicinity of relevant logic B, of this sort. Secondly, even where degeneration occurred and a modal theory was reached, as apparently happened under the dominant position in later medieval thought, the type of logic involved, hardly uniquely determined, would not have been 'normal', but typically somewhere inside the weakest of Lewis systems $S1$. The general drive for strength is, like generalised will-to-power, and other universalization and maximization drives, a modern phenomenon. These drives are now built into the dominant social paradigm; they affect logic like all other intellectual endeavour.

An exasperating feature, from the system-design perspective, which medieval logic shares with much of less simplistic modern philosophy, is the marked lack of systematisation and the extensive qualification of principles and making of fine distinctions. The trouble is, as again with unsystematic contemporary work, that the massive qualification is frequently once only and for the case at hand, and that the distinctions go nowhere theoretically; these things do insufficient work and hardly justify their keep. These are the so-called 'scholastic' elements displayed in much poorer linguistic philosophy, and copiously exhibited in medieval logic in syntactical qualification, over-qualification and hedging of principles. But in part it is because of this resilient complexity of detail that interesting medieval work is now beginning to break out of (early) contemporary Whiggish strait-jacketing and premature classification in available boxes. Thus too the satisfactory excavation and reconstruction of medieval logic calls for considerable patience (by primary researchers). But the effort that has recently been made is beginning to pay off, through the fascinating details, concerning implication especially, now being unearthed and put together. Few recent diggings have been richer than in Abaelard's work.

Abaelard accepted Aristotle's connexive principles AR1 and AR2, making them, along with two equivalent principles, central in his theory of conditionals. Any conditional which can be demonstrated, using acceptable principles, to imply conditionals which contravene them should be rejected. No doubt unqualified Simplification (i.e., $A \,\&\, B \rightarrow A$ and, by symmetry, $A \,\&\, B \rightarrow B$) with arbitrary and perhaps irrelevant or incompatible antecedents conjoined is a candidate for such rejection. For it yields immediately $B \mathbin{\&} {\sim}B \rightarrow B$ and $B \mathbin{\&} {\sim}B \rightarrow {\sim}B$ by unrestricted Substitution, whence given only Adjunction, $(B^\circ \rightarrow B) \mathbin{\&} (B^\circ \rightarrow {\sim}B)$ for $B^\circ = B \mathbin{\&} {\sim}B$, in violation of Abaelard's first connexive principle:

AB1. $\quad {\sim}((A \rightarrow B) \mathbin{\&} (A \rightarrow {\sim}B))$

Abaelard's second principle is likewise a contraposition of Aristotle's, and likewise equivalent using only minimal assumptions, to the first. It is that no statement implies its own negation, i.e.:

AB2. $\quad {\sim}(A \rightarrow {\sim}A)$

As to statements infringing AB2 and implying their own negations, Abaelard remarks:

> No one doubts that this is improper and embarrassing (*inconveniens*) since the truth of one of two propositions which divide truth not only does not require the truth of the other but rather entirely expels and extinguishes it (Abaelard 1956, p.290).

Thus is AB2 underpinned by a cancellation account of negation, with the content of A cancelling the propositional content of ${\sim}A$ (as explained more fully in NC). The response Abaelard could have made to Simplificational strategies directed against his principles, accordingly seem evident. The conditional $A \mathbin{\&} {\sim}A \rightarrow A$ fails because in the antecedent the content of ${\sim}A$ cancels that of A removing the ground from A and so support for the consequent. Or, put differently, the content of the antecedent, since so nullified, no longer includes that of the consequent, as required however for a logical conditional. But though such a response accords well with Abaelard's account of negation and conditionals, Abaelard did not develop any such suggestions. In fact, unqualified Simplication appears not to have been taken very seriously as a general logical principle in the twelfth century; logicians were very ready to surround Simplification with qualifications and to require special justification for its application—Simplication rather than connexive principles, let it be emphasized. Interestingly, Simplification as such does not appear in the twelfth century canon of valid argument. In order to justify it an appeal would have to be made to an appropriate topical principle.

Abaelard thought he could gain control over the wayward connexive principles, and avoid embarrassment, through a tight inclusion account of implication, coupled with a demanding theory of topical rules. A topical relation, or principle, provided a (natural) *connection* between antecedent and consequent elements, which would guarantee a true conditional. Topical rules regulated applications of Simplification and like nonuniversal principles, among other effects. Like Boethius, Abaelard rejects the standard modal account of the truth of a (necessary) conditional, accepting only one half of such necessary and sufficient conditions, namely that 'a necessary condition for the truth of a conditional is that the antecedent cannot obtain without the consequent'.[62] The modal requirement, he argues, is *not* sufficient. Abaelard works his way to alternative more rigorous necessary and sufficient conditions. Under the more rigorous condition for the truth of a conditional,

> not only can the antecedent not hold true without the consequent, but also the antecedent of *itself requires* the consequent. ... nothing can warrant the claim that 'this is antecedent to that' other than 'this *from itself* forces that' (Abaelard 1956, p.283-4, italics added)

Thus Abaelard's tighter conditions give essentially the sufficiency-on-its-own account central to contemporary deep relevant theory (for instance, to the positive theory of RLR I and much of the work on which it was based). Not only are the tighter conditions those of a relevant account; what is more, Abaelard glosses his conditions with a containment requirement: that the content of the antecedent is understood or *contained* in that of the antecedent (Abaelard 1956, p.284; cf. EA, p.393). Exactly such a containment analysis has been much favoured by relevant thinkers, especially those who are (erroneously) prepared to concede a truth-preservation style of analysis to the strict or material opposition. Abaelard was clearly with the forces of truth and justice.

Implicit containment pictures of implication run all the way through logical history, from the Peripatetics well before Abaelard to Mill and Boole long after. But containment may be, and has been, explained in very varying ways, and subject to very varying conditions, from tight sense or property transmission conditions (such as are found in Abaelard and more formally in Priest 1980), though modal analyses (such as Carnap 1956 supplies) to slack material or class calculus ones (such as classical logic offers). Abaelard explained logical conditionals (i.e., entailments) in terms of the consequents deriving necessarily just from their antecedents, and suggested one way of accounting for the necessity involved in terms of intensional containment, that in true conditionals the consequents are contentwise contained, or *understood*, in their respective antecedents. Sometimes (as in Priest 1980),

---

62 Abaelard 1956, p.271. This yields, after some standard modal manipulation contested in later medieval work, Martin's condition 1: for the truth of a conditional it is necessary that it be impossible that the antecedent be true and the consequent false.

this intensional (or sense) containment is taken as primitive, and simply assigned appropriate features; sometimes, as some of Boethius's remarks suggest and later traditional logic made explicit, containment was in turn accounted for through the transmission of suitable properties (for details and explication see RLR I, especially p.216 ff). Like other forms of intensional connection, containment amounts to the transfer or sharing of properties; here sense or content, already implicit in the antecedent, is transmitted to the consequent. Approached slightly differently with the same end result, containment is a partial identity (in one set of equations that were to become about axiomatic in the nineteenth century: $A \rightarrow B$, or $B \subseteq A$, iff $A \,\&\, B = A$, B is the same as part of A). But such partial identity, like identity proper and resemblance and interaction, is intensionally explained through the sharing of suitable properties.

A containment picture of implication, especially when combined with a cancellation account of negation, confirms basic connexive impressions. If $\sim A$ cancels A and vice versa, then A cannot be contained in or contain $\sim A$. So much is evident even on a Venn-style picture of 'cancellation' (where what is cancelled is crossed out, not rubbed out, so remaining in view). Consider $\sim A$'s cancellation of A within A -space. It is evident that A is not contained in $\sim A$, and that $\sim A$ is not contained A, which it erases. In short, $\sim(A \rightarrow \sim A)$ and $\sim(\sim A \rightarrow A)$ where non-implication is represented as non-containment. Similarly, cancellation-wise, as there are no null contents, where $A \rightarrow B$, i.e., B is contained in A, $\sim B$ is not contained in A, i.e., $\sim(A \rightarrow \sim B)$. But these appealing (first degree) pictures were not pulled together into a coherent modelling—not then, and, despite the immense recent increase in logical technology, not very satisfactorily since. Given an implicational setting as strong as that of system E, first degree connexive principles spill over to higher degree connexivism; worse, in much weaker settings they spill into inconsistency. An argument for the first proceeds along those lines: $A \rightarrow B \rightarrow. B \rightarrow \sim A \rightarrow. A \rightarrow \sim A$; $A \rightarrow B \rightarrow. \sim(A \rightarrow \sim A) \rightarrow. \sim(A \rightarrow \sim B)$; whence, as $\sim(A \rightarrow \sim A)$, by the Commutation rule of E, $A \rightarrow B \rightarrow. \sim(A \rightarrow \sim B)$. Arguments for the second were discovered—to the consternation (not of the multitude but) of many—in the twelfth century.

A satisfactory (relevant) account of the conditional is not, in a connexive setting, enough to avoid disaster, as Abaelard realised. Also necessary were further constraints on which materially-supplied conditionals held. Such control was particularly important in medieval theory where systems were always applied ones, with material examples rather freely imported into discussions of logical principles. Abaelard elaborated the theme, drawn from Boethius, that such a conditional is true, or an enthymeme valid, when there is a connection of topics guaranteed by a topical rule (a requirement which, Martin suggests, is rather like the modern idea of an inference ticket). Commonly the topical rules were presented as conditional connections. According to Martin, many logicians of the twelfth century held that there are real connections in the world, to which connections correspond natural conditionals.

Abaelard sought to gain control, then, through a tightening of the topical rules, which justified material conditionals and licensed admitted enthymemes. What, in particular, Abaelard did was to apply his theory of topics to remove potentially damaging conditionals linking opposites, such as "If Socrates is a man, Socrates is not a stone", $A_1 \rightarrow {\sim}B_1$ for short.[63] For if such a conditional as $A_1 \rightarrow {\sim}B_1$ were admitted it could be shown that $A_1 \,\&\, B_1 \rightarrow {\sim}(A_1 \,\&\, B_1)$, contradicting connexive principles. The following embarrassing argument concerning opposites would do the damage:-

$A_1 \,\&\, B_1 \rightarrow A_1$	by admissible Simplification
$A_1 \,\&\, B_1 \rightarrow {\sim}B_1$	since $A_1 \rightarrow {\sim}B_1$, by Transitivity
${\sim}B_1 \rightarrow {\sim}(A_1 \,\&\, B_1)$	contraposing Simplification
$A_1 \,\&\, B_1 \rightarrow {\sim}(A_1 \,\&\, B_1)$	by Transitivity.

This argument Abaelard's restrictive theory of inference licences deflects.

A decisive twelfth century beginning to sociative logic, in connexive form, was not however to be. The control Abaelard managed to achieve was inadequate. By cleverly varying Abaelard's own argument. Alberic produced an extremely embarrassing argument, meeting all Abaelard's requirements. The argument, which led to a counterexample to a connexive principle, and so to inconsistency of certain connexive logics, took the following form:

---

63 Abaelard argued, in effect, that natures are *positively* characterised; it is no part of the nature of a man that he not be a stone. And he contended that maximal propositions of the theory of topics supplying conditionals, are restricted to those connecting natures—a very narrow theory of conditionals. For requisite details of Abaelard's theory, see again Martin, for example EA. One tangential feature, to which Martin does not draw attention, is that Abaelard's treatment of opposites resembles an important move in the theory of objects. Conditionals concerning opposites, which assert that if one of a pair obtains the other does not or is absent, are consistency assumptions, which exclude impossible objects which have both of a pair of opposites.

$A_\circ \rightarrow B_\circ$	*Any* conditional true according to Abaelard (e.g."If Socrates is a man, Socrates is an animal")
$\sim B_\circ \rightarrow \sim A_\circ$	by (rule) Contraposition
$A_\circ \,\&\, \sim B_\circ \rightarrow \sim B_\circ$	by Simplification
$A_\circ \,\&\, \sim B_\circ \rightarrow \sim A_\circ$	by (rule) Transitivity
$\sim A_\circ \rightarrow \sim (A_\circ \,\&\, \sim B_\circ)$	by Contraposition from Simplification
$(A_\circ \,\&\, \sim B_\circ) \rightarrow \sim (A_\circ \,\&\, \sim B_\circ)$	by Transitivity.

Hence $C_\circ \rightarrow \sim C_\circ$, contradicting AR2. Both total-sufficiency and containment analyses of implication appear to straightforwardly validate Transitivity and requisite cases of Simplification; Contraposition (also easily guaranteed) was given in medieval thought. Abaelard was in deep trouble. Abaelard's considered response to this simple and ingenious argument is not known! However Abaelard reacted (the opposition report is that he at once conceded, but more likely he imported further restrictions), the argument was a watershed one. For here was an argument on which everyone interested in the use of conditional—which meant any twelfth century logician—had to take a stand. A conditional apparently satisfying the most stringent requirement for necessity, in that being an animal is part of the definition of being a man, has been shown, by the use of apparently very basic and very plausible principles, to entail the contravention of a principle which, if not Aristotle's own, seems equally certain. It was all extremely embarrassing. Connexive logic had suffered a major fall, from which it has never really recovered.

Many different neutralisations of the embarrassment were soon proposed, some of them anticipating contemporary moves, some of them rather surprising, but none abandoning conditionals as a bad lot or beyond redemption, in the fashion of the narrow contemporary paradigm. Some schools such as the Parvipontani 'essentially gave up on any refinement in the theory of the [demonstrative] conditional at this point' and adopted a modal theory of the broad type that was to become dominant in later medieval thought and is now so fashionable. More relevant, and interesting, are other resolutions, which apparently retained connexive principles as well as Contraposition. Essentially two types of option remain open: some restriction of Transitivity or some further muzzling of Simplification, and both were tried, though not in any thorough-going way. The options may be coupled, of course, with parallel options in the case of implicational paradox arguments.

The Albricini contested, like contemporary nontransitivity positions, Transitivity. But, unlike contemporary positions, they contested it on the grounds that the conditional 'expressed a "casual" relation between properties, and so failed where there was no corresponding property as in this case'. This nontransitivity approach, in denying that inconsistent predicates supply properties,[64] in fact connected with positions qualifying Simplification. For, as Martin has suggested, the idea is that combining the predicate "is a man" with its contrary "is not an animal" destroys the connection required for the natural conditional "if it's a man, then it's an animal" to be true. Thus the idea undercuts conditionals with impossible antecedents. Regulating or removing such conditionals, on one pretext or another, was, and remains, a popular move.

Several objected to the use of impossible conjunctions on the ground (which comes straight from the cancellation account of negation) that such conjunctions could not be posited without destroying part of the conditional (for the support the antecedent offers for the consequent is removed, or the content of the antecedent collapses). The objection appears to imply some kind of (consistency) proviso on Simplification. Such an appealing restriction is ascribed in some texts to Abaelard's followers, the Nominales—but specifically the awkward view that from a conjunction of an affirmation and a negation only the affirmative conjunct follows. This would presumably admit $A_\circ \,\&\, {\sim}B_\circ \rightarrow A_\circ$, in the example, but rule out $A_\circ \,\&\, {\sim}B_\circ \rightarrow {\sim}B_\circ$, thus blocking Alberic's argument, but not others dangerously like it.

Closely related was the resolution of the Porretani, who argued (like contemporary connexivists beginning with Nelson) that Simplification from an arbitrary conjunction was inadmissible. But they claimed this (here in contrast with contemporary connexivists) because such Simplification involves fallacy of non-cause-as-cause, i.e., allows a type of irrelevance.[65] Like the Albricini, the Porretani required at least a casual *connection* between the antecedent and consequent of a true conditional. Thus when one conjunct of an antecedent is not used in obtaining a putative consequent, to take

[64] A similar theme is to be encountered as Australian naturalism, a contemporary school that is very 'scholastic' in its practice, edging crabwise from ad hoc distinction to special counterexample-defeating distinction. This school has not got around however to putting its restrictive theory of properties to work in the logic of conditionals.

[65] It is not clear that the relevance requirement on its own is sufficiently general; for example let $B_\circ$ be a statement duly relevant to $A_\circ$ such as an appropriate compound of it. For connexive success, at least all inconsistent conjuncts $A_\circ \,\&\, {\sim}B_\circ$ need to be excluded from Simplification. But maybe the Porretani intended to exclude all these, and perhaps ban simplification altogether. Then however fallaciousness is no longer apparent in all cases of simplification removed. A revealing discussion of the fallacy of non-cause-as-cause, which helps bring out the relevant use-in-argument requirement it would instate, is given in Hamblin 1970.

the whole conjunction as antecedent, as in unrestricted Simplification, *is* to commit the fallacy. Indeed they generalised their point, claiming that from

> no copulative do either of its parts follow. The reason is because it is a general principle both with regard to consecution and to inference [that it holds] only if the cause of the consequent or the conclusion is advanced for the consequent or conclusion ... What relevance does one of a pair of coupled antecedents have to the consequent when only the other is the cause. Thus in "if Socrates is a man and an ass, then Socrates is a man", isn't "Socrates is an ass", which is not a cause, advanced as a cause. (CL, II. 26, rearranged).

The resolution, because plainly relevantly based, is thus exceedingly interesting for the history of relevant logic.

A much more drastic amputation of consequences of inconsistent antecedents was adopted by the Meliduneses. They claimed not merely that nothing follows from the impossible (a sentiment congenial to connexivist ideas, and rather like modern views of Wittgenstein and others concerning halting at contradictions), but even that 'nothing follows from the false'.[66] Such strong themes need not be quite as restrictive as they at first appear to be, given for example procedures of default or suppositional inference and backtracking. Briefly, a reasoner proceeds from antecedents whose status is unknown but which are supposed possibly true at least until the evidently false is encountered, whereupon backtracking begins. But the approach, though now fashionable in computing theory and theories of theory change, is fraught with difficulties, yet to be decently resolved. Taken straightforwardly such approaches certainly underwrite distinctive features of connexivism. For if A is possible [or true] then A does not imply $\sim A$, else it would not be possible; but if A is impossible [or false], then as nothing follows from A, $\sim A$ does not. So, either way, $\sim(A \rightarrow \sim A)$.

---

66 Parallel themes, not yet taken to be bizarre, are still found nowadays, for example "the false is never a cause", "nothing that is false is a reason". The Melidune theme itself appears to have been adopted by Frege, who held that there cannot be an argument with false premisses! (It is thus easy to see, from one angle, why he was happy with material implication; cases with false antecedents could, as Don't Cares, be treated as convenience and simplicity appeared to dictate). The claim destroyed usual hypothetical proofs; thus reductio proofs, in particular, had to be paraphrased as perfect conditional syllogisms: see Dummett 1981, pp.308-10. Again we owe these details to Martin.

5.4 Later medieval theories of implication: further adjusting 'history'

Although the great twelfth century debate did not persist undiminished into the thirteenth century, but gradually faded away, containment and connexive thinking did not disappear. Kilwardby, writing in the first half of the thirteenth century, appears to have adopted that characteristic combination: a meaning-inclusion view of implication, and a cancellation picture of negation, with a connexive outcome. Thus 'what is understood in some thing or things, follows from it or from them by a necessary and natural consequence; and so of necessity if one of a pair of opposites in repugnant to the premisses ... the other follows from them ... [In particular], if one of the opposites stands, the other cannot', i.e., if $A \rightarrow B$ then $\sim(A \rightarrow \sim B)$ (see Bocheński 1961, p.199). Kilwardby's connexive theory of implication, not so far very thoroughly investigated, perhaps also included some stronger connexive principles, reminiscent of Boethius and flirted with in recent conditional logics, for example the principle $\sim(A \rightarrow B) \rightarrow .\ A \rightarrow \sim B$ (strengthening the material chain principle $\sim(A \rightarrow B) \supset .\ A \rightarrow \sim B$). But whether the final theory was hyperconnexive or not, it contained some serious tensions. For (by contrast with Abaelard and Boethius) Kilwardby also asserted that 'a disjunctive follows from each of its parts, and by a natural consequence; for it follows it you sit, then you sit or you do not sit' (Bocheński 1961, p.199). But, as Kilwardby was well aware, $A \rightarrow (A \vee \sim A) \,\&.\, \sim A \rightarrow (A \vee \sim A)$ counters Aristotle's and his own theme that the same result cannot follow both from a statement and its negative, i.e., $\sim(A \rightarrow B \,\&.\, \sim A \rightarrow B)$.

Kilwardby confronted two waves of objections to Aristotle's theme, objections apparently by that time standard. The first objection was based directly on the positive paradox: that where B is necessary (as for example "God exists") anything whatsoever implies it, in particular both A and $\sim A$ do. Kilwardby easily surmounted this objection, much as Abaelard had done before him, by distinguishing a natural (or natural and necessary) consequence operation, different from "accidental consequence" which is paradoxical. The positive paradox naturally fails for natural consequence. Such a relevant natural consequence connection is again characterised through content containment: the consequent is understood in the antecedent.[67] The second larger wave involved a relevant variation on the first; the argument replaced the arbitrary necessary truth by a specific

[67] This containment account runs right through medieval thought, from Abaelard in the twelfth century, through Kilwardby, Ockham and Strode, to Paul of Venice in the fifteenth century. The story continues, through Javellus and others in the sixteenth century, Leibniz in

one, Excluded Middle in the statement A concerned, i.e., it deployed non-expansive cases of the lattice principle of Addition. This wave may have dumped Kilwardby. For, by contrast with Aristotle, he conceded the parameter-sharing Addition principles as supplying natural consequences. Worse, he proceeded to commit Aristotle to these (apparently) new-fangled lattice principles. Aristotle had meant, Kilwardby maintained, only to deny that the same result could follow naturally in virtue of the *same part of itself* both from a given statement and its negation (see Kilwardby, quoted in Kneale and Kneale 1962, pp.275-6). This amounts to special pleading, geared only to a counterexample based on a disjunctive conclusion like $A \vee \sim A$ where different parts, A and $\sim A$, may serve as antecedent. Kilwardby could more profitably have engaged in further division-chopping, distinguishing a *connexive* natural consequence under which $A \rightarrow. A \vee \sim A$ and its mate fail. An appeal to the ideas of Peripatetic progressive argument, for instance, could have made such a distinction both attractive and historically well-founded. Moreover, such a distinction would have saved his own connexive position.

The vigorous debate of the twelfth century, partly focussed on connexivism, appears to have dissipated in the thirteenth century, when connexivism was slowly routed (with the advent, as above, of lattice connections). By the fourteenth century a strict position had, so it is usually said, become dominant, both as regards the consequence relation and as regards the coupled theory of conditionals. As to the coupling, of some sort, there is little doubt. 'There was always supposed to be a close connexion between the validity of a *consequentia* and the truth of a conditional statement' (Kneale and Kneale 1962, p.293). For example, 'according to Ockham, a consequence is a hypothetical, conditional proposition [and it is true if and only if its antecedent implies (*infert*) its consequent]. That means that a consequence is composed of at least two categorical proposition which are joined by the syncategorematic terms "if-then" or their equivalents... both parts may even be impossible' (Boehner 1952, p.55, insert from Moody 1953, p.66).[68]

What is taken to be the mainstream theory of formal (pure logical) consequence—presented clearly and simply enough by the shadowy Pseudo-

---

the seventeenth, to the present. Moody , who locates it in Strode and Paul, dismisses it, in good logical positivist fashion, as giving a psychological definition (Moody 1953, p.71). But that is to make the crude if common confusion of the more highly intensional with the psychological.

[68] Strictly, the coupling is more complex. Even if the logical truth of an implication (as distinct from that of a conditional) is tantamount to the correctness of a consequentia, the later directly supplies an inference warrant while the former does not. Thus they differ in their relations to the operation of inference, or detachment.

Scotus (sometimes identified as John of Cornubia) perhaps early in the fourteenth century—was, in broad outlines, a strict implicational one; but exact details of the theory were *much* debated. At bottom, Pseudo-Scotus required that for a consequence to be valid it is impossible that the first component (the antecedent) be true and the second (the consequent) be false.[69] Unlike earlier logicians who had used paradoxes of implication to discredit such strict accounts of formal consequence, Pseudo-Scotus tried to make the paradoxes appear inevitable and essential features of the notion of consequence—by establishing them through what amount to the modern 'independent' arguments to them. Such modal views were, however, by no means as widely accepted as the recent conventional wisdom would have.

In a concluding chapter on consequence Ockham sets down general principles for the notion—for *following from*—which were to be repeated with variation, and additions, in many later works on the topic. Unlike some of his commentators (notably Boehner), Ockham did not shrink from setting down, at the end of his resume, the evident paradoxes of strict implication: 'Other rules are given', says Ockham, as if he wanted to dissociate himself from these simple consequences, '(10) *Anything whatsoever follows from the impossible*; (11) *What is necessary follows from anything whatsoever*'.[70] Even so, Ockham evidently felt, like many medieval logicians, considerable reservations about these principles. For he proceeds to comment: 'But such consequences are not formal; neither should they be much used nor are they'. He was not claiming, in contemporary dishonest fashion, that these paradoxes are not or never could be used (but were, so to say, recorded, to complete the books honestly?). Of course they are used: in particular to halt discussion arriving at inconsistency, to exclude logical investigation of what is impossible. Ockham says they should not be used, much (?). More important, he appears, in saying they are not formal, to have been contending that they were not *correct*. For, firstly, '... in medieval latin *formalis* seems to have some of the associations which "proper" has in modern English' (Kneale and Kneale 1962, p.292); but, in addition, Ockham specifically required in

[69] Psuedo-Scotus proposed the following definition of *consequence*: 'A consequence is a hypothetical proposition composed of an antecedent and consequent by means of a conditional connective or one expressing a reason (*rationalis*) which signifies that if they, viz. the antecedent and consequent, are formed simultaneously, it is impossible that the antecedent be true and the consequent false' (Bocheński 1961, p. 190). For a discussion of the basic definition and the careful qualifications Pseudo-Scotus sometimes added, see Kneale and Kneale 1962, pp. 277-288.

[70] The rules of *consequentiae* are from Ockham's *Summa Totius Logique*, iii(iii), 37. As well as being repeated in many medieval texts, they are quoted in several contemporary commentaries, for example Kneale and Kneale 1962, p.291.

a formal consequence some necessary *connection*, for example through a topical rule. According to Kneale and Kneale 1962, 'it is clear that Ockham wishes to call a *consequentia* formal if, and only if, it involves a necessary connection in the stricter sense of which Abaelard spoke' (p.290). Really then, the paradoxes (10) and (11) hold at best for material consequence (such as lax enthymematic arguments), but not for consequence proper. Moreover, the way in which 'the material consequence [is] put aside' in the British medieval tradition 'can only be taken to mean that the material consequence is not considered a consequence in the strict or full sense of that term' (Green-Pedersen 1983, pp.286-7).

To arrive in the thick of what was at issue sociatively, it helps considerably to introduce certain of the basic classifications of types of consequence deployed in medieval theory. One important classification, made in varying competing ways, can be represented thus:

Consequences (inferences, implications)
↙ ↘
Sound *(bona)* | Unsound *(mala)*
↙ ↘
Formal (Literally, defined in terms of form, so substitution for categorematic terms does not affect validity) | Material (Literally, depends also on matter, so affected by content of terms)
↙ ↘
Simple, absolute (hold simpliciter, without temporal condition) | As-of-now (hold when stated)

The distinction between simple and as-of-now conditionals is roughly like that between Diodorean (temporally strict) and Philonian (material) 'implication', at least according to mainstream perceptions (see Moody 1953, p.70 ff). These material types were largely ceded to the forces of irrelevance. At least, there appears to have been fairly wide agreement that what amount to paradoxes of implication hold for material consequence types. What was primarily at issue was whether such paradoxes were *formally* valid, genuinely valid. But even as regards material consequence, simple consequence in particular, there were cross-currents of sociative significance, as a different classification of material consequence starts to reveal.

This interesting cross-classification was 'between consequences valid by reason of some connection of supposition or meaning between a term occurring in the antecedent and another term occurring in the consequent [materially relevant consequences], and consequences valid only because of the falsity or impossibility of the antecedent, or because of the truth or necessity of the consequent [materially irrelevant consequences]' (Moody 1953, p.73). Ockham and successors, such as Strode, considered such relevant consequences *adequate, formal* in their sense. But materially irrelevant consequences are dismissed as 'merely material' by Strode, who illustrates his point with an example which soon became entrenched; 'If some man is a stone, then a stick stands in the corner'. The rules, for merely material consequences are, of course, the paradoxes themselves. Strode too says of such rules, 'But such consequences are not formal, so that those rules are not much used' (quoted in Moody 1953, p.74 ff). Since the principles supposed to govern as-of-now consequences were also essentially material, 'not all the medieval logicians accepted the consequence "as of now" as a valid consequence, and those who did recognise this sense of implication indicated that it was not of much importance for scientific purposes' (Moody 1953, p.79). This scientifically unimportant "implication" they realised could be defined truth-functionally, for example through $\sim(A \mathbin{\&} \sim B)$.

Ockham was far from alone in his reservations and apparent dissent from the modal position, surprisingly veiled given some of his other political activities. Strode, for one, took the distinction between material consequence—which was conceded to be paradoxical, but neither much used or useful—and the stronger formal consequence—which was not paradoxical, but required connection—in much the same fashion as Ockham. But Strode is more precise about the more rigorous connection involved in consequence proper. 'According to him every valid material *consequentia* allows for inference (*illation*) from the antecedent to the consequent, but a formal *consequentia* involves in addition a connexion of meaning or understanding (*intellectio*) between the antecedent and the consequent' (Kneale and Kneale 1962, p.292). The connectional requirements of Abaelard persist, as do connexive features (see Strode, for example f.12rb-va). Such an account of formal consequence, in terms of meaning containment, occurs frequently in British (but not Parisian) works of the fourteenth century (cf Green-Pedersen 1983, p.287). Ferrybridge also appears to have shared Ockham's reservations, but he is less specific than Strode. He does say however, of non-formal 'arguments to the necessary from something not specially relevant', that he does 'not consider that kind of argument valid'. It is interesting that the paradoxes are given in the shape of arguments involving something 'not specially relevant'. Regrettably no information is so far available on how exactly these medieval Englishmen proposed to break the known arguments to the paradoxes; but later commentators on Strode do offer suggestions.

It is not difficult, moreover, to ascertain how Burleigh's theory of consequences apparently avoided the paradoxes (thanks to recent reconstruction). Burleigh, a contemporary of Ockham, who appreciated the logical priority of propositional logic and developed it to a considerable extent, elaborated a system of absolute (i.e., simple) consequence. The system included neither the paradoxes of implication nor (unqualified) simplification principles.[71] As it stands, it is a sociative system almost in direct line of (unhistorical) descent from mainstream Stoic logic (perhaps through twelfth century schools). It is very like Stoic logic as tentatively sketched earlier, but enriched by modal functors and an inclusive disjunction V, (with $A V B \leftrightarrow . \sim(\sim A \circ \sim B)$. Burley's demodalised system, formulated in terms of connectives $\rightarrow$ (for consequence), $\circ$, V, $\sim$, $\leftrightarrow$, with $\therefore$ indicating the rule linkages, takes the following lines (to adapt Boh's 1982 reconstruction, pp.313-4):

$A \rightarrow A$		$A \rightarrow B \therefore B \rightarrow C \rightarrow . A \rightarrow C$
$A \rightarrow \sim\sim A$	(the principles are	$B \rightarrow C \therefore A \rightarrow B \rightarrow . A \rightarrow C$
$\sim\sim A \rightarrow A$	implicit only)	$A \rightarrow B \therefore \sim B \rightarrow \sim A$

The implication-negation fragment thus appears to be exactly the same as that of basic system B of relevant logic. As with Stoic logic the differences (if any) come with $\circ$ and V.

$\sim(A \circ B) \leftrightarrow . \sim A V \sim B$	$A, B \therefore A \circ B$
$\sim(A V B) \leftrightarrow . \sim A \circ \sim B$	$A \rightarrow B, A \rightarrow C \therefore A \rightarrow B \circ C$
$\sim(A \rightarrow B) \leftrightarrow . A \circ \sim B$	(Burley states only indicative special cases of this rule)

(The last left-column principle is especially striking, and part of what justifies taking Burleigh's conjunction as intensional. Had he been working in a

71 The argument, too dependent on omission as opposed to explicit rejection, is hazardous—especially as medieval logicians commonly omitted assumptions they used. And some contrary Burleigh material may, all too easily, come to light! It should be added that there are also some problems with Boh's reconstruction on which the argument tends to rely. For Burleigh made various qualifications to syllogistic and to nonsyllogistic reasoning, as if they satisfied some different principles, which Boh passes over (see also Moody 1953, pp.92-3).

conventionally ascribed strict system his consequence system would collapse into a material one, i.e., absolute consequence would collapse to material, as-of now).

$$A \circ B \rightarrow C \therefore A \circ {\sim}C \rightarrow {\sim}B$$

etc.

(In fact Burleigh only formulates antilogistic principles for *syllogistic* consequence, where it takes literally such forms as $A, B \therefore C \,/\, A, {\sim}C \therefore {\sim}B$ and permutations thereon.)

Such a consequence system is undoubtedly sociative, as it is (properly) contained within the corresponding system R of relevant logic.[72] The argument depends upon the representation of Burleigh's conjunction and disjunction as intensional, and not lattice-theoretic. There is some evidence that an intensional conjunction was in use about that time in England. Strode formulates a rule which is classically invalid, but which can be given good status using intensional conjunction, namely 'If one argues from the opposite of the consequence to one of the premisses, the opposite of the other premiss follows' (see Moody 1953, p.93).

To locate the dwindling medieval mainstream, travel to the neighbouring Continent helps and may be required; perhaps best is a visit to Paris and to the harder-line modal school that formed around Buridan. Buridan shared none of those reservations felt earlier and elsewhere about the paradoxes as formal consequences. He offered the same 'independent' proof of the negative paradox as Pseudo-Scotus, from which he concluded that the consequence was formally valid.[73] Influential logicians taught by Buridan felt none of Ockham's reservations or Strode's disbelief, perhaps as some sort of strict position became established in Paris. Thus Albert of Saxony, building on Buridan, felt sufficiently confident to present the paradoxes as first and second rules of (simple) consequence (see Bocheński 1961, p.200: note again however, the common qualification to *simple* $\approx$ *strict*). But even among such staunch supporters of the mainstream Parisian story as Buridan and as Albert, there were some interesting twists and variations (with truth playing an essential non-redundant role in Albert; see Boehner 1952, pp.71-2). All

---

[72] Burleigh's modal principles can also be included in an appropriate sublogic of $\Box R$. The main principles are the connected group, $A \rightarrow B \,/\, {\sim}\diamond(A \circ {\sim}B)$; $\diamond(A \circ {\sim}B) \,/\, {\sim}(A \rightarrow B)$; $\diamond A, {\sim}\diamond B \,/\, {\sim}(A \rightarrow B)$.

[73] Buridan claimed 'Ad omnem propositionem copulativam ex duabus invicem contradictoris constitutam sequi quamlibet aliam ... consequentia formali' (18, 7a). See also Moody's 1953 discussion of Buridan's emumerations of the types of formal consequence (p.77 especially, also p.88, p.90).

this adds to warranted speculation that the mainstream position, weakened by division and dissension within, held only an uneasy, dubious dominance—at least by comparison with the contemporary scene.

The contraction of the supposed mainstream to the outflow of a Parisian school gives a very different look to high medieval logic. It is worth briefly reassessing the standard contemporary argument, perhaps best assembled in Moody 1953, for the modal picture of medieval logic. What Moody's interesting, but tendentious, depiction of the medieval logic of propositions as a strict implicational system—Lewis system $S3$ so it is claimed[74]—may do some justice to is only this: the consequence system of Buridan and his students at Paris, or more narrowly the views of Buridan and Albert of Saxony. To arrive at his medieval strict system Moody proceeds by the evidently invalid method of selectively amalgamating the statements of several, disputatious, medieval logicians, namely Buridan, Albert of Saxony, Ockham, Burleigh, Strode and Paul of Venice, whose working lives furthermore spanned almost a century. He proceeds to compare 'their' principles with those of Lewis and Langford (and also to a subset of those of *Principia Mathematica*: see p.82 ff). The net result of such an exercise should really be, as it would be if attempted for contemporary logic, the trivial system (in the given morphology), not $S3$ or $S2$. For example, Burleigh's thesis, $\sim(A \to B) \leftrightarrow.\ A$ and $\sim B$, yields classical logic, whereupon Strode's nonclassical principles yield triviality. But for reasons already advanced, Strode, Burleigh and Ockham should be counted out, for they all dissented, in one way or another, from such a modal pasiche; so also no doubt should Paul of Venice (as we shall see, and as Perreiah contends). That leaves Buridan and his former student Albert, i.e., the Parisian school.

It is cvident, then, that rival positions to the dominant strict position were certainly not lacking; and these persisted through into later medieval times. Certainly there were many different positions adopted as regards the issue of correct conditionals, and these conditionals continued to be taken as virtually tantamount to valid consequences. But again appropriate details of many of these positions, who maintained them, the extent of their support, and so on, remain very scanty. Insufficient research into less orthodox positions and thinkers is no doubt one source of the trouble, of this information gap. Fortunately something, to build upon, is known.

---

[74] But it should be $S2$, it seems. The evidence for $S3$ rather than $S2$ (i.e., $A \to B \to.\ \Box A \to \Box B$ instead of $A \to B\ /\ \Box A \to \Box B$) is flimsy at best. The evidence—from Strode and Paul, both to be counted out—suggests the rule form only.

By later medieval times, the number of interpretations of the conditional under discussion, but four or so in ancient times, had expanded to at least ten. In his *Logica Magna*, Paul of Venice listed, and summarily rejected, ten different, apparently extant, accounts of the truth of a conditional (and of the 'meaning of implication' as Bocheński 1961 puts it, pp.145-6), and these accounts do not appear to span all ancient positions. Several of the accounts cited certainly *admit* of sociative elaborations. Whether they actually attained such elaboration is another matter; however it seems likely, given the later logical ubiquity of the paradox arguments, and the apparent resistance to them. But how they were elaborated, or would have been— that sort of logically crucial information is again lacking. In typical medieval fashion Paul of Venice does not even give the names of medieval authors or schools who held the various views he mentions.

The variety of the views does however indicate both considerable dissatisfaction with the mainstream strict style of account, which had not stabilized to invariant agreed form, and also that no consensus as to the trouble with arguments like the paradoxes was emerging, as almost every partway plausible counterview was still being tried for size.

In Paul of Venice's list strict accounts tend to come first, with the first being in fact the first of the strict (modalist) views considered, and also counter-exampled, by Pseudo-Scotus (see Kneale and Kneale 1962, p.286, p.293). According to this account, 'for the truth of a conditional is required that the antecedent cannot be true without the consequent', "without" no doubt being read as "and not" (see the Buridan text elaborating the account, Bocheński 1961, p.196; otherwise, were "without" read intensionally, a very different theory than the mainstream strict one could emerge). Among positions which readily admit of a sociative construal are at least these:

- The third, where 'people have said that for the truth of a conditional it is required that it is not possible that the antecedent of that consequence be true unless the consequent be true':

  and

- The fifth, where 'people say that for the truth of a conditional it is required that if things are as is signifiable by the antecedent, necessarily things are as is signifiable by the consequent'.

Among positions which almost certainly obtained some sociative elaboration are the last three given:

> Eighthly people say that for the truth of a conditional it is required that the consequent be understood in the antecedent...
> Ninthly people say that for the truth of a conditional it is required that the adequate significate of the consequent be understood in the antecedent.
> Tenthly people say that for the truth of a conditional it is required that the opposite of the consequent be incompatible with the antecedent... (Bocheński 1961, p.696).

The tenth position, which superficially resembles that of Chrysippus, is here already distanced from strict interpretations. The eighth position, which resembles that of Abaelard's school, is tantamount to that of Strode. Evidently sociative accounts of implication and conditionality had by no means been entirely overwhelmed, or died out, by the fifteenth century, even if, as contemporary mythologic has it, they had been pushed into minority positions far down the lists.

The position of Paul of Venice, often represented as in the mainstream of medieval logic, is instructive, illustrating something of the complexity of later medieval theory. Having argued against *all* the positions listed, he says that 'my own opinion is that we cannot give any completely general account of what is sufficient or necessary for the truth of conditionals, for they are divided according to a variety of differentiae because of the variety in which they occur'. But his opinion comes close to the tenth position, appearing to amount to a case by case complication of that (and so indeed to be reducible, despite what he asserts, to a very complex set of necessary and sufficient conditions). Moreover, in the *Logica Parva* he characterises a true conditional (and similarly a *bona consequentia*) as that in which the opposite of the consequent is repugnant to the antecedent (p.131, p.167), with such repugnance glossed in terms of *unimaginability* of the denial of the consequent with the antecedent (thus a modal-style reduction of relation *is* reached, but through imaginability or conceivability). Paul of Venice's position if *not* a sociative one, tying the tenth account with connection; but *nor* is it a straightforward modal position by any means, among other things because imaginability differs from possibility (on medieval perceptions those properties only properly overlap one another, cf. p.51). Thus a different perspective would be expected towards implicational paradoxes. And indeed in the *Logica Magna*, he argues against the ninth account, a refinement of the eighth, that the popular post-Ockham conditional, "If you are something other than yourself, a stick is standing in the corner", 'is true and necessary, and yet what its consequent signifies is not thought about in the antecedent because it is not signified by the antecedent or by any part of it'. Plainly the antecedent and consequent are not connected (except in that tenuous modal sense). Though Paul offers a formal deduction of "A stick is standing in the corner" from "You are other than yourself", he apparently does not hold the standard view that every consequence with an impossible antecedent is valid. Such a principle holds good only when "impossible" is taken in a certain strong sense (presumably that of unimaginability), which the example given "You are other than yourself" meets.[75] Elsewhere however (*Logica Parva*, pp.167-8), Paul repeates the familiar paradoxes of implication (statements (10) and (11) of Ockham), and it certainly looks as if they can be derived from principles he gives, if not from the characterisation he offers. That Paul of Venice does not represent the acme of medieval logic is now conventional wisdom; but nor does he signal the later decline. Some of the most interesting medieval developments in sociative logic were still to come.

In the late (perhaps *post*-) medieval work of the Cologne school (active around 1493) the argument to the hard negative paradox through Disjunctive Syllogism was stopped in its tracks, in a way anticipating that of contemporary relevant logical theory. The argument to *ex falso quodlibet* $A \,\&\, {\sim}A \;\rightarrow\; B$ (or inferentially $A, {\sim}A\,/\,B$) was broken, as in relevant logics, by rejection of Disjunctive Syllogism (in inferential form licensing general inference from ${\sim}A$ and $A \vee B$ to B). It was destroyed on the elementary grounds that where both ${\sim}A$ and A are assumed, ${\sim}A$ cannot *also* be legitimately used to rule out A in the disjunction $A \vee B$. Failure to grasp this elementary point, easily explained to novices in logic or volunteered by them, has not merely diverted the whole course of contemporary logic into philosophically disastrous classical ways, but has also helped in substantially distorting the history of medieval logic, with the result that nonclassical positions are not taken very seriously or examined carefully or at all. The procedure of the Kneale and Kneale 1962, who, unlike many (unsympathetic) commentators, do give some small coverage to nonclassical positions, is instructive. They condense the main paradox argument from Pseudo-Scotus in the following tree of derivation:

---

[75] See Hughes' commentary on the *Logica Magna*, attached to his edited translation, from which the unpaginated quotations from Paul earlier in the paragraph of the main text are drawn. George Hughes kindly supplied this material before publication; hence the lack of page numbers. A preliminary comparison of the different theories of consequence of *Logica Magna* and *Logica Parva* is given in Perreiah's appendix to the latter edition, pp.336-7.

Socrates exists and Socrates does not exist		Socrates exists and Socrates does not exist
↓		↓
		Socrates exists
		↓
Socrates does not exist		Socrates exists or a man is an ass
↘		↙
	A man is an ass	

> As the Pseudo-Scotus says, each of the *consequentiae* used in the derivation is formally valid; and so, paradoxical as it may seem, no one can reject it without rejecting something essential to primary logic (Kneale and Kneale 1962, p.282).

After that supposedly compelling demonstration they have a difficult time giving any credence to accounts of consequence which dismiss or try to avoid paradox.

But suppose it holds or is granted (per impossible, but what is impossible can be taken as an hypothesis) that Socrates exists and Socrates does not exist; but it does not hold or is not conceded that a man is an ass (or *Baculus stat in angulo* or such like). Then, applying normal *and* and *or* principles, both "Socrates does not exist" ($\sim A$) and "Socrates exists or a man is an ass" ($A \vee B$) hold, but "A man is an ass" (B) does not. Accordingly, the inference from $\sim A$ and $A \vee B$ to B fails, for it leads from what holds or is granted to what does not. The inference flunks an evident generalisation of the truth-preservation form, as in RLR I). The idea that $\sim A$ eliminates A in $A \vee B$ is based on that assumption that not both $\sim A$ and A can ever hold; but that assumption breaks down, because precisely what has been granted is that both do hold. Thus it is not true that Pseudo-Scotus's argument is formally valid (step by step), unless "formally valid" is assigned, in an issue-begging way, some sort of strict construal. On the holding account indicated it is not valid at the last stage. Rejecting the step involved, as the did, and as relevant theorists now do, is not 'rejecting something essential to primary logic', unless "primary" is given a partisan construal. Rather it is to reject the extrapolation of a principle, good no doubt for negation-consistent situations, beyond its legitimate range of application.

The Cologne commentators, analysizing the paradox argument in the form given, leave no doubt that the last (junky syllogistic) step cannot be accepted. The reason given is that the initial premiss, of the form $P \,\&\, {\sim}P$, can be taken in two different ways: either absolutely, as an asserted contradiction, or hypothetically, for the sake of argument. 'In this proof, it is accepted for the sake of argument, and since both P and $\sim P$ have thus been conceded, one cannot use part of the formal contradiction to deny the other part. That is, $\sim P \mathbin{\&} (P \vee Q) \rightarrow Q$ has to be rejected!' (Ashworth 1974, p.135, symbolism adjusted). The same grounds for rejection of Disjunctive Syllogism—as distinct from the rule γ of Material Detachment which is admissible, granted as always in those days that the actual world is consistent—are equally explicit in the subsequent work of de Soto (of 1529):

> ...A pair of contradictories can be taken in two ways, one way absolutely and without any supposition, and then the one is destructive of the other, that is, the truth of the one takes away the truth of the other. In the other way, they are taken as conceded by someone from some assumption for the sake of argument to see what follows, and then neither destroys the other, for conceding that one is true, it does not follow that the other is false, since both are conceded to be true. So the inference
>
> Peter disputes or a man is a stone
> and Peter does not dispute,
> hence a man is a stone,
>
> if those contradictories are taken absolutely, is valid. But if they are taken as conceded in some initial premise, it is not valid, for from the fact that a negative is true it does not follow by the assumption given that the affirmative is false, since in the cause of disputation both were conceded (f.74va).

The contrast, between obtaining absolutely (or truth) and holding by assumption, is tantamount to that between actual world (world T) semantical assessment and hypothetical (off-T) assessment. A formally valid principle must be more than merely admissible; it must hold for suppositional situations also. Thus Disjunctive Syllogism is *not* formally valid. Observe however that de Soto assumes a cancellation account of negation for absolute cases; indeed he goes further, rejecting implications from the impossible (for example 'God does not exist') to the necessary ('God exists'), apparently on the basis of Abaelard's principle $\sim(\sim A \rightarrow A)$. If so, de Soto is committed to (what is viable enough, though incompatible with relevance logic) a *connexive* relevant framework. But it may be that de Soto intended only to emphasize his repudiation of the paradoxes and of modal construals of the Chrysippean definition of consequence (a consequence is valid iff the negation of the consequent is incompatible with the antecedent) which lead to them. De Soto, like others such as Blanchellus Faventius who adhered to that, was not convinced by the modal reduction of incompatibility. He sarcastically remarked that 'although "the moderns" accept "if God does not exist, then God exists" on the grounds that the impossible leads to anything he could not bring himself to accept an inference of this form'. He also pointed out he was hardly on his own, appealing 'to common usage to support his doubts about the paradoxes: who would say that if you are a stone, it follows both that you are and that you are not?' (see Ashworth 1974, whose account of late medieval activity is heavily exploited, p.127, p.135). Consistently with his stance, he supplied a framework of principles for consequence and conditionality which is appropriately relevant, lacking both paradoxes and their source, Disjunctive Syllogism (p.184). The Cologne framework—although apparently based on a different (through reconcilable) definition of consequence: containment rather than Chrysippean—was also bound to be duly relevant; for those commentators espoused a principle of connection of terms for valid consequences.

While the clear emergence of a nondissembling position is the most exciting sociative development in later medieval logic, it was by no means the only event of relevance.[76] The earlier British tradition, set out in textbook form by Strode, gained a following and obtained different elaborations, a significant one of them plainly nontransitive. Cajetan of Thiene not only, like Strode, adopted an information-inclusion account of formal consequence and accordingly rejected the paradoxes as formally valid; as well he explained that valid consequence implies connection, that the terms involved are *linked*, and went on to discuss several interpretations of Strode's requirement that the consequent be formally understood in the antecedent (for a summary, see Ashworth 1974, p.129). Strode and Cajetan were followed by 'the Cologne commentators, Greve, and John of Glogovic [who] all seem to have accepted the claim that the consequent should be understood in the antecedent without comment; while Major added that a relation of pertinence between the terms was required' (p.130). Here is our first sighting of the connectional requirement explicitly identified as one of relevance.

---

[76] Much else of sociative interest remains to be investigated in the later medieval period; for instance, more precise details of the logical frameworks various schools and authors were committed to. Further logicians whose work may repay study include Blanchellus Faventinus, Fonseca, Kesler, Maiolus, Juvallus, and Regius.

The importance of relevance, and its failure for material and strict consequence, was also clearly recognised by others who adopted a containment story, notably Ferebrich and the unknown author of *Libellus Sophistarum*. According to the latter, in rejecting paradoxes of formal validity,

> There are ... three kinds of materially valid consequence, that whose validity depends on the terms employed, that where the antecedent is both impossible and irrelevant to the consequent, as in "A man is an ass, therefore the stick is in the corner", and that where the consequent is both necessary and irrelevant to the antecedent, as in "You ran, therefore God exists". ... of course ... there were some formally valid consequences whose antecedents were impossible, such as "A man is an ass, therefore a man is an animal", but these held by virtue of some other rule (Ashworth 1974, p.135).

Subsequently some went further; Javellus, himself averse to the paradoxes and committed to a formal containment story, reported that 'realists would not even allow the paradoxes to be materially valid, because they exhibited no inferential link, no dependence and no incompatibility between the antecedent and negated consequent' (p.135). They were not alone; the Cologne commentators required connection of terms for both usual types of material consequence.

Important, and importantly different, elaborations of the formal containment account were also offered, along with (different) suggestions as to what went wrong in the paradox arguments and which consequence were formally invalid. The author of *Libellus Sophistarum* explained formal containment as holding just when in making the antecedent true the consequent is made true or verified, i.e., effectively in truth preservational terms. That text also delivers an elegant if weak logical system, which (by contrast with the disappointing formal theory of Major, p. 179) is duly relevant (pp.182-3, pp.275-6). Given that the system supplies Simplification, Addition, Composition principles *and* Disjunctive Syllogism, but not presumably the paradoxes (which are *not* cited), Transitivity (also not given, though it was standard principle) has to fail, both syntactically, and less plausibly, semantically for verification. For *understanding in*—as distinct from formal or intelligible *inclusion*—eventual failure of transitivity is no doubt to be expected; it is just surprising that it is bound to happen in sequences as short as the paradox arguments. That 's formal containment account *did* lead to failure of Transitivity was explicitly stated by Sermonete in his commentary. He 'said that a consequence of Strode's view was that something could follow formally from the consequent of a valid consequent without so following from the antecedent, since in the case of the paradox "c is understood in b, b in a, and c is not understood in a".'(p.135). The British nontransitivity bent reaches a long way back; but it was then, as now, a minor affair. The moderns of later medievaldom were heavily committed, like the modern moderns of later capitalism, to modal-style theory (as well as to the reference theory). By contrast with present complacency, however, the modal position did not enjoy an easy or stable dominance. While a few logicians did accept the received strict account of consequence validity, 'many felt that it expressed neither the necessary nor the sufficient conditions for validity' (p.126). Even within the ranks of the faithful then, all the details were not merely up for reassessment, but regularly being reassessed.

Even in those medieval centres where a modal position of some sort became dominant after the thirteenth century and the logical canon, and other approaches declined to minority concerns or were abandoned, certain key elements underlying relevant positions and more satisfactory theories of conditionals, such as containment accounts, did not disappear entirely. Rather they continued to play a part, sometimes a significant part, in the ancillary theories that helped to prop up the dominant position; most notably, in the theory of what ensues from, and what commitments are made, with false and even certain impossible assumptions. A contemporary parallel lies in the way the dominant logical position is hedged about by (competing) theory-saving supplements, notably theories of counterfactual and counterlogical implication and recently modal theories of conditionals. But whereas contemporary supplementation of the dominant paradigm takes no due account of connection and relevance, and really has as yet no approved theory for accommodating impossible assumptions, medieval supplementation gave sociative considerations an important place. Moreover, the damaging paradoxes of implication were only incorporated into the later medieval logical canon as regards certain types of consequence and inference, such as material consequence. Not only for many situations, most conspicuously those where impossible assumptions operate, but also for significant types of consequence, including that of valid (*bona*) consequences according to several authors, the strict canon was suspended or ditched. Even so, such a 'high' medieval approach, and onus of organisation, resembled that of liberated contemporary positions rather than a deep approach. It was, after all, variation *from* the strict canons. By contrast, what deep sociative approaches do is to *reverse* all this. A more general theory which can accommodate impossible assumptions is sought from the outset; then special relaxed or shortcut procedures are introduced where consistency or other prized properties (for example truth) can be assumed.

One relevant medieval supplement to the main logical theory, of consequence and of signification (or reference), was that of obligationes (or commitment); another was that of insolubiles. The theory of obligationes, a partly formalised rule-controlled kind of dialogue, guided a method of training in dialectic.[77] The business of obligationes concerned both the theory and practice of dialectical argument; that is, it concerned argument, according to prescribed rules, between disputants. Obligationes discussions—of which there were several types (with rules also varying from place to place)—were closed under restricted (sociative) consequence. A standard introduction to obligationes texts runs along these lines: 'Obligation is a certain art whereby some opponent can bind a respondent to reply at the opponent's pleasure to the obligatory sentence posited to him. Alternatively, an obligation is a sentence by virtue of which someone who is obligated is committed to reply affirmatively or negatively to the given sentence' (Billingham's rather full introduction, quoted by Ashworth 1983, pp.309-310). One important kind of obligationes disputation was that called *positio*, in which the opponent presents the respondent with a simple declarative proposition, the positum, which he is obliged to maintain, while the opponent seeks to draw him

[77] There is now an expanding literature on obligationes, in some of which it is said that the earlier literature on the topic is not very satisfactory, and in much of which it is claimed that the practice is not well understood, a claim given *some* confirmation by the diverse accounts offered of what it is really about. (In this respect it resembles contemporary literature about a main predecessor of obligationes, namely topics.) Thus it has been presented as a set of schoolboy exercises (cf. Stump 1985), as the rules for dialectical exchange, as anticipating the modern axiomatic method (Boehner 1952, p.15), as a theory of counterfactual reasoning (Spade 1982a), as a logic of disputation (Stump 1985). But the prize in these varying presentations undoubtedly goes to Moody, who—after a decidedly misleading description of '*obligationes*... as forms of agreement concerning things to be assumed for the purposes of argument, and the exposition indicates how to state such agreements without having loop-holes which would afford one's adversary the chance of gaining an advantage in the argument'—asserts that 'this topic is not very relevant to logic ... and can conveniently be passed over' (Moody 1965, reissue of a much earlier work, p.294). It seems that obligationes theory can go some way to satisfying several of these varying roles. The theory certainly contained the rudiments of a dialogue logic, and has a good deal in common with modern tableaux theory, tableaux theory with nonstandard rules. And it did investigate what emerged from counterfactual, and sometimes counterlogical, assumptions; and thus insofar as conditionalisation was permitted, it gave a direct guide to which counterfactual conditionals could be sustained. The theory may also have been (as Martin says) the context, otherwise rather mysterious, in which medieval resolutions of semantical paradoxes, naturally arose. Much of the preceding literature on the topic is noted in Spade.

into explicit contradictions. In *positio* the respondent is then required to concede everything that follows, according to the prescribed inference rules, from the proposition to which he or she is committed and anything else already conceded, to deny everything that is inconsistent with all this, and to respond to what is irrelevant as he or she takes it to be the case. In other kinds of obligationes, the respondent has to maintain a proposition about his own response, or what he knows, or some such. Always certain critical factors are involved: a starting *assumption*, or proposition(s) considered, and an investigation by (two) sides, as to what it *binds* holders and opponents to, where it *leads logically*, according to certain predetermined inferential rules. The business plainly differs from the axiomatic method, with which it has recently been assimilated, in that it involves *sides* or participants (characteristically two disputing a single proposition). As each obligation thus takes a certain structured dialogue form, a route to contemporary formalization is evident. Elements of such a theory can be neatly formulated, like parts of dialogue theory which the theory anticipates, in two sided tableaux style, with two columns, one for the respondent and one for the opponent, with the rules figuring rather like tableaux rules.[78]

Obligationes theory intrudes upon connected issues of much interest for sociative logic: firstly, the types of assumptions that can ligitimately be made, and the types of rules that can be applied in such argument situations. It will guickly emerge that because false, and certain impossible, assumptions may be granted, the standard principles of material and formal consequence cannot be applied, but have to be qualified or suspended. It is this adjustment of consequence rules, made in many handbooks on obligationes, that gives the theory its main sociative interest: not what the theory variously says on assumption or on relevance. For there is now—since the Hilbert school, for all its stress on consistency—nothing very striking in the idea of arbitrary assumptions, including those of any grade of inconsistency. The theory of 'natural deduction', for instance, admits *any* well-formed assumptions whatsoever. The worst problems with contemporary logical theory are bound up with where such assumptions lead or are supposed to lead, what they commit adherents to, and so on. On relevance obligationes theory is also a trifle disappointing. For relevance, or 'pertinence' as it is better called,

[78] Such tableaux representation is of course not uniquely determined, even as to format. Rather better than standard two column (O/R) tableaux—which impose excessive constraints on connectives like negation—may be the following scheme of interactive tableaux. Each side, both O(pponent) and R(espondent), is assigned *two* columns, one of which represents what is asserted (or follows from what is granted) and other what is denied (cf. a tableaux formulation of the 'American plan', RLR I, chapter 3).

is defined in such a restrictive way that it is certain irrelevant statements, including relevant impertinent statements, that become of absorbing interest in the development of obligationes dialogues. In that setting B is pertinent to A, $B\ Pt\ A$, iff B follows from A or is repugnant to A, i.e., $\sim B$ follows from A.[79] The relation Pt is reflexive and symmetric, but not transitive. But what discussants, especially those with a sociative bent, want to determine is: what happens when irrelevant statements are infiltrated into dialogues?

The favoured material rules would, if applied to obligationes, have led, like those for mainstream tableaux, to paradox. For instance, in any disputation in which a contingent falsehood is posited, any irrelevant proposition must also be granted! This material paradox derives from the rule for irrelevant propositions, namely, that if a proposition is suitably independent of the positum (i.e., neither follows from, nor is incompatible with the positum, a proposition already granted, or the opposite of any correctly denied proposition), then it must be granted, denied or doubted solely insofar as we know, from outside the dialogue, its truth. The reasoning is like that of familiar paradoxes. Let the positum A be contingent and known to be false, and let B be any irrelevant proposition. Then not-A or B, though irrelevant, is known to be true, and so must be granted, by the rule. As it follows from the two propositions granted, the irrelevant B must also be granted. The upshot is that in any disputation in which a recognised contingent falsehood is posited, arbitrary irrelevant propositions must also be granted.[80] Medieval logicians, such as Burley, were much occupied with such paradoxes. As elsewhere in medieval logic, different rival theories emerged, several of them, like Burley's still unclear. All these theories satisfied, through their qualification or suspension of the standard material rule, *ex falso quodlibet*, a basic requirement for paraconsistency. Not surprisingly some of the theories that were devised anticipated features of contemporary paraconsistent and sociative logics. For example, Swineshed appears to have anticipated the recent non-adjunctive approach. In this theory, while one is sometimes required to concede two contradictory statements in a disputation that begins with a noncontradictory hypothesis, nonetheless, one must always deny the conjunction of these contradictories.

Thus certain standard material principles were suspended in obligationes. For example, Ockham commenting on rules of consequence observes

---

[79] This passage borrows from a communication from Stephen Read on Stump 1985. To Read, who read and commented constructively on this chapter, a considerable debt is owed.

[80] Again by Martin 1986b (whose translation is used), building on De Rijk 1986. For various interesting and curious later restrictions on what could be posited, see Ashworth 1983, pp.314-5.

> That though, in a simple consequence, from something possible something impossible does not follow, nevertheless it sometimes happens that if something possible is affirmed, something impossible has to be conceded and something necessary has to be denied; but this can be done only in the art of *Obligatio* [that is, in the art of purely logical disputation] and only for the course of a given disputation (Ockham 1957, p.88).

While this does not upset the principle under discussion, namely:

$$\diamond A, \sim \diamond B \,/\, \sim(A \rightarrow B),$$

it does overthrow the paradoxes of implication (to which Ockham is reluctantly coming in this passage) given normal logical connections of implication with what is asserted and denied.

What was required of initial assumptions varied, like the rules determining commitment, from university to university and even logician to logician. Although consistency of assumption was *not* in general required, *blatant* inconsistency tended to be excluded, since explicit inconsistency signalled the termination of a game or dialogue. Albert of Saxony, for instance, contended that propositions from which the contradictory opposite immediately followed should not be admitted (Boehner 1952, p.15). Even so *positio* were generally allowed to start from propositions where consistency was in very considerable doubt or had lapsed altogether, such as some important theological propositions concerning the Incarnation and the Trinity. Consider, for instance, the issue of whether the Holy Spirit proceeded from the Son, which involved both their identity and their difference.

As the theory 'developed' it appeared to become progressively more restrictive as to what was admitted in the way of impossible assumptions. In earlier days, before the high, more reactionary, stages of medieval logic were reached in the fourteenth century, assumption was much freer. A most significant text recently unearthed[81] is the *Tractatus Emmeranus de impossibili positione* (TE), apparently produced by the Nominalists, followers of Abaelard, active between 1140 and 1200. There several arguments are advanced to show that an impossible positio should be admitted. The first group moves from an appealing and significant analogy

[81] A notion of pertinence also appears in the medieval literature on insolubilia, where 'a (self-referring) sentence "falsifies itself" just in case it is pertinent to inferring it is false' (see Spade 1982b, p.250, who remarks that 'this notion of "relevance" is a complex one, and needs further study'). But a notion of pertinence directly relevant to contemporary logical concerns *does* appear in late medieval literature.

through to appeal to high authority. 'Just as we assert that what is possible should be conceded in order to see what follows thereupon; similarly we have it from Aristotle that what is impossible should be conceded in order to see what would happen then.' So, as we noticed, Boethius also correctly argued. Next the point is made that we can quite correctly assert things which involve impossibility, for instance that God is a man. Such assertions as these are certainly significant. Furthermore, we can often enough understand what is thus asserted in such impossible assertions; the point is neatly illustrated. But what is asserted or understood can be investigated as a posit, indeed what is understood can in no way be excluded.

Even some highly implicit impossibility is destructive of modal theories of consequence, as the authors of TE proceeded to stress. They contended, invoking the analogy with false positio again, that

> everything that follows from the *positum* is to be conceded, where 'follows' is construed in terms of a proper [*recta*] consequence. And a consequence is proper when the understanding of the consequence is contained in the understanding of the antecedent... [Thus] from an impossible obligation not everything follows. Whence the consequence of the followers of Adam [no doubt the Parvipontani] must not be conceded in this [form of] question in which the understanding of the consequent is contained in the understanding of the antecedent (TE).

Thus proper consequence is supplied through content containment. So far, fine. But they immediately, and too quickly, proceeded to impose heavy connexive requirements on proper consequence. 'Since only such consequences are to be admitted, ... a consequence in which a negation follows from an affirmation is not to be admitted... .' The requirement not only ensures $\sim(A \rightarrow \sim A)$, but takes out *all* such implications of the form $A \rightarrow \sim B$, such as "if it is a man, then it is not an ass". Much of the rest of the tract (TE) is devoted to meeting a series of objections to, and softening, such a problematic outcome of heavy connexivism. The crucial point for present purposes is however that obligationes theory, virtually from its inception, became a stage where a *genuine* consequence could operate, and where the sociative outcomes of false and impossible assumptions could, to a considerable extent, be decently and freely investigated. Unfortunately, later theory became progressively more restrictive, and appears to have degenerated in some centres towards modalism.

But, medieval logic was never, it seems, dialethic; it could not have tolerated some proponent resting comfortably, or however uncomfortably, with an explicit contradiction. Consistency remained a fundamental, and substantially unquestioned, good. But some of the theories, especially parts of obligationes, were paraconsistent; they permitted consequential reasoning from and through inconsistency, even if explicit contradiction could not be a resting stage, but was an endpoint, when the game was up.[82] It is worth observing, however, that obligationes practice could be redesigned so that explicit contradictories are admitted. The game would stop when everything had to be conceded (and so some acknowledged severe falsehood admitted), or differently when some statement was both asserted and rejected.

5.5 Modernity, enlightenment, and logical stagnation

Although the traditional logic was expanded, in sociatively relevant ways, in the long period between the late middle ages and the contemporary period, little advance was made in the fundamentals of sociative logic, indeed comparatively little advance was made in logic at all. 'From the 400 years between the middle of the fifteenth and the middle of the ninetheenth century we have ... scores of textbooks but very few works that contain anything at once new and good' (Kneale and Kneale 1962, p.299). So despite the oft-told greatness of the times, for the rise of science and technology, for mathematics and the arts, neither the Renaissance nor the Enlightenment, nor the early Industrial "Revolution" itself, engendered significant contributions to the growth of logic sociative or not. For the most part, logic stagnated or even declined, not to revive until well into the nineteenth century.

The situation has not been satisfactorily explained. The attractive (if insufficiently specific) suggestion—that it had an intellectual explanation in the rise and domination, at least during the Renaissance and through the Enlightenment, of ideologies and philosophies unsympathetic or even hostile to logic—is rejected by the Kneales. Instead they are inclined towards a more individualistic and superficial explanation, that 'although the subject survived in the elementary instruction of the universities, it no longer attracted the attention of many of the best minds' (p.298, the later point is repeated without the qualifying "many" two pages later). Some pretty good minds were

[82] Even Martin lapses into the commonplace (if convenient) confusion of paraconsistency with dialethism. 'The logic of the Middle Ages were never paraconsistent, but there was an urgent reason for denying that in general everything follows from an impossibility' (Martin 1986a, p.571).

attracted nonetheless, Leibniz for one (as the Kneales themselves observe). Why were other 'best minds' no longer attracted, when certainly they mostly encountered logic? Because, presumably, they came across little they were able to get to grips with intellectually. The time was not ripe, for instance, for the algebraic comparison, which occurred to mathematicians in the nineteenth century, once algebra had been much elaborated (though rudiments of the main idea appear in Leibniz). Because, more important, of the depauperate type of logic they came to encounter. With the changeover in education from scholastic to humanistic grammars, all students, including the best, were directed into humanistic studies. The new humanistic texts 'severed the links with scholastic logic and philosophy [and] ... prepared students for the study of Latin poets and orators'. Thus Ashworth (1974, p.22) who makes 'humanism ... the culprit' for the decline of logic 'because it rendered students unfit for the study of logic, rather than because of its more generally seductive properties.' Such an attractive explanation does too much; the changeover, which could hardly have stopped extra-curricula education, would also have blocked the flourishing of scientific and technological investigations. A more likely explanation for the stagnation of logic lies in a combination of factors: the unpreparedness of the times for certain developments, the rich intellectual opportunities elsewhere, the indifference or hostility of humanism and especially humanistic education to scholastic logic, and the rise to ascendancy of empiricism and idealist-rationalism (later reflected in negative influences of Hume and Kant especially), both of which tended to view logic as a closed and empty arena.

Despite the general logical poverty of these intellectually rich human times, things were happening of much interest for the story of sociative logic (as well as for the history of paraconsistent logic; see PL, chapter 1). Most of these doings concern attempts to expand the Aristotelian synthesis, traditional formal logic (as it was to be called), to encompass further areas of logic, not so far satisfactorily accommodated under the paradigm. Thus the theory of conditionals, the treatment of singular terms and propositions, the logical analysis of complex terms, the quantification of predicates, and so on. Since the traditional logic elaborated, based on syllogistic, was sociative, in a liberal sense, these extensions may be counted as applications of sociative logic.

Of particular note is a theory of conditionals to be found in modern seventeenth century 'traditional' logic. Through the theory is a reductionistic one, treating conditionals as certain enthymematic syllogistic arguments, it is nonetheless a relevant theory. The reason that relevance results, by complete contrast with contemporary enthymematic accounts of hypothetical reasoning and conditionals, is the tight control on terms and connection which syllogistic theory itself provides.

The traditional theory of conditionals, as it is perhaps appropriately called, is much easier to illustrate than to state in complete generality; and most authors who present the theory content themselves with illustrations, often copious illustrations, of the method. The theory is elegantly illustrated in the Port Royal logic of Arnauld or Arnauld and Nicole (Arnauld 1964, pp.224-235), which we shall follow, and is repeated in subsequent logic texts such as those of Whately and Mill. Put roughly, the theory is that a conditional holds good if there is an underwriting valid background syllogism, or chain thereof, into which standard form it can be transformed without loose ends. A main contemporary theory of conditionals is but a relaxation of this theory, obtained by extending syllogistic (as its original sense intended) to a wider range of arguments and by giving fuller scope to enthymematic character.

As is the usual way, cruder reductions were progressively refined 'The work of Wallis ... contains an attempt to bring conditional statements within the Aristotelian scheme by treating them as universal. According to Wallis, "*Si universaliter sumendum est, quasi tantumdem valens atque omni casu quo*"' (Kneale and Kneale 1962, p.306). While this may generously be seen as affording a pleasant anticipation of semantical analyses of "if", as a reduction of conditionals to complete syllogisms, it is too simple, as many examples from Port Royal logic attest. Consider, for instance, negative conclusions drawn from the maxim "Every feeling of pain is a thought", such as "if no animal thinks, then no animal feels pain" and "if some part of man does not think, then some part of man does not feel pain". The first argument compresses—so it is maintained, a syllogism of the form *Camestres*, the second of the form *Baroco* (Arnauld 1964, p. 226). More generally, a conditional compresses part of a syllogistic argument. Consider, for instance, the following conditional (similarly "since" and "because" statements): 'if every true friend must be ready to give his life for his friend, then there are few true friends since there are few friends so devoted'. The single statement compresses the following syllogism:

> 'All true friends must be ready to give their lives for their friends. There are few people ready to give their lives for their friends. Therefore, there are few true friends' (Arnauld 1964, p.225).

A precise and detailed presentation of the initial theory in traditional fashion would be extremely case-ridden because of the numerous forms involved, and appears not to have been attempted anywhere. As it was only an ancillary theory rounding out and helping to complete the established paradigm, and as everyone competent presumed they could see how the additional theory went and deal with virtually any case that arose, there was probably little pressure to present it in proper contemporary detail. Nor shall we attempt to present the theory in full generality, allowing in an unqualified way for complex and relational terms; but we can obtain considerable generality and thereby remove much case structure by taking advantage of contemporary formalisations of syllogistic. Let the terms (perhaps complex but with structure unanalysed) be $T_1, T_2, \ldots, T_n \ldots$, and the term connectors be a, e, i, o. Let variable μ range over term connectors. The *well formed* initial *sentences* (*wfs*) of the syllogistic are of the forms $T_i a T_j$, $T_i e T_j$, $T_i i T_j$, $T_i o T_j$, i.e., $T_i \mu T_j$ for short, with μ variable. Well formed arguments are built up from sequences of wfs in the standard way. These will comprise both familiar syllogisms, consisting of three linked wfs, and chains of three or more linked wfs (called *sorites*), which are in fact resolvable into sets of familiar syllogisms and their validity so determined. Now the traditional theory of conditionals appears to be as follows: If A then B holds if and only if there is a well-formed correct (syllogistic) chain $T_1 \mu T_2$, $T_2 \mu T_3, \ldots,$ $T_{n-1} \mu T_n$ underwriting it, with terms T_1 and T_n at least supplied in A and B, and all terms of A and B included in $T_1, T_2 \ldots T_n$ (i.e., no loose terms). The theory is perforce a first degree one, but permits higher degree substitution instances.

The traditional theory is bound to be relevant because of its common terms and no loose term requirements. Nothing like intuitionism with its free introduction of irrelevant conjuncts is permitted. Consider, what allows similar irrelevant rubbish, the partial-truth-valued theory of conditionals sketched out by the Kneales in substitution for a proper discussion of the third and fourth positions in the Alexandrian debate on conditionals (Kneale and Kneale 1962, p.132). What the Kneales' theory comes to is that "if A then B" behaves like the conjunction, $A \,\&\, B$, when A is true, and (what is often wrong) is not assigned a truth value otherwise, when A is false. In particular then, as with intuitionism, A and B, when *any* truths at all, suffice for "if A then B". For example, "all wild lions are confined now to Africa" is a perfectly satisfactory antecedent for "all straightforward conditionals admit of traditional syllogistic analysis". The traditional theory quite correctly does not validate such irrelevant conditionals, since there is no way of syllogistically linking the terms without loose ends. Thus although the traditional theory is unduly roundabout and has some marked weaknesses (to do as usual with the linguistic forms syllogistic can satisfactorily accommodate), in important respects it represents a significant improvement upon the mainstream contemporary theories of conditionals which have displaced it.

Some further features of the emerging theory beyond relevance are also evident, despite evident limitations of the formulation given. For example, as on recent modal theory, Augmentation will fail, i.e., the following principle does not hold in general:

Aug. if A then B / if A & C then B.

For C may introduce loose terms (in special cases of course the term of C may be packed into terms of the syllogistic chain that underwrites "if A then B"). More generally, any principle introducing new unrelated terms will be in trouble.

By contrast however with most recent modal theories, transitivity *will* hold, i.e.,

Trans. if A then B, if B then C / if A then C.

For the chain underwriting the second premiss can be backed up to that underwriting the first with the coupled chains thereby underwriting the conclusion. To be sure, if some of Arnauld's pronouncements were heeded, and only syllogisms proper used to underwrite conditionals, transitivity would break down (sometimes condensation of chains would be possible, but not in general). Moreover, Contraposition too will hold, also in rule form (as inevitable on essentially first degree theory), i.e.,

Contrap. if A then B / if not B then not A.

The justifying argument, a little hairy, involves antilogising the underwriting chain. Strictly more has to be said as regards what underwriting comprises, a matter that becomes crucial in confirming Modus Ponens:

MP. if A then B, A / B

and in assessing Connexive rules such as:

Con. if A then B / not (if A then not B).

For sufficient generality something like the following is required: A is equivalent to (coentails) a conjunction of wfs of an initial segment of the corresponding chain, and B to a disjunction of wfs of a final segment, with initial and final segments not overlapping. (Furthermore, any wfs which are not derived from preceding members are *taken to be* true; this complication is needed to deal with enthymematic examples, best removed however in an initial attempt to capture the traditional theory). While such a specification removes embarrassment concerning MP, it leaves open the admissibility of Con, and also of reflexivity, i.e.,

Id. if A then A.

Both Id and Con turn on what types of syllogistic chains are admitted. If syllogistic is always, as was usually thought, progressive and never repetitive or re-entrant, then Id is out, though the theory is easily adjusted to readmit it. The issue of Con is more difficult, and turns on such issues as whether syllogistic can lead to both positive and negative conclusions. While it looks as if the simpler examples traditional theory characteristically worked with would rule this out, and confirm Con, as Aristotle did, problems undoubtedly result. Most notable among these is the undermining of the premisses of Eleatic reduction arguments which take the form:

Red. if A then B, if A then not B / not A.

Is this then vacuously admissible? Presumably not, especially as any other irrelevant conclusion would serve as well. But the traditional theory is not clearly enough articulated to give the more decisive answers we might now expect on these issues.

The reduction of all reasoning to syllogistic was pushed further in the eighteenth and nineteenth centuries. The sort of reduction, through compression and underwriting, that the Port Royal logicians applied to conditionals and certain other parts of reasoning, subsequent logicians like Whately, and theoretical hangers-on like Kant, proposed to apply to the *whole* of reasoning, including probabilistic reasoning.[83] The Port Royal logicians themselves seemed prepared to admit, as Leibniz certainly was (cf. Kneale and Kneale 1962, p.322), that not all argument could be brought into syllogistic form. Not so later traditional logicians; the scene was thus set for the downfall of the traditional paradigm, which proceeded to overextend itself. Thus too, given its (then) limited resources, it became an easy target,

---

[83] The idea appears to be ancient. Thus '...Alexander's defence of Aristotelian logic consists largely of attempts to show that all interesting arguments can be represented as categorial syllogisms. A belief in the universality of the categorical syllogism was probably common among the Peripatetics. There are also grounds for thinking that some Stoics held an analogous belief about their logic: every deductive argument can be represented as a propositional one' (Mueller 1969-1970, p.174: the qualifications slid on, in easy fashion, such as "interesting", "deductive", make very significant differences). Whence in part the genesis of the ancient feud between Peripatetics and Stoics (cf. Frede 1974).

open to many counterexamples, for instance, from probability logic and from relational logic (as Leibniz had observed, without observing the damage done to his own metaphysics, and as De Morgan detailed). At least the reduction program pursued by Whately and others had the important merit of making all reasoning relevant—given the unexplicated, but tight and presumably logical, rules of reduction. What a bonus! But the program was never carried through, or even seriously attempted, in a systematic way; it succeeded, while it did, like most other monopolistic paradigms, apart from political strategems, through selective examples, handwaving and bluff. Nor could it have succeeded, without substantial enlargement, as recent developments in syllogistic theory have begun to show (cf. Bacon 1985). Indeed, until recently the (relevant) reduction program to syllogistic appeared to be leading nowhere, but simply impeding the progress of logic (whence heavy, and sometimes justified, criticism from proponents of the new mathematical paradigm).

What proved, of course, a much more fruitful line of development was the adaption of algebraic ideas and techniques to logic, by Boole and others. This line of thought in fact stretches back at least to Leibniz, but it took a long time to activate the algebraic rolling stock properly, having to await further maturation of mathematics. Again the movement began in a sociative way, and only later became derailed. For Leibniz's calculus of containment and content is, so far as it goes in main sketches, a sociative logic, fashioned along algebraic lines. This abstract calculus, presented in various partial stages of development, admits of various partial intepretations—among those noticed by Leibniz a propositional interpretation. In one 'fragment on his project for a calculus of inclusion he suggested that the consequent of a conditional proposition might be said to be contained in the antecedent. ... it is clear that we can interpret his formulae as assertions about the relations of propositions' (Kneale and Kneale 1962, p.344, where other interpretations are tabulated). Leibniz's suggestions give some warrant for symbolising not only the containment relation "is in" by $\leftarrow$, the converse of implication $\rightarrow$, but for representing the conditional in term of the same notation. The result is striking. Proposition 12, for example, of the most developed of Leibniz's many suggestions (conveniently presented in Kneale and Kneale 1962, p.340 ff) becomes, after notational updating, $C \rightarrow B \rightarrow . A \& C \rightarrow A \& B$, i.e., the principle of Factor. Shedding $\rightarrow$-representation of the conditional it is the general first degree principle, if $B \leftarrow C$ then $A \& B \leftarrow A \& C$. The example was chosen for two reasons. Firstly, Factorisation of one sort or another plays a significant role in Leibniz's theorizing. Secondly, the most developed calculus LC, is an apparently relevant factorisation system (an I system), heavy with factorisation features:- not only Factor itself and complications thereof; but the interconnection of inclusion with identity in principles $C \rightarrow B \rightarrow . \; C\,\&\,B = C$ and $C\,\&\,B = C \rightarrow . \; C \rightarrow B$ (Proposition 13 and 14); and the emphasis on identity, $=$, and its substitutivity and other features, as the starting point of a logical calculus (Definitions 1 and 2 and initial Propositions).

As a matter of inspection, this calculus, and $\{\rightarrow, \&\}$ system (without or with a parasitic negation $\sim$), can be represented within an extension of the common first degree system, FD, of entailment. FD is extended by substitutional principles, that upon variables and, more important, that for replacement of equals (substitutivity of provable equivalents). Represent propositions of the form "if α and β, then γ" in rule style as: $\alpha, \beta \,/\, \gamma$. It follows from well-known results that LC, so represented, is weakly relevant.[84]

In the subsequent algebraic elaboration of logic in the ninetheenth century, connection vanished. Though it did not need to vanish at all, it quickly did. The second, algebraic, wave of mathematical logic, called 'the Boolean period' by Bocheński (who presents a neat chronological table, p.279), is said to begin with Boole and De Morgan's publications in 1847. But the logical calculi that Boole and De Morgan somewhat independently devised—though intended, as main applications, to represent both syllogistic forms (by way of familiar dubious extrasystemic translations of A, E, I and O forms)

---

[84] In a fuller investigation of Leibniz's various fragmentary calculi (perhaps, like contemporary attempts of individual researchers at formalising preanalytic notions, only hazardously and ahistorically lumped together), much turns on how negation is accommodated, no straightforward matter given, if Lenzen is believed, Leibniz's 'somewhat uncertain and partly erroneous theory of negation' (Lenzen 1988, p.3). On the treatment of negation (especially), stands the theme, advanced by a bickering phalanx of Leibniz scholars, that Leibniz's ('explicitly given') logic is a strict implicational system, namely $S2°$ or perhaps $S3°$ according to Lenzen's detailed investigations (for example Lenzen 1988, p.24, p.1). Such a modal theme deserves—a bit like the parallel findings planted on the Stoics and medieval 'mainstream'—to be viewed with some scepticism. The arguments to the theme depend upon, what looks methodologically unsound, pooling virtually all of Leibniz's work—which apparently results in an inconsistent corpus (an easy strict route to the sought conclusion)—and then consistencizing in a way that suits sought results. The arguments also depend on strategies such as illegitimately upgrading what are but one-way conditionals (perhaps only rule linkages) to biconditionals or even definitional status (cf. Lenzen's 1988 moves to P3 and P1 on p.4 and to P1S, p.13).

Leibniz's calculus LC will be examined in more detail elsewhere, along with other Factor systems of historic interest, such as that of MacColl in his anticipation of Frege. MacColl's later (1906) sentential logic amounts to $S2$.

and also intermediate inferences—were essentially first degree algebraic systems. They did not admit nesting of the equality relation (or of it with inequality). They did not contain an inclusion relation or an implication. These enlargements, and the nesting of implication and equivalence were to come rapidly enough in the next thirty years, but they were not present at the beginning. Apart from interesting and damaging curiosities like Boole's symbol ν (for representing particularity) which quickly disappeared, the calculi were narrow algebraic systems, fully formalisable (as in RLR I, p.118) *without* propositional apparatus. Though it could accordingly be claimed that these systems are (variously) relevant, because the criteria offered do not apply to them, it would be more accurate to say that the question of relevance does not so far arise. However the systems are hardly neutral under intended applications. Much as $-(x \,\&\, -x) = -(y \,\&\, -y)$ for any classes or conceptions x and y whatsoever, so, for any universal classes or conceptions u and v whatsoever, $u = v$, since $u = 1$ and $v = 1$. So, under application, any two universally true propositions whatsoever are identified, for example "All humans are mortal" says the same as "All kangaroos are herbivores". Content relevance is hardly preserved. Furthermore, Boolean logic, which Boolean algebra straightaway induces (see RLR I, p.119), is irrelevant.

The trouble arose from the introduction, by algebraic analogy, of 0 (read *Nothing*!) and 1 (read (the) *Universe*), conforming, in effect, to the equations $0 \,\&\, y = 0$ and $1 \,\&\, y = y$ (for example Boole 1951, p. 47), with $y = 1$ and $y = 0$ representing, respectively, truth and falsity of corresponding propositions (representations adopted from Boole through MacColl: see Bocheński 1961, p.309). The rather inevitable result, in quite ordinary algebraic settings, is that any designated truth will be assigned to 1 at the top (and designated falsehoods to 0 at the bottom of a Boolean lattice), and so distinct truths will be identified. The fault lay at bottom with the (elective) class model, assumed throughout by Boole, which is entirely inadequate for propositional representation. Mistaken identifications are not the only defective outcome, simply among the worst and the basis of relevance shedding. (Compare the problem with strict implication: because $A \,\&\, {\sim}A = B \,\&\, {\sim}B$, substituting in Simplification gives $A \,\&\, {\sim}A \dashv B$). Another little remarked outcome consists in unsatisfactory interconnections for Nothing and Universe. According to Boole, not-Nothing = Universe, but according to pre-analytic assessments, not-Nothing is something or other; in Boolean symbols, $-0 = \nu$. It took about a century from simplistic beginnings for a satisfactory algebraic treatment of truth and falsity to eventuate, with scrapping of the dud equations, $x = 1$ and $x = 0$, and replacement of 1 and

0 respectively by truth filters, and falsity ideals (c.f. ENT, p.193). Really satisfactory treatments of Nothing, Something, and their discredited mates, will have to wait their time.

With the rise of the crude Boole-Schröder algebraic representation of traditional logic and the ensuing development of classical logic in the late nineteenth century, niceties of connection vanished in favour of mathematical analogies and *calculability*. Relevance changed from being something automatic, that was taken largely for granted, to being something that wasn't guaranteed and didn't matter anyway. Its status changed from a "Not to worry. I'm OK, George" to a "Don't Care. I'm unnecessary, Will". With the ensuing adjustments in logical paradigms, former logical necessities like relevance of conditionals and implications, and good connections in logic, became luxuries and matters of nonstandard concern.

There were other streams of thought in the nineteenth century where connection did matter, or should have, and where sociative logics really were required—not merely in the emerging social sciences, but even to demonstrate the viability of whole ideological movements, most notably German idealism and some of its off-shoots. But although such philosophical enterprise cried out for sustaining logical foundations, of paraconsistent and sociative cast, none of appropriate detail were furnished (despite many texts allegedly about logic). Indeed idealism proved a hostile environment for the growth of sophisticated logic; and most of the development of mathematical logic has occurred in a very different milieu, primarily that of reductionistic mathematics, but with inputs from pragmatism and positivism. Nonetheless parts of idealism were subsequently drawn heavily upon by logical and foundational theory, conspicuously intuitionism, especially elements of the precipitating idealism of Kant (the pivotal figure who also 'began the production of the curious mixture of metaphysics and epistemology which was presented as logic by Hegel and the other Idealists of the nineteenth century', Kneale and Kneale 1962, p.355). Much less momentously according to mainstream history, but more relevant for sociative history, ideas drawn from Kant (and by him from Leibniz and Locke) also played a justificatory role in Parry's elaboration of containment logics.

As the very name "analytic implication" indicates, the origins of systems of analytic implication or logical containment have been speculatively linked with Kant's celebrated explication of analyticity. As Kant explained it, 'the connection of the predicate with the subject is thought through immediately', or, as Anderson and Belnap try to explain Kant, '"S is P" is to be *analytic* just in case the subject is in some sense partially identical with the predicate'

(ENT, p.429). Transcribed as a necessary condition for implication, Kant's dictum then becomes: $A \rightarrow B$ is to be an *analytic* implication only if the consequent B is in some sense partially identical with the antecedent A. So far, if partial identity is taken in an obvious way as symmetrical, this does not swing Kant's dictum in the direction of containment systems, rather than the relevance ones (cf. ENT, p.155). If, however, traditional containment accounts of analyticity (and implication) are recalled, accounts appealed to alternatively by Kant, then partial identity is cashed as containment, and the dictum is swung in the conceptive direction. It remains only a substantive matter, to replace "in some sense" by "as regards sentential variables", a matter that can be argued on the grounds that it is containment of content that counts and that content is carried by variables. The latter claim, though implicit in much logical theory concerning form and content, is highly contestable; for surely such forms $p \& q$ and $p \vee q$ have different content though they have the same variables? Though the distinctive features of analytic implication systems can be tenuously linked with Kant, there is no investigation of any such systems in Kant (who thought logic was a closed shop) or indeed in propositional form until Parry arrived on the Harvard scene earlier this century.

5.6 The contemporary period: the rise of mathematical logic and sociative alternatives to the triumph of irrelevance

In certain respects logic has flourished in the last century as never before. But it has been a strange monocultural, probably unstable, mathematically complex but philosophically poor, mega-development. By contrast with earlier major periods, the classical, degenerate, problem-making theory appeared first, by a good margin (the system of Frege's *Begriffschrift*, the conventional starting point, is dated as 1879, and Russell's equivalent *theory of implication* 1906: on both see Prior 1962, p.301, who follows and reinforces the received history). Naturally the theory did not spring from nothing; the way had of course been prepared through significant earlier development and supplementation, especially in algebraic elaborations exhibiting similar degeneracy of traditional logic. Moreover, some of the most celebrated, but now questionable, developments, such as the theory of quantifiers, were not only long anticipated, but broke through in several places somewhat independently—reflecting a common pattern of human intellectual discovery.

What the history sketched reveals is that classical logic is not in a privileged historical position. It is a rather recent upstart, a quite minor character pushed to prominence by new but crude mathematical technology. Moreover, almost as soon as it was realised what was being claimed as to conditionals and implication under the new 'mathematical logic', there were objections, even outrage, particularly (thanks to Russell's publicizing activities) in England and North America, but also in France. Corresponding with Russell on the subject of nested implications, Couturat wrote: 'These sorts of implications ... are very lacking in evidence, or paradoxical, or even unintelligible ... . And the formulas which I consider as *obvious*, such as: $pq \supset p, p \supset q \,.\, q \supset r :\supset: p \supset r$, are only derivable [in your system] by means of the law of importation, which is again one of these paradoxes ... the equivalence $p \supset q. =. \sim p \vee q$ is yet another paradox' (letter of 8 November 1903). 'In general, [Russell's] correspondence shows that the introduction of $\supset$ was seen as a radical, and counter-intuitive innovation, and not only by Couturat.'[85] While many informed intellectuals (not exactly a large band) protested, few did anything much about it. The first systematic attempts at repairs and improvements were largely modal in orientation (and due principally to MacColl, an underrated researcher, Johnson and Lewis). These attempts *did* offer improvements to the theory of logical implication or entailment—though not that of conditionals—and they were eventually absorbed in the broader classical paradigm, as extensions of classical logic. But these improvements were resisted by the earlier narrower paradigm, whose commitment to extensionality such modal explications did upset; and they still are resisted in logical backwaters.

Lewis's investigations of strict implication were particularly influential, they have been important to all subsequent alternative enterprises, and they have largely withstood subsequent extensionalist attempts to discredit or eliminate them. Much subsequent alternative enterprise has in fact consisted in varying techniques and methods, assumptions and systems worked out for an expanding galaxy of modal logics (as modal logics themselves mostly advanced through adaption or elaboration of methods first devised for, and in the simplified setting of, many-valued logics). Lewis himself engaged in a good deal of variation of system and much experimentation in arriving at his set of strict systems, in the course of this investigation arriving at some systems which collapsed back into classical logic, and some which were in fact relevant (see RLR I, p.356). After his rediscovery of the medieval paradoxes of implication, his investigation narrowed, his position rigidified.

[85] Letter from Alasdair Urquhart (1 Dec 1987), who supplied the quote from Couturat.

Lewis became convinced that there was no paradox-free "promised land" (something logicians, of all people, tend to be *much* too easily persuaded of), that some paradox *had* to be lived with. Such a conviction has not proved highly robust; Lewis's conviction did not transfer even to those who helped him investigate the properties of strict implication systems by such new-fangled techniques as matrix methods, Parry in particular.

In part because of Lewis's influence at Harvard, in part owing to Scheffler and his students, a brief golden age of alternative logic, of which Parry has told (for example Parry 1932, chapter 10), followed at Harvard. Among the fruits of this period were Nelson's rediscovery of connexive logic (a framework Parry sets aside too lightly), and Parry's systems of analytic implication, which represent types of containment logics. Most of this alternative work (theses or equivalents bearing on implication by Bronstein, Emch, Henle, Nelson, Resinger, Shen, Weiss, and others) has however vanished from easy public access.

Parry's work on analytic implication, recorded in an unpublished Harvard dissertation in 1932 (and reported briefly and inaccurately in Parry 1933), itself disappeared from view for more than thirty years until interest in it was revived by Anderson and his students, and their colleagues and competitors. Since then information concerning such conceptivist systems has increased rapidly, with the result that several of the main open questions (for example those raised in ENT, pp.430 ff) concerning Parry's principal system $PA1$ are now resolved or can be straightforwardly solved. As usual, the resolution of such problems has generated another set of problems, concerning a wider class of transcending systems. These include relations with other similar sociative logics, arrived at in slightly different way, such as relatedness logics, problems of quantification, and so on. Now that they have been developed to some extent, many applications for containment logics have been noticed. These typically involve areas of systemic limitations, areas where contexts are limited or information available is restricted in some way. Salient examples include: logics of fiction, and surrounding areas such as ideologies and dreams; dialogue logics; and frame and other problems in artificial intelligence (see further Chapter Seven).

Much of the remainder of the more accessible, contemporary sociative story has also been told elsewhere in sufficient detail for present purposes; and the rest is readily pieced together.[86] Most of the contemporary strands involve

[86] For instance, in ENT, PL. But for proper historical pruposes, it has not been told in nearly enough detail. That however is a task for someone else, more energetic and further removed from the action.

no very organised movement, but, in the fashion of much contemporary alternative activity, simply comprise a few individuals or small groups of individuals writing largely in isolation. Certainly such a pattern obtained with connexive logics, which have been advanced substantially independently by Nelson, Angell, McCall and others. Connexive principles have also been espoused by many others who have done little or no systemic work; connexive principles just refuse to go away easily, no doubt because there is a viable intuitive picture underpinning them (based on a cancellation view of negation, as speculatively explained in NC). Until very recently, containment logics for instance were represented by a one-man band, Parry, playing, so he argued, technical variations on a theme drawn from Kant. Likewise Ackermann, first prime technical mover of the contemporary nondissembling approach, worked entirely on his own (in a high school, having lost Hilbert's patronage). Again with nontransitive logics, there is no orchestrated movement, no school, though those promoting such systems tend to have been British based (not North American as with connexive theories) and to know one another.

By contrast again, both relevance and relational logical approaches have come to involve very loosely affiliated schools. Relevance logic developed around Anderson and Belnap, at first at Yale from 1958, but chiefly at Pittsburg to which they both soon moved and where they gathered around them, especially in the 1960s, an active and intelligent group of graduate scholars (see ENT p.xx). Anderson and Belnap and others *much* elaborated a very fruitful synthesis of the formal and philosophical work which had preceded them: notably, variations upon Ackermann's technical achievements combined with adaptions of methods of Fitch and of Curry, on the one hand,[87] combined with significant explications of the demands of connection and relevance in a theory of entailment urged by a small chain of opponents of strict implication before them, on the other. Different again, relational logics were explicitly launched (though the idea is an older one) in 1979 in *Philosophical Studies*, mainly by a small group of North

[87] For example, only system T (of 'ticket entailment') appears to be strictly due to Anderson and Belnap (and to Belnap precisely), if precision is to be insisted upon in these matters. E was obtained as postulate-chopping variation of Ackermann's II systems; in original intention it was equivalent to them, and it was duly proved theoremwise equivalent to them. Bacon and Prawitz appear to be responsible for the elevation of system R, with later help from Meyer, who pushed NR as an explication of E.

Ticket entailment in general, not as uniquely (but with a certain arbitrariness) encapsulated in system T, has a long, interesting history (some of it traced in Martin 1987). Both Boethius and Abaelard maintained that a conditional is true or an enthymeme valid if there is a topical connection between antecedent and consequent, a connection, guaranteed by a topical rule, very like that of an 'inference ticket'.

Americans, mostly committed at bottom to classical logic, from Wellington, New Zealand, a (declining) centre of modal logics.

In Australia, as in the north, opposition to Russellian logic and insistence upon connection goes back a fair way. But the opposition took a stronger form, which unfortunately helped to delay the development of any sort of contemporary logic in Australia. In the regionally important Sydney sphere of influence, dominated by J. Anderson from the late 1920s, Russellian logic was roundly rejected, at least until the 1960s. In particular, material-implication was dismissed philosophically because it supplied no requisite connection of components.

> ... one reason why the doctrine of 'material implication' is a philosophical blind-alley is just its ignoring of such connection; implication then becomes quite arbitrary or 'magical'—divorced from inquiry .. [as] where reference to pressures is lacking ... the connections between the barometer and the weather would be magical rather than scientific connections (Anderson 1962, p.145, rearranged).

The relevant logic movement, based in Australia, while applauding constructive features of Andersonian criticism of classical logic, is not an off-shoot or continuation of Andersonianism, but an independent development. The relevant movement differs from Anderson and Belnap's relevance enterprise, though it grew up alongside it intellectually, rather as a slightly younger sibling, and borrowed heavily from it as well as contributing to it. (It has benefitted through contributions of a major defector from the relevance enterprise, Meyer, who is not a relevantist either.) Since the movements are popularly confused, it is worth setting out some of the salient differences (these are presented in more detail in RLR I).

- The relevant movement rejects *most* of the sweeping themes of the relevance enterprise, for instance that E is entailment, that it is the logic of relevance and necesity, that R is relevant implication, that E and R (and perhaps T) are the important sentential relevance-preserving logics. According to the movement, the selection of E (and of R) was premature, and mistaken. If there is a unique logic of entailment—increasingly doubtful as evidence accumulates—it lies somewhere among deep relevant logics.
- The movement is committed to paraconsistency, and to taking inconsistent theories, situations and worlds seriously. Thus, for example, the relevant movement courteously disowns Ackermann, whose logics were not paraconsistent. By contrast, the main text of relevance enterprise, ENT,

> is dedicated to Ackermann, 'whose insights' are correctly said to have 'provided the impetus for this enterprise'; and at the very beginning of the text (p.xxx) relevance logic is held up 'as a new branch of mathematical logic, initiated by ... Ackermann'. Ackermann enjoys no place in the history of paraconsistent logic, though some of his technical contributions have, like those of Church, proved valuable in the development of relevant paraconsistent logic. In fact, very much of his activity was directed, like that of other members of the Hilbert school, at establishing consistency, and at exorcising inconsistency—a practice of which Anderson, if not later Belnap, strongly approved (see, for example, his review of Wittgenstein).

To descend again to routine grimy city life and twist the logical scene back into real perspective, it only needs stressing that *all* these alternative logical positions are tiny, with but few genuine proponents, and with at least as many fellow travellers. The overwhelming bulk of contemporary logical theory makes no claim to relevance or connection, the logics involved are not sociative, and most logical practitioners are substantially uninformed about sociative logics of any sort. Moreover logic itself has largely ceased to be an independent discipline. It does not count for much in the contemporary world, and much of what role and support it has derives from its use as a servile means to other ends. In service roles however lies some new hope for sociative logics. Fortunately there are other purer grounds as well.

The ideas of alternative nonstandard logics are, in main relevant cases, obvious, appealing and exciting. So they keep reappearing or being rediscovered, especially on the fringes. Only in intellectually repressive or inactive societies will these ideas be kept from developing and flourishing, and ultimately from pulling down the present anti-pluralistic, narrow and authoritarian, logical establishment and structure. Things logical can change; given a little time and a fair run they will change. Many major happenings in sociative logics, helping to make traditional concerns good, are very recent; whereas ideological evolution remains slow, and challenging ideas still take much time to spread. Despite much internal and ideological opposition, but fortunately for its satisfactory survival, logic has increased enormously in diversity in the last thirty years—in ways even very few intellectuals appreciate. Contemporary relevant logics, in particular, only really began as a distinctly technical enterprise in the 1950s. Semantics and other analyses of certain modal logics, which were to give enormous stimulus to the investigation of a burgeoning range of modal and intensional logics, were only beginning to be worked out at about the same time. Though the prospect of reasoning successfully with inconsistent information and from inconsistency still went largely unrecognised (so Popper could republish his astonishing, but rapidly outmoded, diatribe against against dialectical logics), paraconsistent logics and all that were also beginning. Much more was still to come (for one survey, see PL).

But though logic has come a long way very recently, it has a longer way to go, both in *whom* it involves and *what* it investigates. There are, for instance, virtually no black researchers, and exceedingly few women are engaged; and for all the proclaimed rationality of modern humans and their institutions, logic touches comparatively little human practice. Differently, there remain *many* notions of considerable logical import, *some* of historical significance, of which we lack decent accounts or, sometimes, a clear appreciation. To the satisfactory elucidation of these, sociative logics can make essential contributions.

Part II

Reasoning and Computation

6 On Reasoning: (ponible) reason for (and also against), and relevance

Co-authored with Len Goddard, this essay originally appeared in 1989 as essay 1 of RCR. *It is an amalgam of two papers co-authored with Goddard: Goddard and Sylvan 1989*a *and Goddard and Sylvan 1989b. A third paper was also planned. Draft material for this third paper was found under the title "Non-demonstrative reasoning and non-ponible logics" and has been included in Part III. A fuller treatment of the material dealt with therein was planned, with additional material to be supplied by Goddard, but the paper was never completed. [Editors]*

> Now reasoning is an argument in which, certain things being laid down, something other than these necessarily comes about through them. (Aristotle, *Topica*, 100^a 25)

The assumption is now commonplace that logic, classical logic even, takes care of reasoning, correct reasoning at least. That assumption is very far astray, as we shall try to explain. A corollary is that the fashionable idea of automating reasoning by mechanising proof procedures in classical logic, is misguided (see further the attached Appendix). Despite the long-standing claim of logic to formulate, or to encapsulate principles of correct reasoning, contemporary mainstream logics have much less to do with reasoning than has been supposed.

One source of this failure arises through the historical identification of demonstrative logic with reasoning and the ensuing entanglement of *implication* with *reason for*, a conflation fostered by the presentation of both through conditionals, by an undifferentiated "if". Within the narrow confines of Aristotelian logic, where syllogisms are progressive and sentential connectives do not figure explicitly, such conflation remained invisible. But that entanglement of connectives is not the only source of the confusion; reduction attempts such as those aiming to reduce all correct reasoning to deductive logic, or still worse to first-order classical logic, have also played significant parts.

An early task, then, consists in scraping off the blanketing excrescence of often shoddy logic (while trying to retain worthwhile logical elements) in order to uncover features of reasoning. Such logical stripping is important, not only in exposing the crucial notion of reason, but also for other purposes, such as decisively defeating the charge that logic involves reasoning in a circle. Briefly, contemporary logic is centered on implication, which is reflexive and need not lead to new information, whereas reasoning is normally irreflexive and progressive. Thus conditionality (or, in inferential form, argument) is at least two-way determinable; progressive reasoning takes one determinate route, implication and deduction another. The highly enthymematic implications of mainstream logics (which permit truth suppression) fall under the latter determinate. Some of the story to be elaborated we try to capture in the following elucidatory figure:

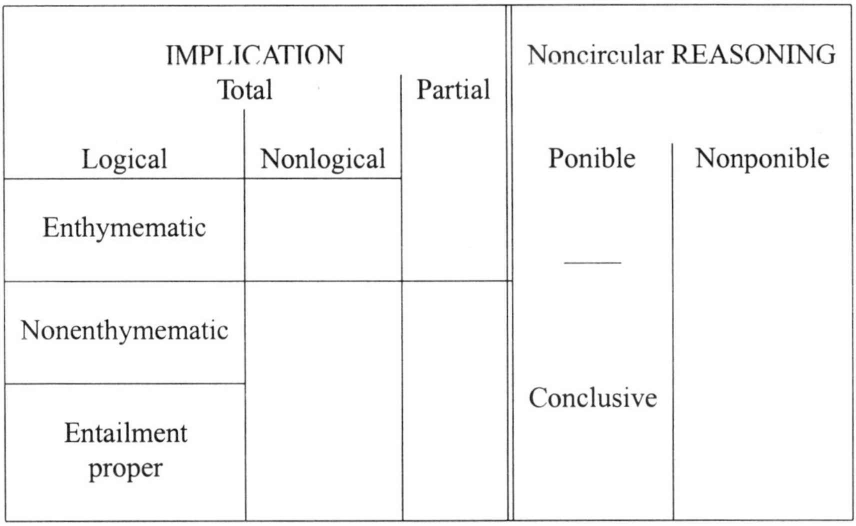

Figure 6.1 ARGUMENT ("syllogism") in conditional, not inferential, form

We do not wish to deny that the investigation of reasoning is, and ought to be, a proper part of logic. It should be; indeed it was once supposed to be the main matter of logic; but it has fallen off the contemporary symbolic agenda. We try here to make a small beginning on restoring the topic to its fundamental position.

6.1 Reason-giving connectives: symbolising some basic connections

Reasoning consists, above all, in the presentation of reasons, reasons for and reasons against. Reasons may be propositional in character, but they need not be. Seeing a green bird is a reason for claiming that the bird is green; feeling flat is a reason for having a drink. More generally, perceiving and emoting can serve as reasons, often fundamental reasons, for or against judgements or actions. Accordingly the grammatical place-fillers in the locution "... is a reason for (against) ..." may signify events, actions, attitudes, states and so on, as well as propositions. For a general theory of reasoning, then, variables need to stretch much more widely than merely propositionally (in the fashion explained in more detail in Chapter Twelve).

Often reasons are *presented* in terms of *if*-constructions, and the locution "if ... then ..." stands in for "... is a reason for ...". Certain kinds of place-holders are then inappropriate, for example "seeing a cat". In this respect the *if*-construction is more limited than the relational form, and unless suitable translations can be made some features of reason-giving are lost by focussing attention on the reason-giving *if*. Even so, the *if*-construction is not limited to indicative sentences as place-holders, and hence not to propositions as relate. For in general, *if*-constructions serve to present reasons of a very broad sort; for instance, ordinarily motives reason-giving *if*-sentences. In fact reason-giving locutions only overlap *if*-constructions. For by no means all *if*-constructions are reason-giving; for instance, concessive conditionals, take-it-or-leave-it conditionals, and the like.

In what follows, we try to elucidate among other things, some of the main features of reason-giving constructions, by analogy with corresponding *if*-constructions. But what we say is by no means confined to *if*-forms. However we do focus on one important kind of reason-giving, namely, as we shall explain, ponible forms. There are, thus, for the present, some limitations to the analysis. We shall also make the further unrestrictive assumption that when an *if*-construction is used to present A as a reason for (or against) B, A always occurs as the antecedent. This condition is not satisfied in general when *if*-constructions are used to present reasons in ordinary discourse; but it is normally a straightforward matter to re-write those which do not satisfy the condition in an equivalent *if*-form in which they do.

Since the logical work is accomplished primarily through specialised forms of reason-giving, we adopt a familiar style of notation in which the connective is embedded between antecedent and consequent. Once again we do not, however, require that the antecedent or consequent should be propositional. The key determinable connections in the bag of notions involved can now be presented and symbolised as follows:

> A is *reason for* B iff A's being the case or occurring counts towards or for the truth of B, i.e., in symbols, iff $A \overset{+}{\supset} B$. Put differently, $A \overset{+}{\supset} B$ iff A counts positively as regards B (or A lends weight to, or strengthens B).
> A is *reason against* B iff A's being so or occurring counts against the truth or occurrence of B, i.e., in symbols, iff $A \overset{-}{\supset} B$ (or, put differently, A weakens B).

Normally, A is a reason against B (pushing B away) when, and only when, A is a reason for the reverse (pulling B in), i.e., for the negation of B. Thus in symbols, $A \overset{-}{\supset} B$ iff $A \overset{+}{\supset} {\sim}B$, given a decent (reversal representing) negation.

> A is (consideration) relevant to B iff $A \overset{+}{\supset} B$ or $A \overset{-}{\supset} B$, i.e., iff $A \overset{\pm}{\supset} B$.[88]

It may be suggested that the *reasons* so far symbolised are rather putative reasons than reasons proper; for nothing requires that A obtains. Thus it is rather that A would be a reason were it to obtain. What is represented, the suggestion accordingly continues, is the hypothetical or counterfactual extension of actual reasons, sometimes also inaccurately accounted reasons in natural discourse. But, to the contrary, there is no inaccuracy. What has happened (which gives some credence to the suggestion) is that there has been an interesting historical shift in the use of terms such as *reason*. It has shifted from a use requiring actual grounds or facts to a modern use allowing in *alleged* fact or *hypothetical* grounds. (A similar shift has occurred with allied terms such as "grounds"; see OED for data.) Of course not any falsehood can serve as a (putative) reason. There are conditions, such as epistemic priority, to be met for something to be a reason; known falsehoods are thereby typically excluded as reasons.

With a view to obtaining a flexible and more workable theory, we shall adhere to the contemporary use. The older rejected use much complicates

[88] This type of relevance is distinguished as *consideration* relevance after an earlier classification proposed by Goddard, because it has A as a *consideration*—or so it is assumed, reason—for or against B. The implicit equation of reasons with considerations is however open to question, short of a narrow redefinition of 'consideration', because reasons are ordinarily a more demanding class than considerations.

Note that here $\overset{+}{\supset}$ and $\overset{-}{\supset}$ are being used as generic symbols, encompassing types of reasoning that will shortly be distinguished. Subsequently, from section 2 on, they become specific, representing forms of putative ponible reasoning.

obtaining a satisfactory theory, as will emerge. It can be conceded, however, to the rejected suggestion that the putative reasons introduced may sometimes be rather weak reasons (almost sliding into motives or excuses); such reason might better be rendered in the terms of the locution "some reason". But plainly increased strength or merit of reason can be indicated by familiar modifiers, such as *good, adequate, sufficient, decisive*, etc. By such modifications we move to stronger connections of various sorts (and we can even move through extensions, which relax normal progressiveness requirements and the like, to entailment, which corresponds to logically sufficient extended reason).

An importantly different style of modification is by terms like *actual* or *evidential*; the point is that it is still a modification, or restriction, of a more comprehensive and more tractable notion. Where A is evidential reason (or evidence) for or against B, the truth (or like) of A is given in, it is part of the data; that is, it follows that A does obtain or occur. For the evidential cases the basic symbolism will be adjusted from "$\supset$" to "$>$" (i.e., straightened out). Thus A is evidential reason for B (A supports B) iff $A \mathrel{+\!\!>} B$, etc. Roughly (at least extensionally), $A \mathrel{+\!\!>} B$ iff $A \mathrel{+\!\!\supset} B \,\&.\, A$, etc. Then the form $A \mathrel{+\!\!\not>} B$ says that A is evidentially relevant to B, or is a relevant evidential consideration for B; the form represents a type of consideration relevance (as contrasted with structural relevance, such as sharing variables or content, or use relevance—all related relevance determinables). The contrast between $\mathrel{+\!\!\supset}$ and $\mathrel{+\!\!>}$ can be coupled with the contrast of information and evidence. In $A \mathrel{+\!\!\supset}$ ... there is information laid, while in $A \mathrel{+\!\!>}$ ... evidence is presented.

A curiosity regarding 'modifiers', from the orthodox standpoint, is that there are clear differences between relational and propositional modifiers, i.e., between *de re* (*latio*) and *de dicto* forms. Suppose the aim is to distinguish conclusive (or total evidence affording) reason, from partial reason. The familiar lines along which entailment is distinguished among implications, as a logical or necessary form, by a modal propositional operator $\Box$, encounters difficulties almost at once. Suppose instead a functor $\boxdot$, for *conclusive* or *certain*, is introduced. The trouble is that "A is a conclusive reason for B", in new symbols $A \boxdot \rightarrow B$, is not approximated by "it is conclusive that (guaranteed that) A is a reason for B", for that says something very different, and a trifle strained. An initial way around this problem is to render the parameters as *suffixes* instead of prefixes. Thus $(A \mathrel{+\!\!\supset} B)\boxdot$ reads: A is a reason for B, (and) conclusively. In those terms $A \boxdot \rightarrow B$, i.e., A is a conclusive reason for B, iff $(A \mathrel{+\!\!\supset} B)\boxdot$. By contrast, A less than conclusively implies B iff $A \mathrel{+\!\!\supset} B \,\&\, {\sim}(A \boxdot \rightarrow B)$. Naturally conclusiveness is not required for all reasoning or conditionals; for example, it is too strong for forms of probabilistic reasoning. Conclusiveness serves primarily in helping to distinguish a major class of conditionals, those we presently know most about.

Conclusiveness, so discerned, may well not afford logical conclusiveness, for that calls for suitable logical reasons. More important, logical conclusiveness of reasoning does not supply, or closely approximate, entailment, any more than conclusive reason delivers a genuine implication. Implication, like entailment meets requirements such as reflexivity and monotonicity which (as further explained below) conclusive reasoning does not. No doubt however there are connections between implication and affirmative reasoning. One bridge is afforded by closure operations. Implication is roughly a closure of (ponible) reasoning under critical operations such as reflexivity, transitivity and monotonicity. That is, implication is a certain *extension* of affirmative reasoning. Plainly this does not offer a particularly neat approach to implication. So if we seek an implication (to be symbolised, $\rightarrow$) to carry the sought logics of reasoning—as we do, for logical ease—it may as well be supplied.

Another major cross-classification of reasons, which applies to both actual and putative reasons and also cuts across conclusive and inconclusive reasons, is that between closed and open reasons. For closed or decisive reasons, A guarantees that B holds or closes its relative truth status, for open reasons, it does not. Where A is a open reason for B, all that is required is that A supports B or lends weight to B. There is of course a gradation from the wide open to the fully closed (conclusive), which in ordinary language is usually marked by modifiers of various kinds. We shall however initially limit attention to the simple dichotomy, and we shall take Modus Ponens to mark the distinction between open and closed reasons. That is, reasons which fail to satisfy Modus Ponens are open; those which satisfy it are, in one way or another, closed. We also call the latter *ponible reasons*. Thus, a ponible reason for B guarantees that B holds, while a non-ponible reason for B, though relevant to the assessment of B, does not guarantee B. The distinction is evidently related to the difference between deductive or demonstrative reasoning and inductive or evidential reasoning.

To begin with, we shall limit attention to this single characterizing difference between open and closed reasons, even though the difference in forms matters not merely for detachment, but for other principles such as transitivity. In fact our main aim in this essay is to gain semantical control over ponible reasoning, leaving for later discussion the more exacting non-ponible form, as well as the different varieties of ponible reasoning, for example transitive and not, conclusive and not, and so on. The ponible-*if* is the closest obvious relative among the reasongiving *if*'s to the familiar truth-preserving *if*'s of standard logics; that is one reason why we select it as the first target.

But although our aim is to get a grasp on ponible reasoning substantially independently of nonponible reasoning and, only subsequently to investigate nonponible reasoning, there are situations in which both need to be considered together. For we may have, say, a certain amount of inconclusive evidence in favour of B but then discover a decisive reason against it, i.e., in symbols $A \mathrel{+\!\!\supset}^{\circ} B \,\&.\, C \mathrel{-\!\!\supset} B$ for some A and C. It would then be pointless, if C is true, to look for further evidence in favour of B even if there is such and it would have supported B had C not been discovered. Thus questions of epistemic priority arise in determining whether or not something can be properly counted as a reason. The eventual goal, then, must be to develop a 'mixed' system which allows for both open and closed reasons. A less difficult problem arises from the fact that both open and closed reasons satisfy a number of similar conditions: for example, A can never be a reason, whether open or closed, for itself. We indicate this in what follows by putting the superscript $^{\circ}$, which indicates openness, in parentheses showing that it may be kept or dropped *uniformly*.

The difference between $\mathrel{+\!\!\supset}^{(\circ)}$ and $\mathrel{+\!\!>}^{(\circ)}$ is that in the case of $\mathrel{+\!\!>}^{(\circ)}$, but not $\mathrel{+\!\!\supset}^{(\circ)}$, the truth (or like) of the antecedent is given in as part of the data; that is, it follows from $A \mathrel{+\!\!>}^{(\circ)} B$ that A does obtain or occur. We take this to be characteristic of the concept of evidence. That is, where A is offered as evidence for B, the presumption is that A holds; where the presumption is defeated, A is not acceptable as evidence. Thus the contrast between $\mathrel{+\!\!\supset}^{(\circ)}$ and $\mathrel{+\!\!>}^{(\circ)}$ can be coupled once again with the contrast between information and evidence. Similar remarks hold for the difference between $\mathrel{-\!\!\supset}^{(\circ)}$ and $\mathrel{-\!\!>}^{(\circ)}$. The connexion between $\mathrel{+\!\!\supset}^{(\circ)}$ and $\mathrel{+\!\!>}^{(\circ)}$ is given, at least extensionally, by the relation $A \mathrel{+\!\!>}^{(\circ)} B$ iff $A \mathrel{+\!\!\supset}^{(\circ)} B \,\&.\, A$. Similarly, $A \mathrel{-\!\!>}^{(\circ)} B$ iff $A \mathrel{-\!\!\supset}^{(\circ)} B \,\&.\, A$. Then the form $A \mathrel{\pm\!\!>}^{(\circ)} B$ says that A is evidentially relevant to B, or is a relevant evidential consideration for or against B.

As already indicated, intended interpretations of these connectives can often be given through requisite uses of *if* which present the antecedent as a reason for or against the consequent. In the case of $A \mathrel{+\!\!\supset} B$, the interpretation is simply "if A then B"; in the case of $A \mathrel{+\!\!\supset}^{\circ} B$, also construed using *if*, a more explicit ordinary-language presentation would normally be in terms of an explicitly weakened *if*, for example "If A, then it is likely (possible, probable) that B". For $A \mathrel{+\!\!>} B$, the interpretation "if A then B" is less appropriate; since the truth or obtaining of A is given in, $A \mathrel{{}^{+}\!\!\rhd} B$ can however be read, at first approximation, "since A, B" (sometimes *as* or *because* can replace *since*); in fact in the strong case, $A \mathrel{{}^{+}\!\!\rhd} B$ can almost be read "A, therefore B ". In the case of $A \mathrel{{}^{+}\!\!\rhd}^{(\circ)} B$, suitable weakening modifiers may again be introduced; for example "since A, possibly (probably) B".

But although $\mathrel{{}^{+}\!\!\supset}^{(\circ)}$ can be read as *if*, it is clear that none of the standard logics which contain connectives intended to be interpreted as *if* will suffice as the logic of $\mathrel{{}^{+}\!\!\supset}^{(\circ)}$. In the case of $\mathrel{{}^{+}\!\!\supset}^{(\circ)}$, this is immediate since it fails Modus Ponens. The required logics are accordingly more like J-systems (in the sense of Arruda-da Costa). In the case of $\mathrel{{}^{+}\!\!\supset}$, the material implication $\supset$ of classical logic, though ponible, is not a substitute for $\mathrel{{}^{+}\!\!\supset}$. As already observed, other mainstream connectives do not fare significantly better, such logics give little guidance as to what *reason* is all about.

Mainstream connectives, such as material, intuitionistic and strict implications, satisfy core requirements for deduction. So do the arrow connections (symbolised $\rightarrow$) of relevant logics. These requirements are (semantically) those of universal truth preservation, or equivalently syntactically those of reflexivity, monotonicity and portability (for this important equivalence see RLR I, for example p.432). By contrast, the general logics of $\mathrel{{}^{+}\!\!\supset}^{(\circ)}$ and $\mathrel{{}^{+}\!\!\rhd}^{(\circ)}$ are highly nondeductive. For, unlike the standard implicational connectives, they exhibit the following unorthodox features:

- They cast doubt upon Modus Ponens: in particular, A and $A \mathrel{{}^{+}\!\!\supset}^{(\circ)} B$ together do not ensure B. Similarly for $\mathrel{{}^{+}\!\!\rhd}^{(\circ)}$.
- They all flunk Identity: thus $\dashv A \mathrel{{}^{+}\!\!\supset}^{(\circ)} A$; ditto $\dashv A \mathrel{{}^{\pm}\!\!\supset}^{(\circ)} A$. Indeed they fail it in a strong way, since we can take as a postulate irreflexivity, $\sim(A \mathrel{{}^{+}\!\!\supset} A)$. A itself is not a reason for A; nor is $A \mathbin{\&} A$ nor $\sim\sim A$.
- Both $\mathrel{{}^{+}\!\!\supset}$ and $\mathrel{{}^{+}\!\!\supset}^{(\circ)}$ fail Augmentation (Monotonicity): for where $A \mathrel{{}^{+}\!\!\supset}^{(\circ)} B$ it is by no means guaranteed $(C \mathbin{\&} A) \mathrel{{}^{+}\!\!\supset}^{(\circ)} B$. For C may undermine A and remove its reason-giving force.

There are of course variants of Augmentation which need to be considered. Thus the following rule, which imposes a relevance requirement on C,

$$A \mathrel{{}^{+}\!\!\supset}^{(\circ)} B,\ C \mathrel{{}^{\pm}\!\!\supset}^{(\circ)} B \ /\ A \mathbin{\&} C \mathrel{{}^{+}\!\!\supset}^{(\circ)} B$$

might seem to be admissible. But relevant information can still be negatively bearing and undercut support. Indeed C may be relevant to B just because $C \mathrel{{}^{+}\!\!\supset} B$ holds; and C may in fact be conclusive evidence against B, in which case $A \mathbin{\&} C$ can hardly be reason for B.

Even less acceptable, of course from the point of view of the basic theory, are implicational forms of relevance-qualified Rule Augmentation, such as $A \supset_{+}^{(\circ)} B \,\&\, C \supset_{\pm}^{(\circ)} B \rightarrow . \; A \,\&\, C \supset_{+}^{(\circ)} B$ and $A \supset_{-}^{(\circ)} B \,\&\, C \supset_{\pm}^{(\circ)} B \rightarrow . \; A \,\&\, C \supset_{-}^{(\circ)} B$, and parallel implicational forms for $>^{(\circ)}$. Less problematic, it might seem, are some signed augmentation-style rules, such as the &-Composition Rule, and their implicational mates: for example $A \supset_{+}^{(\circ)} B, C \supset_{+}^{(\circ)} B \,/\, A \,\&\, C \supset_{+}^{(\circ)} B$. However, there are situations in which composition increases the strength of a reason. Thus, though A and C may each, separately, be quite open reasons for B, $A \,\&\, C$ may be a ponible reason for B since C adds to A in such a way that the two together constitute a closed reason. In this case $A \supset_{+}^{\circ} B, C \supset_{+}^{\circ} B \,/\, A \,\&\, C \supset_{+}^{\circ} B$ fails; rather, for such A, B, C we have $A \supset_{+}^{\circ} B, C \supset_{+}^{\circ} B \,/\, A \,\&\, C \supset_{+} B$. The proper investigation of such relationships evidently calls for a 'mixed' system; but we can for the present avoid these and related difficulties by limiting the basic theory to ponible reasons. Then, we can reasonably expect $A \supset_{+} B, C \supset_{+} B \,/\, A \,\&\, C \supset_{+} B$ to hold.

Naturally, given suitable restrictions, the rejected augmentation forms—and others equally questionable—are admissible. But we shall regard them as optional (but restrictive) extras, and omit them from the basic logical theory. There is a large class of such extras. Imposing some such structure leads into *reclassification* of types of reasoning, for example augmentable reasoning, enthymematic reasoning, etc. Thus, given that principles can be reintroduced for types of reasoning for which they *do* hold, we can be free and properly uninhibited in our rejections of erroneous or questionable orthodoxy.

In general, the crucial distinctions between $\supset_{+}^{(\circ)}$ and mainstream implicational connectives supposed to approximate a reason-giving *if*, can be summarized by saying that the latter conform to truth-preservation features, many of which $\supset_{+}^{(\circ)}$ sheds. Even the ponible $\supset_{+}$ does not satisfy all the standard conditions for truth-preservation (for example reflexivity, augmentation); and by all usual over-powered logical standards it is incredibly weak. But we want to dispell straightaway the idea that the sort of theory of ponible reasoning we are working our way towards is bound to be hopelessly weak. On the contrary, it corresponds logically to causal reasoning, and it appears to provide a framework for the complete unreduced classical theory of syllogism, the argument-representational capacities of which are considerable. Syllogistic involves progressive demonstrative reasoning; it affords a wealth of examples of widely deployed ponible reasoning and it is intended as a major application of the present theory. Syllogistic in turn appears to offer some welcome constraints on ponible reasoning; for it seems that what a logic of demonstrative reasoning requires in the way of logical principles is like what would yield unreduced syllogistic theory (with duly restricted terms). But it would not require vastly stronger principles, such as those Łukasiewicz and others have foisted on Aristotle.

6.2 Ponible reasoning: some features of the zero-order logics

Thus far too much of our wealth lies in idling connectives. Accordingly we turn to further formal features of ponible reasoning, of $\stackrel{+}{\supset}$ in combination with truth-functional connectives, &, $\vee$, $\sim$. To carry the logic deductively, we deploy at least initially an implication, $\rightarrow$, as well. To be suitably deductive, $\rightarrow$ should satisfy the requirements of reflexivity, portability and monotonicity; indeed the $\rightarrow$, &, $\vee$, $\sim$ sublogic should be a relevant logic, which satisfies at least the postulates of first degree entailment, as argued in detail elsewhere (especially RLR I). In fact, for semantical ease, we shall take the underlying logic to be a weaker relevant affixing system (a deep relevant logic in the vicinity of basic system B of RLR I). For reasons we shall soon reach, this carrier logic should not be conspicuously stronger than it need be.

Primitives: &, $\vee$, $\sim$, $\stackrel{+}{\supset}$, $\rightarrow$

Defined: $A \stackrel{+}{>} B =_{df} A \stackrel{+}{\supset} B \,\&.\, A$

$A \stackrel{-}{\supset} B =_{df} A \stackrel{+}{\supset} \sim B$; similarly $\stackrel{-}{>}$

$A \stackrel{\pm}{\supset} B =_{df} A \stackrel{+}{\supset} B \vee.\, A \stackrel{-}{\supset} B$; similarly $\stackrel{\pm}{>}$.

There are significant questions about the extent of well-formed formulae, both those already alluded to and others, in particular, to what extent higher degree wff make sense? A simple resolution (explained in detail in Chapter Twelve) is to admit higher degree formulae as well-formed, and to classify them as nonsignificant if need be. For a start, counting out implication wff, there will be few higher degree theorems, except substitution instances of lower degree wff. But substitution itself is problematic: it seems to be *field* restricted, in a way formalisation seldom or never captures. To skirt the problem, a field restriction will be taken as a background (contexual) assumption; but in this way it won't effect the formal workings.

In our first attempt at writing down postulates for $\stackrel{+}{\supset}$, on a pre-analytic approach, we arrived ineluctably at implicational transitivity and irreflexivity, i.e.,

ITrans. $(A \stackrel{+}{\supset} B) \,\&\, (B \stackrel{+}{\supset} C) \rightarrow.\, A \stackrel{+}{\supset} C$

Irref. $\sim(A \stackrel{+}{\supset} A)$

An immediate theorem is $\sim(A \ni B \,\&. B \ni A)$, whence $A \ni B \nrightarrow B \ni A$; reason-for is not symmetrical. But in revision of the postulates we were led to reflect that such transitivity may be too strong, because there appear to be cases where $A \ni B \,\&. B \ni A$. For instance, "Felix is a black cat and Fluffy is a black cat and Minou is a black cat [and ...] $\ni$ all cats are black", and also "all cats are black $\ni$. Felix is a black cat". On the other hand, it is easy to point to missing premisses such as that Felix is a cat in the last connection. But such a suppressed premise is a common phenomenon (and a feature of mainstream logics which are highly enthymematic), which presumably does not tell against being a reason for something? In the light of such (inductive) counter-examples, we are inclined to retreat—in revision leading perhaps towards a nonponible structure—to a rule form, namely:

RTrans (Rule Transitivity). $\quad A \ni B, B \ni C \,/\, A \ni C,$

which, so far at least, does seem correct. At least it seems correct so long as it does *not apply* to any examples there may be of symmetrical reasoning; else reflexivity is restored. There are, as it happens, other reasons for retreating to a rule form; namely, apparent counter-examples, involving non-conclusive prudential reasoning, to implicational transitivity. Consider a situation where avoiding something (for example, a spanking) is a reason for going to bed now, and where it is also true that going to bed now is a reason for escaping tiredness tomorrow. But avoiding such a thing as a spanking is scarcely a reason for escaping tiredness tomorrow.

While much stronger transitivity principles than ITrans are offered on the logical market, under such tempting labels as "noncircular reasoning", we find little evidence, or argument, for supposing that they hold for ponible reasoning. Consider the difficult pair:

ESyll. $\quad A \ni B \ni. B \ni C \ni. A \ni C.$
$\quad B \ni C \ni. A \ni B \ni. A \ni C.$

While we can transliterate these, into a sort of English, it is difficult to know what to make of them; but linguists would very likely reject the English renditions as deviant. Insofar as we do get a grasp of them (by casting them along such lines as "A's being a reason for B is a reason for B's being a reason for C being a reason for A's being a reason for C") we doubt that they are generally true, any more than:

Com. $\quad A \ni (B \ni C) \ni. B \ni (A \ni C).$

(Reasons for rejecting all these third degree and beyond principles as correct for much *stronger* notions, like implication and deducibility, are given in RLR I, chapter 2). Certainly such principles as ESyll do not appear in Aristotle's work, and there is no substantial reason for believing he, preeminent authority on noncircular reasoning, would have endorsed them. An appeal to Aristotle's authority thus provides no support for such principles.

It soon emerges, then, that the logics of reasoning will be rich in rejections. To obtain expected rejections as theorems, adjoin the following rule: From $\sim C$ and the information that C is substitution instance of D, infer the rejection of D, i.e $\dashv D$. Then, for example, because purely redundant 'reasons' are no reasons and because reflexivity thus fails: $\dashv A \& B \overset{+}{\supset} B$; $\dashv A \& B \overset{+}{\supset} B$; $\dashv A \overset{+}{\supset} A \vee B$; $\dashv B \overset{+}{\supset} A \vee B$. Indeed virtually all lattice properties appear to be rejected; for example, we should reject Distributivity, i.e., $\dashv A \& (B \vee C) \overset{+}{\supset}.\ A \& B \vee.\ A \& C$ because $\sim(A \& (A \vee A) \overset{+}{\supset}.\ A \& A \vee.\ A \& A)$. A similar series of rejections follow for $\overset{+}{\rhd}$, since $A \overset{+}{\rhd} B \rightarrow A \overset{+}{\supset} B$, whence also $\sim(A \overset{+}{\rhd} A)$. Given extensional linkage we do obtain as theorems however $A \overset{+}{\rhd} B \rightarrow.\ A$; $A \overset{-}{\rhd} B \rightarrow.\ A$; $A \overset{\pm}{\rhd} B \rightarrow.\ A$.

If A cannot count as a reason for itself, then, for similar reasons (whatever they are), A presumably cannot count *against* itself. For reason then, a further postulate is $\sim(A \overset{-}{\supset} A)$, i.e., $\sim(A \overset{+}{\supset} \sim A)$, Aristotlc's principle holds. Thus reasoning is connexive,[89] as well as irreflexive, and so decidedly nonclassical. In such a framework, furthermore, $\sim(A \overset{\pm}{\supset} A)$, whence $\dashv (A \overset{\pm}{\supset} A)$, i.e., consideration relevance cannot be reflexive. (Some doubt is cast upon the linkage of reason and relevance by the way however in which non-circularity of reason leads to irreflexivity of relevance. Some doubt too is cast on traditional connexive principles by Reductio, which requires for its application cases where $A \overset{+}{\supset} \sim A$.)

Many old favourites, plausible for implication, fail for reason. An interesting case (related to Reductio) is Counter-example, $A \& \sim B \rightarrow \sim(A \overset{+}{\supset} B)$, which appears implausible. A mere truth-functional combination does not bear on reasons for, even less on consideration relevance. And, more positively, A can be a reason for B even if, in the end, B turns out to be false. Another old favourite of relevance theory, Assertion, this time in the form $A \& (A \overset{+}{\supset} B) \rightarrow B$, is also in deep trouble. A may only be an insubstantial (or material) reason for B, in which event it will not carry an implication from A through to B.

89 But more powerful connexive principles such as $A \overset{+}{\supset} B \rightarrow \sim(A \overset{+}{\supset} \sim B)$ are not thereby underwritten. They are optional, and difficult, extras.

A crucial issue for any logic with higher degree parts is the extent of substitutivity of equivalents. Reasoning connectives do *not* in general guarantee replacement, but $\rightarrow$ does, for instance (for suitable sentence contexts):

SE. $A \rightarrow B, B \rightarrow A \,/\, \ldots A \ldots \rightarrow \ldots B \ldots$

But forms with implicational premisses, as distinct from such coimplicational forms, are not so evident. Nonetheless for implication we are inclined to plump for affixing principles, i.e.,

RA. $A \rightarrow B, C \rightarrow D \,/\, B \rightarrow C \rightarrow .\, A \rightarrow D.$

Arguments for this 'extended transitivity' principle, which at once give prefixing and suffixing principles as special cases, are the standard ones (as in RLR I). Indeed we shall take it, for similar previously developed reasons, that a sufficiency implication, $\rightarrow$, conforms to a deep relevant logic; for example it meets distributive lattice logic principles. Then from such coimplications as $A \,\&\, A \leftrightarrow A$ and $A \leftrightarrow A \vee A$, with $A \leftrightarrow B =_{df} (A \rightarrow B) \,\&\, (B \rightarrow A)$, we can derive by SE little theorems like $\sim(A \,\&\, A \not\supset A)$ and $\sim(A \not\supset A \vee A)$.

Again derived evidential principles are not so neat, confirming impressions that evidential notions afford a more difficult and messy starting point.[90] We gladly avoid the area, in favour of semantical exploration.

6.3 Analyses and semantical recipes for demonstrative reasoning: failure and apparent success

We scramble upwards through failed approaches. Failure is instructive, not merely or at all because our first attempts at semantics, along more or less received lines, were failures, but, more important, because failure can indicate improved or correct directions, and also because what suggestions there are

---

[90] While, given $\boxdot$ principles and the definition $A \rightsquigarrow B =_{df} (A \not\rhd B)\boxdot$, $A \rightsquigarrow B \rightarrow A$ and $A \rightsquigarrow B \rightarrow .\, A \rightarrow B$, and also (where $A \,\&\, (A \rightarrow B) \rightarrow .\, B$) $A \rightsquigarrow B \rightarrow .\, B$, expected principles, such as $(A \rightarrow B) \,\&\, A \rightarrow .\, A \rightsquigarrow B$, appear to fail. Since $A \rightarrow B \,\&\, A\boxdot \rightarrow .\, A \rightsquigarrow B$, an alternative somewhat superior definition is suggested, viz. $A \rightsquigarrow' B =_{df} (A \not\supset B).\boxdot \,\&\, A$, i.e., $(A \rightarrow B) \,\&\, A$. As trivially, $(A \rightarrow B) \,\&\, A \leftrightarrow .\, A \rightsquigarrow' B$, $A \rightsquigarrow' B \rightarrow .\, A \rightarrow B$. Note that where it is given that $A \rightsquigarrow B \leftrightarrow .\, (A \not\rhd B) \rightarrow B$, as for instance as postulate, then $(A \not\rhd B) \rightarrow B \rightarrow .\, A \rightarrow B$ and $(A \rightarrow B) \,\&\, A \rightarrow .\, (A \not\rhd B) \rightarrow B$, by substitution.

We are again bordering the territory of *therefore*. Another initial stab makes "A therefore B" true iff $A \rightsquigarrow' B$ is true. But any purely conditional account fails to capture the crucial detachment feature of inferential connectives like *therefore*.

for semantical analysis of reason in the literature generally fall into failure classes. The biggest trap is undoubtedly probability; it has been death to many a notion or proposal to have fallen within the precincts of probability theory, especially classical probability logic.

A first familiar idea is that $A \supset B$ holds, or is true, iff A increases the likelihood of B , i.e., so it seems, iff $\mu(A \,\&\, B) > \mu(B)$, where likelihood μ is treated relevantly to avoid classical pitfalls (otherwise decidedly false or inconsistent information could never serve as reasons). Under this proposal, A undoubtedly does have several of the right associations, for example connections with enhanced credence. Also the proposal appears to do many of the right sorts of things. For example, $A \supset A$ does not hold, because $\mu(A \,\&\, A) = \mu(A) \not> \mu(A)$. However central elementary principles such as Rule Transitivity break down :- Where $A \supset B$ and $B \supset C$, $\mu(A \,\&\, B) > \mu(B)$ and $\mu(B \,\&\, C) > \mu(C)$, but it does not follow, without additional dubious assumptions, that $\mu(A \,\&\, C) > \mu(C)$, as required for $A \supset C$.

Where probability appears to fail there is a batch or loosely connected notions nearby to occupy the breach, for instance, real possibility, credibility, plausibility, even preferability. On most of these notions we still lack much of a logical grasp (despite the work of modern masters like von Wright and Rescher); some of them remain neglected by mainstream logical enterprise, but no matter. Let's try to cash $A \supset B$ semantically as: $s(A)$ increases or enhances $s(B)$, where $s(C)$ is variously the *standing* of C, real possibility of C, plausibility of C, prospects of C, etc. This sort of approach straightaway links $\supset$ with discussive logics, where for example, $A \supset B$ is true iff $\diamond A \rightarrow \diamond B$, and enhancement is construed (no doubt less than satisfactorily) through implication $\rightarrow$. Variants, under the discussive approach (for more on which see OP), are $A \rightarrow \diamond B$, with $\diamond$ a real 'maybe' or possibility, and also $\diamond(A \rightarrow B)$. But such discussive approaches not only seem intuitively wrong—what have these *modal* things to do with *reason for*—but also accomplish the wrong things. For instance, they deliver transitivity in excessively strong forms and also, in contrast to s-enhancement, reflexivity. Consider instead the received discussive analysis which would make "reason for" dissolve along the following unlikely lines: if ... then plausibly/presumably/(really) possibly ... . But then, $A \rightarrow \diamond B$ yields $A \supset A$. Also again it renders Rule Transitivity dubious: it becomes $A \rightarrow \diamond B, B \rightarrow \diamond C \,/\, A \rightarrow \diamond C$; so it requires $B \rightarrow \diamond C \,/\, \diamond B \rightarrow \diamond C$, or (for example) $\diamond B \rightarrow \diamond\diamond C \,/\, \diamond\diamond B \rightarrow \diamond C \,/\, \diamond B \rightarrow \diamond C$, neither of which have much appeal under intended interpretations.

Enhancement of some sort is no doubt part of the reasoning story, though in exactly what form is less clear. Nor is all that it represents adequately taken up in the implicational connection, which is no doubt also part of an account of ponible *reason for*. All this suggests a two component account, and that previous attempts (such as the discussive effort) go down because they include only one of two interconnecting components or try to merge the two in one recipe.

Such a duplicate account of reasoning is also suggested by a process analysis. Reasoning is, as Aristotle said, an argument, in which things come about; that is, it is a type of process. Accordingly, reasoning is certainly a process, and so should reflect the double aspects of process, a relation and a superposed direction. The relation involved in *ponible* reasoning is some sort of appropriately *weak* implication, while the direction is some sort of priority ordering or enhancement of standing. *However* these components are filled out, the general form is evident:

R1. $A \mathrel{\supset\!\!\!\!{\scriptscriptstyle +}}\, B$ iff $A \rightarrow B \,\&\, s(A) < s(B)$.

The form, already familiar from other analytic exercises such as that of causal implication, leads in turn to other interpretations. It leads in particular to the idea that the priority involved is primarily *epistemic*, as opposed for instance to causal or temporal or logical; naturally it also reflects enhancement. In the setting of reason-giving, the assumption that what gives a reason has epistemic priority over what is given a reason for is both appealing and plausible. It surely can't hold in the reverse direction, so we shouldn't expect both $A \mathrel{\supset\!\!\!\!{\scriptscriptstyle +}}\, B$ and $B \mathrel{\supset\!\!\!\!{\scriptscriptstyle +}}\, A$; and it must be progressive, so $A \mathrel{\supset\!\!\!\!{\scriptscriptstyle +}}\, A$ fails because A has no priority as regards A. Upon symbolising "A is prior information to B" by $A \multimap B$ (thus approximating a sideways appropriately directed P), the priority analysis takes the form

R2. $A \mathrel{\supset\!\!\!\!{\scriptscriptstyle +}}\, B$ iff $A \rightarrow B \,\&\, A \multimap B$.

The two analyses are elementarily connected by construing $s(A)$ as the epistemic content, or information of A, i.e., $A \multimap B$ iff $s(A) < s(B)$.

These analyses of ponible reasoning suggest in turn analyses of reason more generally. Evidently, implication, no matter how enthymematically diluted, is too strong, guaranteeing ponibility. It is implication that has to be weakened, in generalising the process account. The general form thus takes the following shape:

R1G. $A \mathrel{\supset\!\!\!\!{\scriptscriptstyle +}}\, B$ iff $A \mathrel{\circ\!\!\rightarrow} B \,\&\, s(A) < s(B)$,

where s is as before but $\circ\!\!\rightarrow$ (or $\rightarrow^{(\circ)}$) replaces $\rightarrow$; plication (p-ization) replaces implication. Plying expands upon implying, generalising over such propositional support relations as plausibilifying and probabilifying. Its explication, perhaps best approached in a case by case way with *implying* a first determinate, should generate much future logical ado. For the present, we stick, however, with that suggestive determinate and accordingly with ponible reasoning.

A worthwhile analysis often yields an accompanying deeper-reaching semantical theory. In the case of ponible reasoning, appropriate semantical rules fall out by situational relativisation of R1, or R2. So results the interpretation principle:

I$\ni\!\!\!\supset$ $\quad I(A \ni\!\!\!\supset B, a) = 1$ iff $I(A \rightarrow B, a) = 1 = I(A \multimap B, a)$,
i.e., iff $(b, c)(Rabc \,\&\, I(A, b) = 1 \supset I(B, c) = 1) \,\&\, s(A, a) < s(B, a)$,

set within the framework of an enlarged relevant logic semantics. In logic-school English, the first line of I$\ni\!\!\!\supset$ reads: that A is a ponible reason for B holds at situation a iff both that A implies B and that A is prior enhancing information to B hold at a. Of course *reason for* is commonly enthymematic. But that's all right, that feature can be logically matched. The implication mainstream logics do, ad lib introduction of irrelevant components. Rather it will be relevant enthymematic, after the fashion of favoured relevance logics such as systems E and R.

With but few further assumptions, of considerable initial plausibility, the analysis confirms preanalytic impressions, delivering for the most part expected themes and effecting anticipated rejections. We can verify, in particular, that the ponible reason-for relation $\ni\!\!\!\supset$ is a connexive implication satisfying Irreflexivity, Aristotle, Modus Ponens, Modus Tollens, Rule Transitivity, Composition but failing Identity, Simplication, Addition, Substitution, Contraposition, etc. To derive these results, it will be assumed—what can be cogently argued in demonstrative cases—firstly that implication ($\rightarrow$) conforms at least to the principles of first degree relevant implication (thus it may be enthymematic and it could be a strict or other irrelevant implication), and secondly that the order $<$ is transitive and irreflexive (and so, antisymmetric). Call ponible reasoning so far delimited *demonstrative* reasoning. This corresponds roughly to Aristotle's account of reasoning, where what comes about comes about necessarily—plainly, necessity linkage is not, on the face of it, an essential feature of all reasoning, certainly not of non-ponible, non-conclusive reasoning.

Proposition 1. Demonstrative reasoning connective $\supset$ is an implication satisfying Rule Transitivity, Irreflexivity, Modus Ponens, Modus Tollens but failing Identity.

Verification uses R1 and properties given.[91]

To proceed the further intuitive (but nonetheless more controversial) distance, it will be assumed, thirdly, that at least for extensional connectives in the same components, priority is determined component-wise, that is, $s(\sim A) = (A)$, $s(A \,\&\, A) = s(A)$, $s(A \vee A) = s(A)$. This leaves open, so far, what is said as to $s(A \,\&\, B)$, and so on, for different components.

Proposition 2. The reasoning connective $\supset$ is a connexive implication satisfying Aristotle, but failing Contraposition, Simplification, Addition (and elementary lattice conditions generally), and very important, failing substitutivity of equivalence for demonstrative reasoning.

Verification is again elementary, but some examples are instructive. Consider Aristotle, which expands to $\sim(A \rightarrow \sim A) \vee \sim(s(A) < s(A))$, and is carried by the latter disjunct. If Simplification held, then $A \,\&\, A \supset A$ would hold, implying $s(A) < s(A)$. Rule Contraposition fails because $<$ is not symmetric. Thus, appealing as the component-wise assumption may seem, it puts us out of line with logical tradition. A long tradition, stretching from Aristotle through Boethius to 12th century logicians like Abaelard and beyond, accepted both connexive principles and contraposition (though, be it noted, for conditionals). Moreover, as the example of Aristotle reveals, some very curious negative statements are also verified. Let $\Delta_1(A)$, $\Delta_2(A)$ be any extensional functions just of A. Then $\sim(\Delta_1(A) \supset \Delta_2(A))$. The component-wise information assumption is too simple. Connectives *do* carry information; $\sim A \,\&\, A$ does not convey the same as $\sim A \vee A$. The assumption will need upgrading, once it has served its purpose of enabling logical take-off.

To fill out the first degree picture, it remains to impose appropriate connections between $s(A \,\&\, B)$ and the pair $s(A)$, $s(B)$. An obvious move is to let information sum, so the information of $A \,\&\, B$ is simply that of A and of B, i.e., $s(A \,\&\, B) = s(A) \cup s(B)$, where $\cup$ is some sort of union operation (a semi-lattice operation suffices). Then, since any truth functional combination $\delta(A_1, ..., A_n)$ can be defined in terms of just connectives $\&$ and $\sim$, $s(\delta(A_1, A_2, ..., A_n)) = s(A_1) \cup s(A_2) \cup s(A_2) \cup ... \cup s(A_n)$. For instance,

[91] *For details see Chapter Twelve. [Editors]*

$s(A \vee B) = s({\sim}({\sim}A \,\&\, {\sim}B)) = s({\sim}A \,\&\, {\sim}B) = s({\sim}A) \cup s({\sim}B) = s(A) \cup s(B)$ (given semi-lattice conditions and either suitable definitions of truth-functions or the principle $A \leftrightarrow B \,/\, s(A) = s(B)$ for A and B truth-functional). It follows, what may be more than a little surprising, that information reduces to that of sentential variables, at least in truth-functional cases.[92] Plainly similar distribution principles fail, or should fail, for such functors as the tense-logical F. it will be the case that: for $s(A) \neq s(FA)$ for many A.

Proposition 3. Connective $\supset\!\!\!\!\cdot\;$ satisfies various Composition principles, conforms to Augmentation only in restricted form, and fails Factorisation principles unless also restricted.

Verification and elaboration. Consider for instance &-Composition. Assuming $A \supset\!\!\!\!\cdot\; B$ and $A \supset\!\!\!\!\cdot\; C$ it follows that $A \to B$, $A \to C$, $s(A) < s(B)$, and $s(A) < s(C)$. Then, by the first pair $A \to (B \,\&\, C)$, while by the second $s(A) < s(B) \cup s(C) = s(B \,\&\, C)$, whence, assembling, $A \supset\!\!\!\!\cdot\; (B \,\&\, C)$. Factorization by contrast fails, for there is no guarantee that when $s(A) < s(B)$ that $s(A \,\&\, C) < s(B \,\&\, C)$, i.e., $s(A) \cup s(C) < s(B) \cup s(C)$. What happens with Augmentation, which similarly fails in general, is interesting. Consider thc form: $A \supset\!\!\!\!\cdot\; B \,/\, A \,\&\, C \supset\!\!\!\!\cdot\; B$. The critical issue, assuming implication sustains Augmentation, is when $s(A) < s(B)$ guarantees $s(A) \cup s(C) < s(B)$. Presumably generally only if $s(C) < s(B)$; roughly C is content enhancing for B. In short,appropriate content enhancing additions can be tacked on, but not arbitrary ones.

While we shall begin further semantical elucidation with the componentwise assumption, which facilitates reduction of extensional information to parametric components, we shall go on to abandon that restrictive assumption, but not the assumption $s(A \,\&\, B) = s(A) \cup s(B)$ that information sums.

6.4 Ponible reasoning systems: axiomatics and semantics

We begin from the semantical end, which we have already reached, in main essentials, by straightforward situational relativisation of the process

---

92 Here information is looking remarkably like, indeed too like, significance. Superior accounts of information and content appear in the relevant logic tradition, where (at the first degree) $s(A \vee B) = s(A) \cap s(B)$, $s({\sim}A) = Ns(A)$, with N De Morgan complementation (see for example RLR I, p.211 ff). The theory of reasoning will be gradually amended to allow for these connections, the analytic potential of which is far from exhausted.

analysis. We commandeer semantics for relevant logic, formulated in terms of connective set $\{\rightarrow, \&, \vee, \sim\}$. Though we view the semantics as fashioned for some deeper relevant system, such as that of deducibility or conditionality, they will serve for a variety of other systems such as the multitude of strict systems. Accordingly, we do not prejudge questions as to what sort of logic reasoning is coupled with, though we prefer relevant linkages.

A relevant frame (or m.s., i.e., model structure) $\langle T, K, O, R, *\rangle$ comprises a set K of situations (or worlds), a subset $O \subset K$ of regular situations (where theorems hold), a base (actual) situation T in O, a three-place relation R on K (modelling implication) which collapses to a two-place relation in irrelevant logics, and a reversal operation $*$ (which takes care semantically of negation). As it happens, the regular situations will come to play a significant role in relevant reasoning, as in deeper relevant logics. We add to such relevant frames (which are further explained in RLR I), a set $S(a)$ of *information* for each situation a in K, and an operation $\cup$ for summation of information, on each $S(a)$. The operation is commutative, associative, idempotent, i.e., each $\langle S(a), \cup\rangle$ is a semi-lattice. The information of set $S(a)$ (which will be construed 'canonically' just in terms of equivalences of wff, whatever these objects really do) admits of a variety of other interpretations, for example for different readings of "A", as grounds, moments, times. Order relations are defined for α, β in $S(a)$ as follows: $\alpha \leq \beta$ iff $\alpha \leq \beta = \beta$; $\alpha < \beta$ iff $\alpha \leq \beta \,\&\, \alpha \neq \beta$. It is the priority ordering of information that matters, but $\cup$, which facilitates the definition of order, also has the role of enabling the representation of summation of information. Semi-lattice properties used subsequently follow in the standard way; in particular, where $\alpha \leq \gamma$ and, $\beta \leq \gamma$ then $\alpha \cup \beta \leq \gamma$, and both $\alpha \leq \alpha \cup \beta$ and $\beta \leq \alpha \cup \beta$, i.e., $\cup$ provides a least upper bound. For instance, $\alpha \leq \alpha \cup \beta$ since $\alpha \cup (\alpha \cup \beta) = (\alpha \cup \alpha) \cup \beta = \alpha \cup \beta$.

An $R1$ *frame* $\langle T, K, O, R, *, S, \cup\rangle$ thus adds to a relevant frame just a function S from situations to sets of information and a semilattice operation $\cup$.[93] An $R1$ (basic reasoning) *model* adds to an $R1$ frame two valuation functions, I (or ν) and s. Valuation I, from sentential parameters and situations to acceptability or holding values 1(on) and 0(off), subject to

---

[93] It will be evident (at least to connoisseurs of relevant logic) that some conceptual economy could be effected by switching to an operational semantics for the carrier relevant logic, with situations in K construed (after Urquhart's proposal) as comprising bits of information. Then $S(a) = a$, so S can be elided; and the 'same' semilattice operation used both within a and on K, so relation R can be dispensed with in favour of $\cup$. In short, an $R1$ operational m.s. can reduce just to a relevant m.s. For the present exercise we will forego this pleasant if deceptive economy.

a hereditariness requirement, is as in relevant semantics. Valuation s is also from sentential parameters and situations but maps to situationally-associated information, i.e., $s(p, a)$ belongs to $S(a)$. Thus $s(p, a)$ represents the information of p at a.

Both the frame and model take various, weaker or stronger forms, depending upon whether ponible reasoning is to be fully relevant or not and component-wise constrained or not. We present the stronger, more simplistic form first, and then indicate some of the variations demanded by relevance and adequacy. These variations, to come shortly, will involve both relaxing valuational conditions on information and decoupling some of the excessively rigorous semi-lattice requirements.

Both I and s are extended from initial sentential wff to all wff inductively. The evaluation clauses for I, as applied to connectives &, $\vee$, $\sim$, $\rightarrow$, are just those of relevant semantics. The further clauses are as follows:

$$s(A, a) = \sum_{j=1}^{n} s(p_j, a), \text{ i.e., } s(p_1, a) \cup ... \cup s(p_n, a)\ ,$$
$$\text{for } p_1, ..., p_n \text{ exactly the sentential parameters of } A$$
$$I(A \multimap B, a) = 1 \text{ iff } s(A, a) < s(B, a)$$
$$I(A \overset{+}{\supset} B, a) = 1 \text{ iff } I(A \rightarrow B, a) = 1\ \&\ I(A \multimap B, a) = 1$$
$$\text{i.c., iff for all } b, d, (Rabd\ \&\ I(A, b) = 1 \supset .\ I(B, d) = 1)$$
$$\&\ s(A, a) < s(B, a).$$

To make axiomatisation and demonstration of its adequacy run smoothly priority is equality expanded. The connective "$\overline{\multimap}$" symbolises equality expanded priority, and "$\multimapboth$" the corresponding equality. These are subjected to hereditariness requirements and evaluated in the expected way: $I(A \overline{\multimap} B, a) = 1$ iff $s(A, a) \leq s(B, a)$; $I(A \multimapboth B, a) = 1$ iff $s(A, a) = s(B, a)$. Validity and other semantical notions are defined as usual for relevant logics.

The matching initial stronger logical theory $R1$ is adequately axiomatised by some adequate postulate set for a carrier relevant logic in connectives $\{\rightarrow$, &, $\vee$, $\sim\}$ together with the following postulates in "$\overline{\multimap}$" (equality expanded priority) where, by definition, $A \multimapboth B =_{df} A \overline{\multimap} B\ \&\ B \overline{\multimap} A$ and $A \multimap B =_{df} A \overline{\multimap} B\ \&\ \sim(A \multimapboth B)$:

1. $A \overset{+}{\supset} B \leftrightarrow .\ (A \rightarrow B)\ \&\ (A \multimapboth B)$; Analytic linkage, i.e., P1.
2. $(A \overline{\multimap} B)\ \&\ (B \overline{\multimap} C) \rightarrow .\ A \overline{\multimap} C$; Order, i.e., $\overline{\multimap}$ is transitive.
3. $(A \overline{\multimap} C)\ \&\ (B \overline{\multimap} D) \rightarrow .\ A\ \&\ B \overline{\multimap} C\ \&\ D$; Semi-lattice, i.e., $\overline{\multimap}$ fully composes.

4(I). $C \rightarrow . A \mathrel{\vec{\circ}} B$, where the sentential parameters of A are a subset of those of B. i.e., in symbols $v(A) \subseteq v(B)$.

The irrelevant form 4(I) conveniently supplies further partial order and semi-lattice principles, such as:

$$A \mathrel{\vec{\circ}} A;\ A \mathrel{\vec{\circ}} A \,\&\, B\ ;\ B \mathrel{\vec{\circ}} A \,\&\, B,$$

as well as the more dubious lattice connection:

5. $A \mathrel{\vec{\circ}} B \leftrightarrow . A\&B \mathrel{\vec{\circ}} B \,\&. B \mathrel{\vec{\circ}} A \,\&\, B$.

For, by 4(I), $A \mathrel{\vec{\circ}} B \rightarrow . B \mathrel{\vec{\circ}} A \,\&\, B$. and also $A \mathrel{\vec{\circ}} B \rightarrow . B \mathrel{\vec{\circ}} B$. As $A \mathrel{\vec{\circ}} B \rightarrow . A \mathrel{\vec{\circ}} B$, by 3, $A \mathrel{\vec{\circ}} B \rightarrow . A \,\&\, B \mathrel{\vec{\circ}} B$. One half of 5 then results by composition. For the converse, apply 4(I) to obtain $B \mathrel{\vec{\circ}} A \,\&\, B \rightarrow . A \mathrel{\vec{\circ}} A \,\&\, B$. Then by relevant logic and 2, the result follows. Postulate 4(I) in fact immediately yields the pleasanter:

4. $A \mathrel{\vec{\circ}} B, v(A) \subseteq v(B)$

along with much irrelevant junk (such as $q \rightarrow . p \mathrel{\vec{\circ}} p$) which makes proofs simpler and (sometimes) possible. In irrelevant logics which offer some control (over classical licence), C of 4(I) can be restricted in form, for instance to $\Box C$ in modal logics and to $C \rightarrow C$ in intuitionism (and in the I systems of RLR I); and those logics of course enable the derivation of 4(I) thus restricted from 4. It is the quest for relevance, and thus for replacement of 4(I) and its variations by 4, that first leads beyond $R1$ to weaker and deeper axiomatisation of ponible reasoning.

The soundness and completeness of the axiomatisation of $R1$, with respect to the semantics given, builds in familiar ways upon corresponding adequacy results for relevant logics. Details are essentially the same as those already given for causal implication (in Chapter Twelve). In any case, the details sufficiently resemble those for improved system $R2$.

Logic $R2$ abandons 4(I) in favour of 4, and, what 4(I) supplies, the lattice connection 5. But, short of amending the carrier logic, disposing of 4(I) properly involves weakening the previously assumed semi-lattice structure. For consider some semi-lattice principle, such as $A \mathrel{\vec{\circ}} A$. By the valuation rule for s, this holds at arbitrary world a iff $s(A, a) \leq s(A, a)$, which with a semi-lattice structure in place is always. Thus $A \mathrel{\vec{\circ}} A$ holds everywhere, whence, by the rule for $\rightarrow$, anything implies it. So irrelevant principles like $B \rightarrow . A \mathrel{\vec{\circ}} A$ hold. Thus more general structures than semi-lattices are required for nonregular situations, because at them reflexive and other lattice properties may, like all logical principles, fail. An appropriate structure at each such a is an *order-structure* $< S(a), \leq, \cup >$, or $< S(a), \cup >$ since order can be defined in the usual way: for α, β in $S(a)$, $\alpha \leq \beta$ iff $\alpha \cup \beta = \beta$. The relation $\leq$ is semi-lattice-*like* in that it is transitive on $S(a)$, composes fully, and satisfies the (definitional) condition $\alpha = \beta$ iff $\alpha \leq \beta \,\&\, \beta < \alpha$. It follows at once that $(\alpha \cup \beta \leq \beta) \,\&\, (\beta \leq \alpha \cup \beta)$ iff $\alpha \leq \beta$ (corresponding to lattice connection, 5). Given as well that where a is regular, $< S(a), \cup >$ is a semi-lattice, an order-structure, or *mi-lattice*, is, so to say, a *relaxed* semi-lattice. Along with such relaxation the semantics for $R2$ also reduces the component-wise linkage. While $R2$ frames are the same as $R1$ frames, for $R2$ valuations, $s(A, a) = \sum_j s(p_j, a)$ for a in O, but $s(A, c)$ is assigned arbitrarily otherwise. That is, s is only determined through sentential variables at regular situations; otherwise $s(A, a)$ is assigned initially, like $s(p, a)$ as some element of $S(a)$. Adequacy details for $R2$ can in turn be piggy-backed upon corresponding details for a different improvement, $R3$.

Logic $R3$, while retaining the lattice connection (i.e., 5), dumps the component-wise linkage (i.e., 4) altogether.[94] As a result, $R3$ also loses Aristotle's and Abelard's distinctive nonclassical theses (i.e., $\sim(\sim A \ni A)$ and $\sim(A \ni \sim A)$), but they can be recovered in an expanded $R3$ (where $s(\sim A, a)$ and $s(A, a)$ are disjoint at a).

System $R3$ accordingly adds to the carrier logic the following postulates:

1. $A \ni B \leftrightarrow . (A \to B) \,\&\, A \multimap B$ Analytic linkage

2.1 $(A \mathrel{\bar{\circ}} B) \,\&\, (B \mathrel{\bar{\circ}} C) \to . A \mathrel{\bar{\circ}} C$ Transitivity

2.2 $A \mathrel{\bar{\circ}} A$ Reflexivity

3.1 $(A \mathrel{\bar{\circ}} C) \,\&\, (B \mathrel{\bar{\circ}} C) \to . A \,\&\, B \mathrel{\bar{\circ}} C \,\&\, D$ Praeclarum

3.2 $A \mathrel{\bar{\circ}} A \,\&\, B;\ B \mathrel{\bar{\circ}} A \,\&\, B$ Lattice bound

5. $A \mathrel{\bar{\circ}} B \leftrightarrow . A \,\&\, B \mathrel{\bar{\circ}} B \,\&. B \mathrel{\bar{\circ}} A \,\&\, B$ Semi-lattice suppression.

Adequacy theorems for $R3$ and $R2$. A wff of $R3$ is a theorem of $R3$ iff it is $R3$ valid. Similarly for $R2$.

Proof: Both parts build upon adequacy results for relevant logics. In *soundness*, the following properties drawn from relevant semantics simplify (to the verge of triviality) verification of postulates in $\bar{\circ}$:- $I(C \to D, a) = 1$ iff, for every situation a, where $I(C, a) = 1$ then $I(D, a) = 1$. Similarly,

94 The previous over-rigid theory of causal implication of Chapter Twelve admits of improvement along similar lines.

$I(C \leftrightarrow D, a) = 1$ iff, whatever relevant situation a, $I(C, a) = I(D, a)$. Scheme 1 is then immediate from the clause $\mathrm{I}\overline{\supset}$.
ad 2.1. It suffices to show where $I(A \multimap B, a) = 1 = I(B \multimap C, a)$ that $I(A \multimap C, a) = 1$; that is, where $s(A, a) \leq s(B, a)$ and $s(B, a) \leq s(C, a)$ that $s(A, a) \leq s(C, a)$. But that follows by virtue of the fact that $S(a)$ is a mi-lattice, and so $\leq$ a partial order and transitive in $S(a)$, for each a.
ad 2.2. Similar, using reflexivity of $\leq$ on $S(a)$, for a in O.
ad 3.1. As in 2.1, it is enough to show for arbitrary situation a that where $s(A, a) \leq s(C, a)$ and $s(B, a) \leq s(D, a)$ then $s(A \,\&\, B, a) \leq s(C \,\&\, D, a)$. As $s(A \,\&\, B, a) = s(A, a) \cup s(B, a)$ and $s(C \,\&\, D, a) = s(C, a) \cup s(D, a)$ the requisite result follows from mi-lattice properties.
ad 3.2. Similar, using upper bound lattice properties, for regular situations a in O.
ad 5. Suppose firstly $s(A, a) \leq s(B, a)$; to show $s(A \,\&\, B, a) \leq s(B, a)$ and $s(B, a) \leq s(A \,\&\, B, a)$. Now $s(A \,\&\, B, a) = s(A, a) \cup s(B, a) \leq s(B, a) \cup s(B, a) = s(B, a)$. And $s(B, a) \leq s(A, a) \cup s(B, a) = s(A \,\&\, B, a)$, without any assistance from the supposition. The converse reverses what is supposed and what is to be shown. But then $s(A, a) \leq s(A \,\&\, B, a) \leq s(B, a)$, as required. The remaining soundness detail concerns postulate 4 of $R2$.
ad 4. Suppose $v(A) \subseteq v(B)$. Then, for a in O, $s(A, a) = \sum s(p_A, a) \subseteq \sum s(p_B, a) = s(B, a)$, where p_A ranges over the variables of A and p_B over those of B. Hence, for every a in O, $I(A \multimap B, a) = 1$. So $A \multimap B$ is $R2$ valid.

In completeness, the canonical m.s. of the carrier logic (as defined in RLR I chapter 4) is enriched by a canonical order-structure, obtained by a ready adaptation of standard algebraic methods. For each situation a, a quasi-equivalence $\tilde{a}$ on wff is defined: $A \,\tilde{a}\, B$ iff $A \multimap B \in a$ and $B \multimap A \in a$. Then the relation $\tilde{a}$ is symmetric by definition and transitive by partial order postulate 2.1; it is reflexive where a is regular. Now define $|D|_a$ as the quasi-equivalence class of wff D under $\tilde{a}$, i.e., $\{C\colon C \,\tilde{a}\, D\}$. Then $S(a)$ is the class of all these classes, i.e., $S(a) = \{|D|_a\colon D \text{ is a wff}\}$ for every a in K. The following definition of $\cup$, admissible by virtue of 3.1, completes the definition of the canonical $R3$ m.s.: $|C|_a \cup |D|_a = |C \,\&\, D|_a$.

- $S(a)$ supplies an order-structure for each a in K, and a semi-lattice for each a in O.

The central step in showing this, and other vital steps in completeness, is:

† $\quad |C|_a \leq |D|_a$ iff $C \multimap D \in a$.

But $|C|_a \leq |D|_a$ iff $|C|_a \cup |D|_a$, by definition of $\leq$.
iff $|C \,\&\, D|_a$, by definition of $\cup$.
iff $C \,\&\, D \;\tilde{a}\; D$, by canonical definitions.
iff $C \,\&\, D \,\bar{\circ}\, D \in a \,\&\, D \,\bar{\circ}\, C \,\&\, D \in a$, by def.
iff $C \,\bar{\circ}\, D \in a$, applying 5.

Postulate 5 is critical in this final step. Using the definition of $\cup$ it is now a straightforward matter to verify that $S(a)$ is order-structure. Consider to illustrate, the analogue of 3.1; i.e., where $\alpha \leq \gamma$ and $\beta \leq \delta$, $\alpha \cup \beta \leq \gamma \cup \delta$. For a given a, let $\alpha = |A|_a$, $\beta = |B|_a$, $\gamma = |C|_a$ and $\delta = |D|_a$. Then where $|A|_a \leq |D|_a$ and $|B|_a \leq |C|_a$, $A \,\bar{\circ}\, D \in a$ and $B \,\bar{\circ}\, C \in a$, by †. Thus by 3.1, $A \,\&\, B \,\bar{\circ}\, C \,\&\, D \in a$, whence by † again, $|A \,\&\, B|_a \leq |C \,\&\, D|_a$, so $|A|_a \cup |B|_a \leq |C \,\&\, D|_a$, as required.

Canonical valuations are defined as follows for $R3$: $I(p, a) = 1$ iff $p \in a$, for every sentential variable p and every a in K. $s(A, a) = |A|_a$, for every wff A and every a in K. For $R2$, the characterization of s is varied for a in O, where $s(p, a) = |p|_a$ for every variable p. Given the canonical definition of s, the main further work for $R3$ is elementary.

- $I(A \,\bar{\circ}\, B, a) = 1$ iff $A \,\bar{\circ}\, B \in a$,

For $I(A \,\bar{\circ}\, B, a) = 1$ iff $s(A, a) \leq s(B, a)$
iff $|A|_a \leq |B|_a$
iff $A \,\bar{\circ}\, B \in a$, by †.

For system $R2$, it has also been shown:

- $c(A, a) = |A|_a$, for a in O.

Now:

$$\begin{aligned} c(A,a) &= \textstyle\sum_{j=1}^{n} c(p_j, a) = \sum_{j=1}^{n} |p_j|_a \text{, for } p_1, ..., p_n \text{ the variables of } A \\ &= |p_1 \,\&\, ... \,\&\, p_n|_a \text{, iterating } |q \,\&\, r|_a = |q|_a \cup |r|_a \\ &= |A|_a \text{, for } a \text{ in } O. \end{aligned}$$

For, as $v(p_1 \,\&\, ... \,\&\, p_n) = v(A)$, by postulate 4, $p_1 \,\&\, ... \,\&\, p_n \,\bar{\circ}\, A$ and $A \,\bar{\circ}\, p_1 \,\&\, ... \,\&\, p_n$. As a is in O and contains all theorems, both these wff are in each a in O. Hence $p_1 \,\&\, ... \,\&\, p_n \;\tilde{a}\; A$, whence $|p_1 \,\&\, ... \,\&\, p_n|_a = |A|_a$.

System $R3$, which is partly tailored to the adequacy argument, is stronger in certain respects than it strictly needs to be (for example lattice axiom schemes 3.2 and 2.2 could be let go, with semi-lattice structure at regular situations), but is weaker in other perhaps expected features (for example, as to full intersubstitivity of biconditionals). More important, because of the suppression features incorporated in principle 5, $R3$ is not quite the sought basic system; Shangri-la is a little further on. Another logic $R4$ is somewhat nearer, and better as a basis. In $R4$, the need for postulate 5 is avoided by weakening other semi-lattice postulate to rule forms. Specifically 2.1 and 3.1 are replaced respectively by the rules:

R2.1 $\quad A \mathrel{\bar{\circ}} B, B \mathrel{\bar{\circ}} C \,/\, A \mathrel{\bar{\circ}} C$

R3.1 $\quad A \mathrel{\bar{\circ}} C, B \mathrel{\bar{\circ}} C \,/\, A \,\&\, B \mathrel{\bar{\circ}} C.$

Thus $R4$ adds to the carrier logic, postulates 1, R2.1, 2.2, R3.1 and 3.2.

These minor, axiomatic adjustments call for rather more sweeping changes to the semantics. The frame is adjusted: $S(a)$ is only a semi-lattice for a in O. And the interpretation rules are amended: $I(A \mathrel{\bar{\circ}} B, a)$ is cut loose from $s(A, a)$ and $s(B, a)$ except for a in O. That is, for a in O, $I(A \mathrel{\bar{\circ}} B, a) = 1$ if $s(A, a) \leq s(B, a)$, but otherwise $I(A \mathrel{\bar{\circ}} B, a)$ is assigned arbitrarily. Definitional linkages involving $\mathrel{\bar{\circ}}$ are not interfered with.

Adequacy theorem for $R4$. A wff of $R4$ is a theorem iff it is $R4$ valid.

Proof varies that for $R3$. To validate the rules R2.1 and R3.1 it suffices to show that when the premisses hold for arbitrary a in O so does the conclusion. But this is guaranteed by the evaluation of $\mathrel{\bar{\circ}}$ and semi-lattice conditions at a. In completeness, O is defined as the class of regular theories closed under rules (R2.1 and R3.1) for $\mathrel{\bar{\circ}}$ (sometimes a problematic procedure). $S(a)$ is defined for a in O, and results previously established for a in K are restricted to a in O. In particular, $S(a)$ is a semi-lattice for each a in O and $|C|_a \leq |D|_a$ iff $C \mathrel{\bar{\circ}} D \in a$, for each a in O. Use of principle 5 is avoided in the initial last step (of the proof of † for $R3$) as follows: Suppose $C \,\&\, D \mathrel{\bar{\circ}} D \in a$ and $D \mathrel{\bar{\circ}} C \,\&\, D \in a$, then $C \,\&\, D \mathrel{\bar{\circ}} D \in a$. But $C \mathrel{\bar{\circ}} C \,\&\, D \in a$ by 3.2 since a is in O. Then by R2.1, $C \mathrel{\bar{\circ}} D \in a$. Suppose conversely $C \mathrel{\bar{\circ}} C \,\&\, D \in a$. Then $D \mathrel{\bar{\circ}} C \,\&\, D$ already belongs to a, and similarly so does the theorem $D \mathrel{\bar{\circ}} D$. Hence by R3.1, $C \,\&\, D \mathrel{\bar{\circ}} D \in a$. Finally, for a not in O , $I(A \mathrel{\bar{\circ}} B, a) = 1$ iff $A \mathrel{\bar{\circ}} B \in a$, i.e., the canonical connection is obtained for nonregular situations by stipulation, not honest toil.

Systems $R3$ and $R4$ admit of much variation and experimentation, for example by way of content linkages between formulae and subformulae. But these Ri systems are themselves, as is becoming clear, part of a much bigger thing. Instead of messing locally now with $R3$ and $R4$, let's try to make a closer surface approach to this iceberg, glimpsed thus far only from afar.

6.5 On double-barrelled analyses and semantical elucidation

Many implicational relations succumb to analyses in terms of other implications and additional components. (Similarly for other intensional connectives.) Some important classes of analyses are beginning to emerge. One class, of considerable antiquity, is that of *enthymematic implication*, the logical potentialities of which are far from exhausted. Such systemic reductions characteristically represent a 'stronger' system through a tighter system, for example mainstream classical, modal, and intuitionistic logics through a relevant system. Another class of analyses, which typically operate in a reverse direction, consists of *sieves or strainers*, which capture a tighter connection through controlled cases ('sieving') of a slacker one; such for instance are repairs of Lewy's account of (first degree) strict entailment in terms of certain contingent substitutions upon classical tautologies.[95] As well as these specific types of analyses, there are of course many styles of mappings of one system or class of systems into another, for example modal renderings of intuitionistic logics.

Yet another major class comprises all those notions which involve a double analysis. The analysis and semantics of reasoning offered belong, like those for causal implication which they parallel, to this class: the extensive class of double-barrelled or Doppelganger implications. It is decidedly worthwhile displaying the wealth and variety of this major class:

1. *Reason-giving logics*, and the logic of reason more generally. In these systems conditional-type relation ARB, for example, is analysed in terms of a pair of connections AIB, of implication form, and APB. Similarly, generalizing, for *plication*. Moreover, the example we concentrated upon, that of reasoning, belongs to a wider class, which includes also Peripatetic logics (see RLR I, chapter 11). We worked in part by analogy with a second class closely related to that of reasons, namely causes, which permit of representation through causal implication.

2. *Causal implication*, and *cause* generally. Indeed all these notions fall under that of cause in Aristotle's generous sense, with Aristotle's determinates (final, efficient, formal and material) picked out by types of priority and associated modifiers. By displacing priority by the quite different relation

[95] For references see RLR I, p.161 ff. Analyses of this sort have a substantial Cambridge pedigree, going back at least to suggestions of Russell early in the century for getting around the manifest shortcomings of material-implication. In the main, such strainers have been sought with the object of bringing potential rivals within the classical orbit, thereby presupposing that the classical paradigm is correct.

of containment, the next group of double-barrelled notions, takes us beyond the reason-cause nexus.

3. *Containment logics*, including *analytic implication* (and other Parry logics). An early example goes back, like much else in contemporary logic, to Gödel, who proposed that "p analytically implies q" might be exactly interpreted as "p strictly implies q (i.e., in his terms, q is deducible from p and the logical axioms) and q contains no concepts other than p" (see ENT, p.431). This proposal was progressively sharpened up—with concepts or content reducing to variables—by that remarkable trio, Dunn, Urquhart and Fine. That work has culminated (so far that is) in proofs that for Parry's system and others in the neighbourhood, A analytically implies B iff A strictly implies B and $v(A) \subseteq v(B)$. Here $v(C)$ supposedly represents the concepts, or content, of C.

A striking feature of these analytic implication systems is that variables inclusion can be defined *within* them, for instance in Parry's system, $v(A) \subseteq v(B)$ iff $B \rightarrow . A \rightarrow A$. Whatever one's view of such interconnections, they enable an elegant demonstration of the adequacy of the pure Parry systems unencumbered by additional connectives (see Fine 1980). Our view is, of course, that such interconnections depend on a batch of manifestly defective implication principles beginning with Mingle, $A \rightarrow . A \rightarrow A$, and running through such implausible principles as $f(A) \rightarrow . A \rightarrow A$, with $f(A)$ any complication of A. But defective or even simpleminded conceptions often pay off in pretty proofs.

For reason, or for that matter for relevant containment, there appears to be no such representation of variable inclusion (or what it approximates) within the systems involved. Determination of the logic of reason, without parasitic use of relevant implication, will require more circuitous approaches.[96]

4. *Analytic expansion* (and reverse Parry logics). Whereas an analytic implication $A \rightarrow B$ requires the content of B to be included in that of A, an expansion supposes the information of A to be part of that of B. (Thus a reverse Parry logic would validate $A \rightarrow . A \vee B$ and reject $A \,\&\, B \rightarrow A$.) Lewis took expansion for granted in his argument for positive implicational paradoxes, based on the strict expansion principle $A \rightarrow . A \,\&\, (B \vee {\sim}B)$. The initial logics of cause and reason amount to analytic expansion with however implication relevant and containment proper.

Under symmetrisation (achieved also by contraposition, which can thereby be restored) both requirements are imposed, two-way containment,

[96] *Cf. Chapter Twelve on the pure logic of causal implication. [Editors]*

perhaps reducing to an equivalence of some sort. This leads directly to connexive logics. It also suggests, through forms like $A \rightarrow B \,\&\, c(B) \subseteq c(A) \,\&\, s(A) \leq s(B)$, multiple-barrelled analyses. Why stop at 2 when the going is good? Or 3? In progressive generality, $A \rightarrow_R B$ iff $A \rightarrow_S B \,\&\, R_1(A,B) \,\&\, \ldots \,\&\, R_n(A,B)$, and $A\, C^2_R\, B$ iff $A\, C^2_{S1}\, B \,\&\, \ldots \,\&\, A\, C^2_{Sm}\, B \,\&\, R_1(A,B) \,\&\, \ldots \,\&\, R_n(A,B)$, and similarly for n-place connectives.It is obvious however that such multiplicity adds nothing in fire power over double-barrelled analyses. For let $R(A,B) = R_1(A,B) \,\&\, \ldots \,\&\, R_n(A,B)$ and $A \rightarrow_S B = A \rightarrow_{S1} B \,\&\, \ldots \,\&\, A \rightarrow_{Sm} B$; etc.

5. *Connexive logics.* McCall's first-degree connexive system, in $\rightarrow$, quickly succumbed (see ENT, p.450) to a double-barrelled analysis; namely, Meyer's 'poor man's connexive implication' which used but $S5$ strict implication $\dashv$ combined with material equivalence: then $A \rightarrow B$ iff $(A \dashv B) \,\&\, (A \equiv B)$. Some much more interesting connexive logics can certainly be generated by doubling implication with containment connections, a doubling which has a long historical record. The doubling approach also much increases the prospect of satisfactory semantical elucidation of more appealing systems of connexive logic, such as Angell's $PA1$ and relevant-based analogues thereof, which have hitherto resisted such analysis.

6. *Relatedness logics.* What all these double-barrelled analyses have in common is a resultant implication, or connective, characterized in terms of another of a similar type together with a relation. What are called relatedness logics are of precisely this sort, only the starting implication is typically restricted to material-implication and the relation is not further diagnosed. There is no reason for these restrictions.

A generalized relational logic can be relevantly carried. To hook directly onto what has been done, suppose first the relation is (rule)transitive and reflexive, i.e., yields R2.1 and 2.2. Then a somewhat hollow version of the adequacy theory for $R4$ can be replicated. This is, if you like, the semantical elucidation of a (weak) partial order in a relevant setting. Next, suppose the relation R fails one of the order requirements. Then an equivalence class elucidation can proceed on the basis of its transitive and reflexive closure R^+. Usually R is taken to represent content (or topic) relevance, and so assumed to be reflexive and symmetric but not transitive.

7. *Non-transitive logics.* These include not only the classical relational logic of Woods and Walton, and others, but much earlier systems delivered by sieve methods. The analyses again take the form, $A \rightarrow B$ iff

$A \dashv B \,\&\, R(A, B)$, with R reflexive but not transitive, and in classical form, with $A \dashv B$ collapsed to $A \supset B$.

8. *Gricean conditionals.* The account, when decontextualized and boiled down, goes like this: $A > B$ iff $A \supset B \,\&\, \vdash A \supset B$, where $>$ is the conditional concerned and $\vdash$ is an assertability connective. No doubt Grice would not commit himself to anything so crudely exact, so we advisedly say 'Gricean'. For some accounts, by his followers and former adherents of what he says, reduce rather quickly to this. And it no doubt is of interest, particularly if $\vdash$ is given some appropriate properties of a necessity sort. For it is the relational recipe over again, with $R(A, B)$ rendered modally as $\vdash A \supset B$. Now symmetry must fail, and transitivity likely breaks down, but reflexivity is to be expected; so the conditional system elaborates a standard relational logic. Though the Gricean story is supposed to reinstate the material conditional, $\supset$, as a satisfactory formal analysis of 'the' conditional, it is evident, once it is boiled down, that it does not do that at all. It offers another account, certainly one parasitic on material-implication but also involving essentially a further intentional functor. Though the account is inadequate, it admits some ready upgrading, in the same way as other double-barrelled analyses.

9. *Metavaluational methods.* There are many ways the coupled relation can be filled out. An ingenious elaboration (pursued further by Meyer than anyone else, primarily in unpublished work) introduces proof theory into the analysis. Then $A \to_R B$ iff $A \to_S B \,\&\, \vdash A \to B$. or in a first 'semantical' reformulation $I(A \to_R B) = 1$ iff $I(A \to_S B) = 1 \,\&\, \vdash A \to B$. Situational relativisation is straightforwardly introduced. But in its fruitful applications, the provable implication is nothing other than $\to_R$ itself, i.e., as an analysis the method would be circular.

6.6 Brief glimpses beyond: applications, and further ado

An important exercise for showing that the reasoning enterprise has some teeth, consists in demonstrating the adequacy of the framework outlined to represent the main substance of Aristotelian logic. For syllogistic offers a rich array both of proofs concerning demonstrative reasoning and of applications of it. By contrast with grossly enthymematic procedures of mainstream logics, Aristotelian logic appears to need only principles of ponible reasoning. What is involved, then, in demonstrating adequacy is essentially an appropriate reworking of Łukasiewicz's presentation of syllogistic on a more demanding reasoning basis instead of classical (or many-valued Ł) logic. The result will be, it is conjectured, a much better fit with Peripatetic thinking, a fit which makes room for non-classical theses (such as Aristotle and Irreflexivity), while not ascribing to classical thinkers third degree principles which there is no evidence that they accepted.

Traditional logic included not only a theory of demonstrative argument; it came to include also some theory of fallacies. But the theory, going back to Aristotle's sophistical refutation, has remained in decidedly unsatisfactory shape.[97] The development of miniature logics like those for reason-for enables an improved account of fallacies, beyond recent improvements. We begin from what Hamblin 1970 presents as the almost universal account of what constitutes a fallacy: 'a fallacious argument ... is one that *seems to be valid* [to many] but is not so' (p.12). Our aim is to begin to explain why fallacies are not any old invalid or incorrect reasoning but a *pervasive* type. The idea is this: they persist, and seem valid, because there is a closely related reasoning mode for which their procedure is correct. For example, denying the antecedent and rejecting hypotheticals (i.e., where the antecedent is not given) are commonly regarded as fallacious, because they *are* incorrect for conditionals. But they are apparently *correct* for what is closely related, evidential reasoning. Our 'improved' account of fallacies accordingly takes off along the following lines: a fallacious argument is one that is not relevantly correct but which seems to many to be correct because of similar argument pattern which is correct. The account (ultimately a double

[97] The best modern attempts available in English (in book form, Hamblin 1970, and Woods and Walton 1982) follow Hamblin's lead in resetting the full theory of fallacies within dialogue logic, thus reestablishing the medieval approach. But these efforts still have a long way to go; they fall down especially badly on fallacies of relevance, of equivocation, and of reasoning in a circle (*petitio principii*). They fail, in large part because they remain logically conservative, continuing to operate essentially within the classical paradigm. The move to dialogue expands what can be accomplished while not contesting classical logic: for instance, 'classical logic ... is the basic logic of dialogue' (Walton 1984, p.301). By contrast, the early medieval theory of posing correctly tried to incorporate some paraconsistent features.

Relatedness logic is a first step in the right direction to handle fallacies of relevance (and is introduced in Woods and Walton 1982 for that overt reason), but it too remains classically based. Because of their undue attachment to classical logic, Woods and Walton have no decent way of accommodating the fallacy of reasoning in a circle. In that circular arguments are 'not fallacious' (p.190; of also p.295) or at least do not constitute 'a serious logical fallacy' (p.91).

The systematic treatment of equivocation eludes modern work (cf. Hamblin's concluding remarks, Hamblin 1970, p.303). It will continue to do so while the unwarranted restriction that Woods and Walton try to insist upon, that formal logic 'eliminate equivocation altogether' (for example Woods and Walton 1982, p.175), remains in force.

system analysis) largely circumvents the need to shift to a dialogue model to try to account for fallacies. But of course it leaves open, to be filled however in known ways, a conspicuous modern problem, notably explaining requisite similarity—rather than just enumerating cases as on standard treatments of fallacies. Now reasoning in a circle is a fallacy because it would incorrectly imply some statement or claim is a reason for itself, an incorrect impression (induced even in connoisseurs like Walton) because reflexivity is correct for a similar mode of argument, rigorous deduction.

Avoidance of the circle fallacy is crucial in obtaining a nontrivial formulation of the principle of sufficient reason. Within the theory sketched this important traditional principle obtains a particularly simple formulation: $(UB)(PA)(A \supset B)$, i.e., for every B there is some A which is a reason for B, or, loosely, everything has some reason for it.[98] (Even in a protothetic extension of the logic of reason-for which includes full comprehension principles, this does not appear to be logically vindicated: it looks contingent, and arguably, contingently false.) Such principles stress the need to enrich the logic by some apparatus for quantification, or the equivalent.

After the introductory stages, we focused upon ponible reasons for, setting aside the rich variety of reasoning contents where ponability fails. In sequels we shall investigate various types of non-ponible reasoning, probabilistic, plausibility, evidential, and other types of reasoning. This leads us through many little investigated logical labyrinths, through significant parts of scientific, legal and practical reasoning. We shall encounter many vexing notions, such as the accumulation of evidence, and interesting functors, such as "it is beyond reasonable doubt that", and we shall be led to a fuller conception of rationality. Plainly rationality is intimately tied to reason-giving; to be rational necessarily involves preparedness to give reasons for or against.

6.7 Appendix: Critical notes on some kinds of standardized reasoning: natural, artificial, automated

Perhaps not all reasoning is the giving of reasons, for or against (this or that, justifying this opinion or conclusion, that action or course). It depends how far 'reason' and 'reasoning' are stretched, whether they include making comparisons, drawing analogies, and so on. But certainly such giving-

---

[98] *As mentioned in Chapter One, the existentially-neutral quantifiers U and P replace the more usual (and, in Sylvan's view, existentially-loaded) $\forall$ and $\exists$ respectively. [Editors]*

of-reasons reasoning for or against is a large, significant, yet logically neglected, part or reasoning. Such reasoning not only comprehends a variety of argumentation procedures, as in dialogue, discussions, cross-examinations, material at the margin of contemporary logic and scorned by most logicians; it also includes the myopically-focussed part looked at by contemporary schools, the substantially context-independent drawing or deriving of conclusions and consequences from premisses or assumptions—what might be called *standardized* or school reasoning.

Standardized reasoning is subjected to various cross-classifications, which have however largely outgrown their usefulness. One formerly important type of cross-classification is that, hardly well-characterised or exhaustive, into deductive, inductive, and perhaps analogical/abductive/... (a classification, already bothering Joseph and others at the turn of the century). A very different classification, in increasing use, is not in terms of *content* or style of the argumentation connections at all, but rather through the means by which such argumentation is accomplished. Such means divide roughly into natural reasoning (often erroneously equated with that carried out by humans) and artificial, with automated or machine reasoning the main division of the artificial (the division sometimes made into machine-assisted or not, though fuzzy, is certainly exhaustive).

From this angle the fashionable terms "natural-deduction" and "natural logic" are entirely anomalous. On the one hand, natural-deduction may perfectly well be automated, and so not be natural. On the other hand, natural-deduction with its usual highly artifical (not to say precious and precarious) rules and restrictions is quite far removed from most of the reasoning carried out by natural creatures, bearing only a casual resemblance to some of it.

Theorising humans have a special propensity for claiming as the whole, or essence, of some enterprise what is only a part, often an atypical part, of that enterprise. Reasoning and logic have been frequent casualties. It is now a joke that Kant, Whately and others contended that logic was a complete, closed and largely exhausted subject. Natural-deduction and natural-logic are not yet jokes; nor are the rival claims of set theory and of category theory to wrap up mathematics treated as simply amusing (you have to get serious to take them as ludicrous). Still less is what goes by the name of "automated reasoning" on the way to becoming a joke; on the contrary, it is treated with deadly seriousness and solemnity, and frequently at such a pace that it is difficult to observe how *little* of reasoning gets into it.

Automated Reasoning (AR) is thus far not the *automation* of *reasoning*, but (isolated exceptions apart perhaps) the automation of fragments of classical logic, of certain (first order) classical theories. The whole great sweep of *reasoning* is passed by—as for instance in Wos *et al.* 1984, where AR just is the automation of machine confinable parts of classical logic.[99]

True, the techniques for automating classical logic—a logic which lends itself especially well to machine transcription, since the machines are presently made in its image—can *sometimes*, usually with considerable difficulty, be adapted to nonclassical frameworks. A main reason for obstruction is that many classical tricks and shortcuts, such as normal forming techniques, do not adapt at all well, or at all, to intensional logics, and still less to the largely unformalised expanse of reasoning beyond contemporary little deductive systems.[100]

---

99 In a parallel fashion, Tennant 1978 calls a whole text which scarcely moves out of the narrow confines of mainstream (and mostly classical) logic, Natural Logic. Nothing as to naturalness is included. The sole pretext for this outrageous scoop appears to be that parts of the book use mainstream 'natural-deduction' techniques. Tennant is by no means the only or the first offender, see further Sylvan and Günther 1987.

100 We are much indebted to Newton da Costa, Laurence Goldstein and Arnold Günther for comments on an earlier version of this essay.

7 Relevant Containment Logics and Certain Frame Problems of AI

This essay originally appeared as Sylvan 1988. [Editors]

On many occasions there appears to be a real need for a relevant implication relation which does not introduce or carry superfluous or irrelevant information. In an implication connection $A \rightarrow B$, B contains superfluous content if B contains content not represented in A. An implication meets a *tight* containment requirement if it does not sanction the introduction of such superfluous content. The standard relevant logics do not meet such a containment requirement (though they can satisfy *other* plausible accounts of containment; see ENT, p.155). They fail it because, in particular, of principles like Addition, $A \rightarrow. A \vee B$ and $B \rightarrow. A \vee B$, which tack on disjunctively what may be fresh or extra information.

Nor is an obvious alternative to a relevant logic, a Parry logic, or analytic implication system, altogether satisfactory. For while such containment logics, also weakly relevant, do exclude additive expansion, they do not exclude other types of damaging (and implausible) explosion, notably (but unsurprisingly given their modal bases) of necessary and of inconsistent information.[101] For instance, They underwrite implicative explosion in such forms as $A \rightarrow . A \rightarrow A$, $A \rightarrow . (A \rightarrow A) \rightarrow (A \rightarrow A)$, ..., in demodalised systems, or similar chains with antecedents such as $A \rightarrow B$ or $B \rightarrow A$ in modalised versions. Nor do they entirely exclude additive effects; for Addition is available in rule form, thus apparently violating adequacy conditions (as Kielkopf has shown; see further RLR I, p.101). Indeed the systems are worthless for major paraconsistent purposes, since they spread contradictory information everywhere, by virtue of implications like $(A \mathbin{\&} {\sim}A) \mathbin{\&} B \rightarrow. B \mathbin{\&} {\sim}B$ and $(A \mathbin{\&} {\sim}A) \mathbin{\&} (B \vee {\sim}B) \rightarrow. B \mathbin{\&} {\sim}B$, direct

[101] The trouble with Parry logics by no means end here. For a detailed critique of these logics and their motivation, see RLR I, especially p.96 ff, where such logics are called conceptionist (and indexed thus). It needs stressing that standard relevant logics straightforwardly satisfy important containment requirements, such as that $A \rightarrow B$ holds true iff the content of B is included in that of A; see for example RLR I, pp.216-7.

consequences of the relevant *bête noire*, Disjunctive Syllogism, $A \,\&\, ({\sim}A \vee B) \rightarrow B$, rightly excluded in main relevant and paraconsistent logics (cf. Deutsch 1979, p.139). Since one of B or ${\sim}B$ holds true, an isolated contradiction, $A \,\&\, {\sim}A$ say, gets spread everywhere (at least where B is determinate), to arbitrary contradiction $B \,\&\, {\sim}B$.

It looks as if the best of both relevant and containment logics, without the defects, can be had by combining the two, essentially product-wise. So result, in one way or another, relevant containment logics. But, as will emerge, amalgamation is by no means straightforward (especially once the simple deceptions of modal confines are left behind); and combinations are far from uniquely determined.

7.1 Reasons for the enterprise and for dissatisfaction with modally-based approaches, such as analytical implication

There are systems which satisfy containment requirements, and which, as a corollary, are relevant; namely, Parry's system of 'analytic implication' and systems in its vicinity. But though some of these systems have pleasant properties, such as neat semantics and algebraic analyses, they nonetheless include a certain amount of junk. The reason for this is that they amount to restrictions of normal modal logics (or even their degenerate limit, classical logic), and carry over the junk from these systems, some of it already exhibited. Other undesirable junk includes 'relevantised' versions of all the noxious paradoxes of implication. By contrast, the aim of the present enterprise is to start from systems from which modal and classical junk has *already* been appropriately removed, such as certain (deeper) relevant logics, and to add containment requirements in one way or another to such systems.

Put differently, one main reason for proceeding beyond logics like Parry's is that such logics offer inadequate *control*. For example, Parry's logic spreads identity about in a quite profligate way, similarly inconsistency. A mere *filter* on strict implication is far from satisfactory; the rot has already set in extensively with strict implication. As a result, systems like Parry's are not particularly useful for several of the main applications envisaged; and some requisite work has already in effect gone into modifying Parry logics with a view to reducing some of these problems (for example Deutsch 1979, Daniels 1986).

The sorts of applications that have been considered, in one way or another, or that should be, include these. They fall into two overlapping groups, roughly:-

1. Cases and logics involving circumscribed contexts, where additional (irrelevant) information is, in some way, inappropriate:

- logics of fiction and stories, where additions are in serious doubt. According to the story story, stories don't introduce new content (see for example Daniels 1986, pp.221-2).
- communication networks, and polylogue theory. For communication and dialogue, like stories upons which they generalise, are characteristically information circumscribed. (For the many places in polylogue theory where relevant containment logics tend to enter, see Sylvan 1985.)
- meaning, inclusion of meaning, synonymy, analysis, aboutness. For example, even if inclusion-of-meaning theory and Tarski-Davidson meaning-through-truth theory are respectively rectified by setting the theories on relevant logic foundations (see RLR I for the first and Priest and Crosthwaite 1988 for the second), odd puzzles at least remain. For, primarily by virtue of Addition, relevant logics permit the introduction of superfluous meaning, through such biconditionals as $p \leftrightarrow .\ p \,\&\, (p \vee q), p \leftrightarrow .\ p \,\&\, (p \vee q) \,\&\, (p \vee r)$ ..., etc., where q, r, ..., may have nothing to do with p. But these relevantly equivalent expressions don't have the same meaning. So in particular a 'meaning-giving' T-schema of the form, s is true $\leftrightarrow p$, should not deliver, s is true $\leftrightarrow .\ p \,\&\, (p \vee q) \,\&\, (p \vee r)$. By switching to relevant containment logics (which are derivationally adequate for the purposes at hand), *these* difficulties are avoided.
- dependence, functional and predicate dependence, 'relevant' or 'genuine' predication. Genuine functions and relations depend directly and explicitly on their arguments; genuine dependence and predication are neither vacuous nor ensured by entirely remote, irrelevant, content (as Dunn 1987 explains, though he proceeds to pass over an evident relevant containment resolution. For the remarkable large bundle of philosophical issues intertwined with dependence, see the conclusion of Dunn 1987.)
- frame problems. In important respects, frame problems (investigated in more detail below) typify this whole class of problems.

2. Cases and logics where tight relevance is required, or at least paradoxes and like puzzles are resolved through such relevance considerations. As useful surveys of such cases are available elsewhere (see especially Weingartner and Schurz 1986), there is excuse for even briefer listings:

- doxastic logics, assertion logics, epistemic logics and so forth, where again superfluous additions are in much dispute. Belief and assertion functors are normally closed under some relevant operations significantly tighter than implication. A suggestion to toy with, seriously, is that some of at least of the main intensional functors involved are closed under relevant containment.

- deontic logics, preference logics, and volitive logics (with operators like "desires that", "wishes that", "feels that"). Again something is contained in the judgements made, from which something relevant emerges logically; some relevant commitments are made.
- legal reasoning where (Ross's 'paradox', has some exposure), and relevant reasoning more generally (cf. Bollen 1985 for what is to be varied).
- logics of explanation and confirmation, where superfluity again generates several puzzles (see Weingartner and Schurz 1986, whose classical tack-on resolutions are *not* however endorsed, for reasons essentially given in Kielkopf 1977 and RLR I).

Most of these problems—those that are genuine, that is, for not all are—can be tackled, in a fairly uniform fashion, through (deep) relevant containment logics. Some admittedly can be dealt with in Parry logics; but most cannot, so uniformity of treatment would be lost. To be sure, there are economic advantages that logics in the vicinity of Parry's do have, namely that variable inclusion (*if* that is what is sought) can be represented *within* the logics, for example through formulae like $B \rightarrow . A \rightarrow A$ and $B \rightarrow B \rightarrow . A \rightarrow A$. A perhaps surprising feature of relevant logics is that there is (so it is conjectured) no way of representing such variable inclusion within the logic. It is possible to represent *overlap* of variables, for example $A \odot B$ iff $A \rightarrow B \vee . B \rightarrow A$. But to define inclusion in terms of overlap requires (presumably) what amount to propositional quantifiers, for example $C \sqsupseteq D$ iff (for all B)(if $B \odot D$ then $B \odot C$).

To couple inclusion notions with relevant (first order) logics, rcquires then further logical apparatus. Specificially, it calls for an appropriate connective $\sqsupseteq$, to symbolise inclusion of content or information. This content-inclusion connective, obtainable perhaps by definition, should—apart from satisfying expected inclusion conditions such as transitivity—hold *where* variable inclusion obtains (*but* not necessarily conversely); i.e., if the variables of B are included among those of A then $A \sqsupseteq B$. Nor, however, is variable inclusion, though governing $\sqsupseteq$, always enough for containment implication, as implication explosion reveals. A good underlying implication connection, such as deep relevant logics furnish but modal logics do not, is an essential prerequisite.

Relevant containment logics combine then relevant logics with a further inclusion requirement, in product fashion. A relevant containment logic is one which includes an implication $\twoheadrightarrow$ which is both relevant and satisfies tight containment, symbolized through connective $\sqsupseteq$. Relevant containment $\twoheadrightarrow$ accordingly is defined: $A \twoheadrightarrow B$ iff $A \rightarrow B \,\&\, A \sqsupseteq B$.[102] The inclusion of content concerned would ordinarily be presented more explicitly through a content function, c say, and an inclusion relation on terms, i.e., $A \sqsupseteq B$ would expand to $c(B) < c(A)$. But for the present, c functions have conveniently been absorbed. The combined effect of relevant implication *and* containment is to peel off such orthodox principles as $A \rightarrow .\ A \vee B$, $A \rightarrow .\ A \rightarrow A$, $A \leftrightarrow .\ A \vee (A \,\&\, B)$, etc., principles which in many circumstances prove an embarrassment, because they can introduce superfluous content. For $A \twoheadrightarrow .\ A \vee B$, $A \twoheadrightarrow .\ A \twoheadrightarrow A$ and the like do not hold. The first of these fails because content inclusion breaks down, as with analytic implication; the second is invalid for relevant reasons (see the discussion of Mingle in ENT and RLR I).

A *categorematic* account of content inclusion—where connectives (and wider context) do not contribute essentially to content—has already been insinuated. But, despite its historical sponsorship, such an account is in no way dictated, and indeed is decidedly dubious. There are (out there in *aussersein*) other, less restrictive and more satisfactory, theories of content and content inclusion. Content is a determinable with many determinates, which vary considerably in character and quality. Nonetheless, the initial logical theory advanced will begin with categorematic theories of content inclusion; they are places to start from that have much to offer.

7.2 Some details of initial relevant containment logics and their semantics

To begin with, there is an underlying relevant logic, L say, the *carrier* logic, with conventional connective set $\{\rightarrow, \&, \vee, \sim\}$ for instance (as in RLR I). Later, the carrier logic may be removed, with implication $\rightarrow$ supplanted by containment connective $\twoheadrightarrow$. But, for the present, containment logic is superimposed upon the carrier logic. This is done by adding to L, as on the surface in relational logic,[103] a relation r of the right type, a two-

102 In effect this adapts to relevant settings an old idea revived by Gödel (and worked out by Fine) for Parry's system *Al*. In essence the idea is ancient, reaching back perhaps to Boethius but certainly to Abaelard.

103 The approach differs from the relational logic of the literature (presented in most detail in a special issue of *Philosophical Studies* 36 (1979), which is criticized in OP. For here (as in effect in Deutsch 1979) variable-sharing is not taken as a sufficient condition for relevance, and

place connective, with the following features:- By virtue of its grammatical categorization, where A and B are wff, so is $r(A, B)$. The chief logical feature required of r is that it enables the definition of a certain order relation, a (lattice-ordered content) *containment* relation, $\sqsupseteq$, on wff.

Relation $\sqsupseteq$ is required to be transitive (achieved if necessary by taking the transitive closure of r) and is variable governed, and it meets appropriate semi-lattice conditions, for example it composes fully. Specifically:

$\sqsupseteq$1. $A \sqsupseteq B \,\&\, B \sqsupseteq C \rightarrow . A \sqsupseteq C.$

$\sqsupseteq$2. $A \sqsupseteq B$ where $\vee(A) \sqsupseteq \vee(B)$, with $\vee(A)$ representing the set of variables of A.

$\sqsupseteq$3. $C \sqsupseteq A \,\&\, D \sqsupseteq B \rightarrow . C \,\&\, D \sqsupseteq A \,\&\, B.$

$\sqsupseteq$4. $A \sqsupseteq B \leftrightarrow . A \,\&\, B \sqsupseteq A \,\&. A \sqsupseteq A \,\&\, B.$

Scheme $\sqsupseteq$4, though dubious in a relevant setting, supplies the standard lattice linkage, of order with identity. For it reduces to $A \sqsupseteq B \leftrightarrow . A \,\&\, B \equiv A$, where $C \equiv D$ is defined as $(C \sqsupseteq D) \,\&\, (D \sqsupseteq C)$. Together with the vital connection:

$\sqsupseteq$0. $A \twoheadrightarrow B \leftrightarrow . A \rightarrow B \,\&. A \sqsupseteq B,$

this completes the axiomatisation of the basic initial logic CLI. The axiomatisation admits of much tampering; for instance (given replacement) the awkward lattice principles of $\sqsupseteq$4 can be broken down into the perhaps more amenable trio $A \sqsupseteq B \rightarrow . A \,\&\, B \sqsupseteq A$, $A \sqsupseteq B \rightarrow . A \sqsupseteq A$, $A \sqsupseteq A \,\&\, B \rightarrow . A \sqsupseteq B$.

The axiom schemes in $\sqsupseteq$ can be argued to independently of their role for the semantical theory.[104] But in fact the schemes reflect exactly what is required for a modified equivalence-class completeness proof on inclusion of content, with contents getting modelled canonically through equivalence defined classes of wff. But, just as it proves possible to provide Lindenbaum algebras for logics where appealing equivalence connections are lacking, so it is possible to provide somewhat messier completeness proofs where various underwriting schemes in $\sqsupseteq$ are missing. Certainly, as we shall see, it is not too difficult to peel off the less convincing semi-lattice linkages imposed in

accordingly transitivity is not abandoned. Furthermore main paraconsistent purposes are not sacrificed, as in standard relational logics which, like their close relatives Parry logics, simply tack relations onto irrelevant logics, of strict or classical varieties. Thus the main objections lodged in OP, pp.137-8, against relational logics are avoided on the approach taken here.

[104] *See Chapter Twelve. [Editors]*

$\sqsupseteq$4. However the conditions are not demanding by conventional standards, and transitive closures of more familiar variable-connecting relations r will meet them.

More or less any adequate semantics for the carrier logic L can be used, so long as it affords a way of separating out regular theories or worlds or structures, that is those where all theorems of CLI hold. Here we shall suppose that something like the usual relational frames or model structures, of form $M = < T, O, K, R, * >$, for relevant logics are used, with $O \subseteq K$ the class of regular situations, and actual situation T in O.

Interpretations on model structures for the carrier logic are also as for relevant logic (for details see RLR I); whether wff A holds at a, symbolised $I(A, a) = 1$, is defined for every situation or world a in K. What further semantical apparatus is required in order to represent content inclusion? The requisite additional apparatus can be reached in a straightforward naive fashion. Since, naively, $A \sqsupseteq B$ holds true just where the content of A includes that of B, i.e., $I(A \sqsupseteq B, T) = 1$ iff $c(B, T) < c(A, T)$, then, by straightforward world relativisation,

$$I(A \sqsupseteq B, a) = 1 \text{ iff } c(B, a) < c(A, a), \text{ for every } a \text{ in } K.$$

The (categorematic) content of A at a is in turn governed by (the sum of) that of its sentential parameters, as follows, for regular situations a in O:

$$c(A, a) = c(p_1, a) \wedge ... \wedge c(p_n, a),$$

where $p_1, ..., p_n$ are precisely the sentential parameters of wff A. In regular situations A says what its parameters jointly say. In regular situations, moreover, the operation $\wedge$ amounts simply to a (semi-)lattice operation on contents, given for instance by the least upper bound on the content structure ordered by $\sqsupseteq$.[105] (For semi-lattice properties, see Curry 1963, p.131 ff.) For each situation a not in O, $c(A, a)$ is assigned content arbitrarily from $C(a)$, the information or content at situation a, subject only to the restriction that $c(A \,\&\, B, a) = c(A, a) \wedge c(B, a)$. For such nonregular situations more general structures than semi-lattices are required, because at them reflexive and other lattice properties may, like all logical principles, fail. An appropriate structure at each such a is an *order-structure* $\langle C(a), <, \wedge \rangle$ or $\langle C(a), \wedge \rangle$ since order

[105] As it happens, operations can be taken either as a meet (l.u.b.)—more natural for the intended summation of contents story—or as a join (g.l.b.), how it may look. For all that is required in the semantics is that $\wedge$ is commutative, associative and idempotent. Then whether $\wedge$ is a meet or join will depend upon which order is adopted: $\beta < \gamma$ iff $\gamma \wedge \beta = \beta$, or the converse: $\gamma < \beta$ iff $\gamma \wedge \beta = \beta$.

can be defined in the usual way: for α, β in $C(a)$, $\alpha < \beta$ iff $\alpha \wedge \beta = \beta$. The relation $<$ is transitive on $C(a)$, composes fully, and satisfies the (definitional) condition $\alpha = \beta$ iff $\alpha < \beta \,\&\, \beta < \alpha$. It follows immediately that $(\alpha \wedge \beta < \beta) \,\&\, (\beta < \alpha \wedge \beta)$ iff $\alpha < \beta$. It is an elaborate story for a structure which underwrites no theses!

To pull all this together, the structure required is given by a functor C supplying a set $C(a)$ of contents or bits of information, at each situation a in K, with the set $C(a)$ ordered by $<$ with meet operation $\wedge$. It is enough to add C and $\wedge$ to L model structures (m.s.) to obtain CLI m.s.. Thus a CLI m.s. is a structure $< T, O, K, R, *, C, \wedge >$, with $\langle C(a), \wedge \rangle$ an order-structure for each a in K.

A CLI (mark 1 containment logic) *model* adds to a CLI m.s. two valuation functions, I (or ν) and c. Valuation I, from sentential parameters and situations to holding values 1(on) and 0(off), subject to a hereditariness requirement, is as in relevant semantics. The hereditariness requirement and several of the rules resemble those for intuitionistic semantics. Hereditariness is extended to include the further connective $\sqsupseteq$ of CLI; specifically, wherever $a \subseteq b$ and $I(A \sqsupseteq B, a) = 1$ then $I(A \sqsupseteq B, b) = 1$. Valuation c is also from sentential parameters and situations, but maps to situationally associated content; i.e., $c(p, a)$, with c representing the information or content of p at a, belongs to $C(a)$. Both I and c are extended from initial sentential wff to all wff inductively. The evaluation clauses for I, as applied to connectives $\&, \vee, \sim, \rightarrow$, are just those of relevant semantics. The further clauses for c are just those already given.

7.3 An outline of soundness and completeness arguments

In the *soundness* argument, the following properties drawn from relevant semantics much simplify verification of postulates:-

> $I(C \rightarrow D, T) = 1$ iff for every situation a, where $I(C, a) = 1$ then $I(D, a) = 1$.

Similarly

> $I(C \leftrightarrow D, T) = 1$ iff for every situation a, $I(C, a) = I(D, a)$.

These connections enable immediate validation of $\sqsupseteq$0, given the evaluation rule for $\twoheadrightarrow$ at an arbitrary world simply conjoins those for $\rightarrow$ and $\sqsupseteq$.

ad $\sqsupseteq 1$. By the foregoing it suffices to show that where $c(B,a) < c(A,a)$ and $c(C,a) < c(B,a)$ then $c(C,a) < c(A,a)$ for arbitrary a in K. But this is a result of transitivity of $<$ in $C(a)$.
ad $\sqsupseteq 2$. Suppose $\vee(A) \sqsupseteq \vee(B)$. Then for a in O, $c(B,a) = \Sigma c(p_B, a) < \Sigma c(p_A, a) = c(A,a)$, where p_A ranges over variables of A and p_B over those of B and Σ indicates $\wedge$ summation. Hence, for a in O, $I(A \sqsupseteq B, a) = 1$, whence as T is in O, $A \sqsupseteq B$ is valid.
ad $\sqsupseteq 3$ and $\sqsupseteq 4$. By virtue of the simplifying features and the connection $c(A \,\&\, B, a) = c(A,a) \wedge c(B,a)$, these reduce to order-structure properties at arbitrary a in K.

Completeness is established through a canonical modelling. To the canonical m.s. for the carrier logic, defined in a familiar way, are added further details for C and $\wedge$. The requisite canonical order-structure is arrived at by adapting standard methods for Lindenbaum algebras. A quasi-equivalence on wff, $\tilde{a}$ is defined for each situation a thus: $A \,\tilde{a}\, B$ iff $A \equiv B \in a$, i.e., iff $A \sqsupseteq B \in a \,\&\, B \sqsupseteq A \in a$. Then $\tilde{a}$ is symmetric and transitive, and reflexive for $\tilde{a} \in O$. Let $|A|_a$ be the quasi-equivalence class of A under $\tilde{a}$, i.e., $\{B : B \,\tilde{a}\, A\}$. Then $C(a)$ is the class of all these classes, i.e., $C(a) = \{|A|_a : A \text{ is a wff}\}$; and correspondingly $|A|_a \wedge |B|_a = |A \,\&\, B|_a$. By virtue of $\sqsupseteq 3$, $|A \,\&\, B|_a$ is suitably independent of the choice of wff A and B.

- $\langle C(a), \wedge \rangle$ is an order-structure, for each a in K. Requisite properties follow from the CLI axiom schemes using connections established below.

Canonical valuations are defined as expected: $I(p,a) = 1$ iff $p \in a$ and $c(p,a) = |p|_a$, for every sentential parameter p and every situation a. These interconnections are extended inductively or definitionally to every wff A involved. The definitional element is the general stipulation, for a not in O, $c(A,a) = |A|_a$. Then requisite details, beyond those for the carrier logic (which are as in RLR I), are these:

- $c(A,a) = |A|_a$, for every wff A and situation a.

The generalising step, for a in O, is as follows, where $p_1, ..., p_n$ are all the sentential parameters of A:

$$\begin{aligned} c(A,a) &= \Sigma c(p_j, a) && \text{by its evaluation rule} \\ &= \Sigma\, |p_j|_a && \text{by the given basis} \\ &= |p_1 \,\&\, ... \,\&\, p_n|_a && \text{by iteration of } |p \,\&\, q|_a = |p|_a \wedge |q|_a \\ &= |A|_a . \end{aligned}$$

It is in this final step that restriction to situations in O is crucial. By $\sqsupseteq 2$, both the following results are derivable: $p_1 \,\&\, ... \,\&\, p_n \sqsupseteq A$ and $A \sqsupseteq p_1 \,\&\, ... \,\&\, p_n$. Since a is regular, both those results belong to a. Hence $p_1 \,\&\, ... \,\&\, p_n \,\tilde{a}\, A$, vindicating the final step.

- $c(A \,\&\, B, a) = c(A, a) \wedge c(B, a)$. By the preceding argument and definition of $\wedge$.
- $I(A \sqsupseteq B, a) = 1$ iff $A \sqsupseteq B \in a$.

Because of the evaluation rule for $\sqsupseteq$, it suffices to show $c(A, a) < c(B, a)$ iff $A \sqsupseteq B \in a$. Now:

$$
\begin{aligned}
c(A,a) \; < \;& c(B,a) \text{ iff } |A|_a < |B|_a \text{, as above} \\
\text{iff } \;& |A|_a = |B|_a \wedge |A|_a \text{, by order-structure definition} \\
\text{iff } \;& |A|_a = |A \,\&\, B|_a \text{, by definition of } \wedge \\
\text{iff } \;& A \,\tilde{a}\, A \,\&\, B \text{, by canonical definitions, symmetry, transitivity} \\
\text{iff } \;& A \sqsupseteq A \,\&\, B \in a \,\&.\, A \,\&\, B \sqsupseteq A \in a \text{, by canonical definitions} \\
\text{iff } \;& A \sqsupseteq B \in a \text{, by } \sqsupseteq 4.
\end{aligned}
$$

The connection also enables validation of hereditariness as regards $\sqsupseteq$. For $I(A \sqsupseteq B, a) = 1$ and $a \leq b$ supply $A \sqsupseteq B \in a$ and $a \subseteq b$, which guarantee $A \sqsupseteq B \in b$, whence $I(A \sqsupseteq B, b) = 1$. And that finishes the outline of the completeness argument.

As usual, the logical theory can be strengthened by further postulates and corresponding modelling conditions. Let us consider just one important case, that of $\leftrightarrow$ replacement, a case which reveals as well just how precious the basic logical theory *CLI* is algebraically (for example it doesn't permit replacement of $C \,\&\, C$ by C everywhere). The replacement rule, SE. $A \leftrightarrow B \,/\, D(A) \rightarrow D(B)$, reduces in the setting of relevant affixing logics to the cases for $\sqsupseteq$, i.e., to:

$\sqsupseteq$SE. $\quad A \leftrightarrow B \,/\, A \sqsupseteq C \rightarrow.\, B \sqsupseteq C$
$\qquad\quad\;\; A \leftrightarrow B \,/\, C \sqsupseteq A \rightarrow.\, C \sqsupseteq B$

The corresponding modelling condition, which interconnects valuations I and c, is a trifle curious. It is in effect this: If $I(A, d) = I(B, d)$ for every d in K, then $c(A, a) = c(B, a)$.

7.4 Upgrading the logical theory

The logical system CLI presented, while decidedly appealing as regards the $\rightarrow$ and $\twoheadrightarrow$ theorems it delivers and even more for what it fails, still has noticeable shortcomings. For it still yields such dubia as $A \sqsupseteq B \rightarrow . A \sqsupseteq A$. And though it properly avoids such atrocities as $A \rightarrow . A \sqsupseteq A$, and worse $B \rightarrow . A \sqsupseteq A$, it does so at a cost in complexity of specially contrived semantical apparatus. Without that, without substitution of order structures for semi-lattices at nonregular worlds, it would validate those and other noxious results, by virtue of the evaluation rule $I(A \sqsupseteq B, a) = 1$ iff $c(A, a) < c(B, a)$, whence $A \sqsupseteq A$ and other lattice themes would hold at all situations supplied. To make requisite room for impossible situations where lattice linkages fail, a better procedure is to modify the evaluation rule, for instance as follows:

> for a in O, $I(A \sqsupseteq B, a) = 1$ iff $c(A, a) < c(B, a)$; and otherwise, i.e., for a in $K - O$, $I(A \sqsupseteq B, a)$ is assigned arbitrarily, except so far as constraints like hereditariness are to be met. In short, for $a \notin O$, $I(A \sqsupseteq B, a)$ is treated like $I(p, a)$, as an initial assignment.

Such an adjustment also enables dubious lattice connections to be removed, in particular the ungainly $\sqsupseteq$4. However an obvious way in which to accomplish this feat—by weakening $\sqsupseteq$1 and $\sqsupseteq$3 to rule forms—does not prove robust enough to carry the style of completeness argument so far used. Fortunately an alternative is ready to hand, which averts difficulties; namely, make the semi-lattice conditions, what they seem to be, enthymematic. For this purpose, appeal is made to the standard constant t of relevant logic, with t construed contentwise as a zero-place connective. Constant t, which serves to represent the conjunction of theorems, satisfies the two way rule: $A // t \rightarrow A$. Given t the new semi-lattice conditions are the following enthymematic forms:

$t \sqsupseteq$1. $t \,\&\, A \sqsupseteq B \,\&\, B \sqsupseteq C \rightarrow . A \sqsupseteq C$, i.e., $A \sqsupseteq B \,\&\, B \sqsupseteq C \supset . A \sqsupseteq C$, where $\supset$ is an intuitionistic-style connective defined $D \supset E =_{df} t \,\&\, D \rightarrow E$.

$t \sqsupseteq$3. $t \,\&\, C \sqsupseteq A \,\&\, C \sqsupseteq B \rightarrow . C \sqsupseteq A \,\&\, B$.

Note that since t is a theorem, rule analogues are immediately forthcoming. Postulates $\sqsupseteq$2 and $\sqsupseteq$0 of this revised, mark II, containment logic $CLII$ are as before. An analogue of $\sqsupseteq$4 is no longer required; so far as such lattice connections are required they can be derived.

Some theorems applied in showing completeness are next recorded.

T1. $A \sqsupseteq A$, by $\sqsupseteq$2; whence $A \twoheadrightarrow A$.

T2. $A \& B \sqsupseteq A$; $A \& B \sqsupseteq B$, by $\sqsupseteq$2.

T3. $A_1 \sqsupseteq A_2 \& t \rightarrow. A_1 \& B \sqsupseteq A_2 \& B$.
For, as $t \rightarrow. A_1 \& B \sqsupseteq A_1$ from T2, $A_1 \sqsupseteq A_1 \& t \rightarrow. A_1 \& B \sqsupseteq A_2$. But, by T2, $t \rightarrow. A_1 \& B \sqsupseteq B$ also, whence by $\sqsupseteq$3, $A_1 \sqsupseteq A_2 \& t \rightarrow. A_1 \& B \sqsupseteq A_2 \& B$.

T4. $B_1 \sqsupseteq B_2 \& t \rightarrow. A \& B_1 \sqsupseteq A \& B_2$. Similar to T3.

T5. $A_1 \sqsupseteq A_2 \& B_1 \sqsupseteq B_2 \& t \rightarrow. A_1 \& B_1 \sqsupseteq A_2 \& B_2$. By T3 and T4.

The model structure is like that for CLI *except* that $\langle C(a), \wedge\rangle$ can now simply revert everywhere regular to more amenable semi-lattice form. That is, operation $\wedge$, defined (if you prefer) in $C(a)$ for every a in K, is *within* O a semi-lattice operation: commutative, associative *and* idempotent. Soundness and completeness arguments for $CLII$ vary those given already for CLI. Soundness is much as before, and verification of $t \sqsupseteq$1 will indicate all main points. The previous argument as to $\sqsupseteq$1 will work when duly restricted to arbitrary a in O. But that a is in O is guaranteed by the adjunction of t to the antecedents. For by the relevant rule for t, $I(t, a) = 1$ iff $a \in O$. Furthermore, it is evident why $\sqsupseteq$1 itself, by contrast with $t \sqsupseteq$1, breaks down. For $a \notin O$, $I(A \sqsupseteq A, a)$ can be assigned value 0 when the antecedents are all assigned 1.

For completeness, the definition of $|A|_a$ is upgraded, inasmuch as that, for $a \in O$, standard equivalence-class requirements are now met. For instance, as $t \rightarrow. A \sqsupseteq A$, $\tilde{a}$ is reflexive at each a in O. For $a \notin O$, $|A|_a$ is virtually a Don't Care: any passable definition will do, as nothing extra definitional has to be established. For definiteness let $|A|_a$ be $\langle A, a\rangle$ for $a \notin O$. Other crucial new details concern the canonical modelling of $\sqsupseteq$, i.e., proof that $I(A \sqsupseteq B, a) = 1$ iff $A \sqsupseteq B \in a$. For a not in O, this is a matter of stipulation for initial wff, and hereditariness follows in the usual way. Suppose then a is in O. The argument is as before except for the final step,

$$A \sqsupseteq A \& B \in a \&. A \& B \sqsupseteq A \in a \text{ iff } A \sqsupseteq B \in a,$$

which previously applied $\sqsupseteq$4. Suppose now $A \sqsupseteq A \& B \in a$. As $a \in O$, $t \in a$ and $A \& B \sqsupseteq B \in a$ because all theorems do. But a is closed under adjunction and provable implication, so by $t \sqsupseteq$1, $A \sqsupseteq B \in a$. Suppose conversely, $A \sqsupseteq B \in a$. As $a \in O$, $t \in a$ and $A \sqsupseteq A \in a$, so by $t \sqsupseteq 3$, $A \sqsupseteq A \& B \in a$. Also as $a \in O$, $A \& B \sqsupseteq A \in a$, completing the argument.

With a batch of logics in hand, which yield through one of their primary implication connections, namely $\twoheadrightarrow$, only tight (nonsuperfluous) relevant consequences, we are better prepared for applications to typical problem areas. Applications to issues which call for removal of $A \twoheadrightarrow. A \vee B$ (such as Ross's 'paradox' in deontic logic) are immediate. Let us consider then a less immediate and much more ambitious application.

7.5 An application to certain frame problems

The main problem, as often portrayed, consists in suitably confining the informational and inferential frame of an automaton or suchlike artefact—in particular, so it can act in requisite time on *relevant* information and not be sidetracked on *irrelevant* inference-making. The very general setting of the problem is this:- In the course of planning and acting, an automaton needs both a representation—model or some such—of its environment and also to be able to update that representation as circumstances change, for instance because of its own changing position and impact on the local environment. Several of the problems in the bundle of problems that go under the head of 'the frame problem' are then but pointed instances of much more general problems such as, the problems of excess information and of requisite pruning of information bases and restriction of search spaces, and also those problems bound up with suitable formalisation of nondemonstrative inference.[106]

But a major problem, sometimes accounted a 'control problem', just is that of relevance of what is inferred. The logical apparatus of an automaton typically enables it to make (potentially) infinitely many inferences from the information it has stored. But much of that information and many of the inferences it might make will, even if modally valid, be irrelevant to solving specific issues confronting it. A significant part of the problem reduces to that of limiting inferences to those relevant to the issues at hand.

A relevant containment implication which does not allow the iterated introduction of extraneous material can accomplish much of what is sought. For such problem solving situations the automaton's logic is a relevant containment one. Thus it does not derive extraneous material, or waste important time piling up irrelevant consequences of the problem data. This

[106] The many and various problems that now go under the designation 'the frame problem' are distentangled in RCR, essay 2 The proposal made in the text for a logical way of solving the main sort of frame problem discerned by Dennett was put to Dennett back in 1981.

internal resolution of the logical problem can be contrasted with Dennett's external tagging suggestion. Dennett's designer tried to teach their newest and best model '*the difference between relevant implications and irrelevant implications* ... and to ignore the irrelevant ones. So they developed a method of tagging implications as either relevant or irrelevant to the project at hand.' But this robot-relevant-deducer spent its time listing implications and tagging as irrelevant those it should ignore, when urgent evasive action was required. An automaton programmed with a relevant containment logic would spend no time on such irrelevant details. Given relevant inputs (a significant control matter), it would arrive, and arrive directly, only at tightly relevant conclusions.

It is not pretended that such relevant programming would solve 'the' frame problem; of course it does not. But it could chip a worthwhile piece off one of the conglomeration of problems involved. How then is tight relevance programming to be implemented; how, in particular, is relevant theorem proving to be duly mechanised?

A promising approach reformulates relevant containment logics as subscripted tableaux or natural deduction systems. These systems work with expressions of the form A_c, where A is a wff and index c is a situational subscript. The tableaux formulation in fact corresponds directly to an operational reformulation of the semantics, and the natural deduction and tableaux formulations are effectively transformations of one another (on both points see RLR I, chapter 11). The idea is that *subscripted* tableaux or natural deduction theorem proving procedures can take advantage of very many of the features already worked out for corresponding unsubscripted procedures.[107] But that idea has yet to be followed through; the hard work at work-stations lies ahead. Any relevant workers about?

[107] There is also a straightforward way here, with no hidden obstacles, to programming full relevance logics, such as *RQ*. That way combines subscripted tableau (or natural deduction) proof methods with a subscripting of tableaux procedures already worked out for (classical) quantifiers, in a now well-known alternative to resolution methods.

8 Computability is Logic-Relative

Co-authored with Jack Copeland, this essay is an edited and somewhat expanded version of a typescript that Sylvan produced in November 1995 (some seven months before his death). Sylvan's typescript was an early draft and Copeland has reorganised and abridged it, and has in several places attempted to sharpen and expand the arguments (even where he does not agree with them). Nevertheless the words remain for the most part Sylvan's own. While the typescript drew, in part, on Copeland's contributions to their discussions and correspondence, he acknowledges that the work is largely Sylvan's alone. In particular, the material on (what Copeland has subsequently termed) paraconsistent computability theory is mostly Sylvan's. Had Sylvan lived he would no doubt have enlarged considerably upon his remarks on what he called 'dialethic computing machines'. It is hoped that the sketch that he has provided (section 4) will inspire others working in the paraconsistent field to pursue his idea. (A related idea is described by Teixeira and Sarmento 1997). [Editors]

8.1 Introduction

Once upon a time, back in the golden age of recursive function theory, computability was an absolute. Further back still, things were different. While computability (by human hand) was, when it was thought about carefully, taken to depend on the resources available in a given framework or system, it was assumed that there was no upper bound to resources available. Indeed, absolutely sufficient resources were already available within classical logic, so some imagined. Within its comprehensive framework all problems were supposedly solvable mechanically, all mathematical questions were in principle decidable. As we are all supposed now to appreciate, these classical enthusiasts overreached themselves, ascended too high and sailed too near the classical sun, were badly burnt and took a stunning fall.

But in turn the new limitationists, rising out of the ashes of the first Hilbert program, have proceeded also to overextend their claims, claims to absolute results.[108] And they seem to have made a similar mistake: that of assuming that they had built their unsolvability and limitation results on an entirely absolute foundation. Serious cracks in such absolutism have begun to appear, from at least two important directions:

- from *analogue procedures*.

These appear to be able to accomplish more than had been allowed for in the previously 'general' theory of computability (see Copeland and Sylvan 1999 for a survey). The threat to absolutism can be seen as coming even from the old classical physics, especially electrodynamics, theoretical home of many analogue machines.

- from *newer physics*.

Different physics offer different resources, and so different hardware. Quantum computability, for example, diverges somewhat (disappointingly little so far) from classical computability. It was even thought, for a short time, that quantum computability might 'solve the halting problem', a problem that in effect imposes a major limitation upon classical computability. But the buzz of excitement that surged though small circles of cognoscenti soon subsided as it began to appear that quantum computability as so far muddily conceived would do no more in this direction than classical.[109] A different idea takes these suggestive lines: even small chinks of infinity are death to classical limitation results (see Gandy 1980, p.146). For example, a Turing machine that performs infinitely many operations in a finite time will compute functions that are classically uncomputable (see Copeland 1998b for a discussion). General relativity theory may offer alternative frames of reference which provide the effect of infinite time, or other damaging chinks of infinity (see Hogarth 1992, 1994).

Troubles with absolutism are deeper and more extensive than these cracks so far reveal. For one thing, computability is relative not simply to physics, but more generally to *systems* or *frameworks*, which include or contain underlying logics. The idea that computability is relative is familiar from *degree theory*, the branch of recursive function theory that studies the classes of functions which result when the universal Turing machine of traditional recursive function theory is equipped with 'oracles' of varying sorts (a

[108] Turing himself was not among them; see Copeland 1997, 1998a.

[109] On quantum computability see Deutsch 1985. In the end Deutsch's exercise is uninteresting for present purposes, because even though quantum computers may be able to perform some feats of speed that ordinary Turing machines cannot replicate, the class of recursive or technically computable functions does not vary.

notion which, like most good things in computability theory, originated with Turing)—but there the logical setting is usually of a classical nature.[110] We are taking a wider look around. There is a third source of cracks:

- from *radically non-classical logics*, which may carry nonstandard physical theories.

Thus two (overlapping) sorts of relativity can be distinguished:

(i) relativity in procedures, methods or in the devices available for computing; in brief, relativity in the *resources* that are available for computing.

Against a fixed classical logical background, it may make a difference whether computing methods are, for instance, purely manual, or digital mechanical, or analogue, or other. For example, if the resources run to a device that returns values of the halting function, then the classical predicate calculus is decidable.

(ii) relativity in logical systems.

It is this broader *logical* relativity (*not* a logical relativism) which will be the main concern in what follows. The celebrations that there have been for the golden year of 1936, when everything came together and absolutes were forged by human hands (through a sort of American-engineered Gentlemen's Agreement), may in retrospect appear premature.

Let us detour briefly to inspect evidence of absolutist presumptions. Evidence that an absolute conception has become established is not difficult to find, rather is hard to avoid: it is everywhere in the literature. Gödel became almost effusive on the matter:

> the great importance of the concept of general recursiveness (or Turing's computability) ... is largely due to the fact that with this concept one has for the first time succeeded in giving an absolute definition of an interesting epistemological notion, i.e., one not depending on the formalism chosen. (Gödel 1946, p.84)[111]

Church was just as dogmatic. Post tells us that

---

110 There are interesting exceptions, for example Martin Hyland's work on the effective topos.

111 The acclaimed reason: 'By a kind of miracle it is not necessary to distinguish orders, and the diagonal procedure does not lead outside the defined notion'. How this miracle was pulled off will become apparent below: the diagonal procedure does not lead outside the

> The phrase "absolutely unsolvable" is due to Church who thus described his problem in answer to a query of the writer as to whether the unsolvability of his elementary number theory problem was relative to a given logic. (Post 1965, p.340)

Post adds

> True, with a specific criterion of solvability under consideration, say recursiveness, the unsolvability in question, as in the case of the famous problems of antiquity, becomes merely unsolvability by a given set of instruments. ... The fundamental new thing is that ... the given set of instruments is in effect the only humanly possible set. (Ibid.)

The 'given set of instruments' is the paraphernalia of traditional recursion theory, which includes, of course, classical logic.

Such absolutist ideas and claims have been with us ever since, peddled and pushed in the orthodox textbooks. For instance Rogers insists, echoing Gödel, that the 'class [of partial recursive functions] is one of the few *absolute* mathematical concepts to orginate in work on foundations of mathematics' (Rogers 1967, p.19). In an attached note he adds two glosses, one reflecting Gödel's formalistic rendering of systemic independence. The other—which should have given away the orthodox game once and for all (it didn't)—is this: 'where *absolute* means "existing apart from and largely [?] independent of particular symbolic formulations". Quantificational provability (for classical two-valued logic) is another such concept.' (!)

What is more, the absolutist theme became entrenched upon the discovery that several different analyses of computability provably coincide. There is much to be said in criticism of this extraordinarily influential argument for the absoluteness of computability (see Copeland and Sylvan 1999). The point we wish to emphasise here is that these analyses all took logic as a given, not to be tampered with or varied. (In any case, there are grounds for suspecting that the famous convergence may have been, in part at least, contrived. Those attempting later analyses knew where they were going, and what equivalences would be nice to prove and would be duly applauded. The vaunted *independence* of these analyses is something of an illusion. Several of the analyses were pursued even in the one small place, Princeton.)

---

defined notion only because the procedure was adjusted—fixed—to that end.

8.2 Inside and outside classical settings

In undermining absolutism, one easy strategy starts with what classical absolutism does not really dispute, but for the most part simply ignores, namely weak logical systems where *less* is computable than classically. Examples are afforded by systems where not all of the usual range of recursive or primitive recursive functions are definable, such as classical systems with arithmetics weaker than Robinson's Arithmetic.

Admittedly, that initial play does not put much of a dent in absolutism. But there are more substantial and established settings where *less* is computable than is classically claimed, for instance *constructive* systems (of various sorts furthermore). The following function f, one among many such, is classically computable but not constructively computable. f is defined thus on the set of natural numbers:

for all n, $f(n) = 1$ if the Continuum Hypothesis (CH) is true and
$f(n) = 0$ if CH is not true.

(CH is classically undecidable within ZFC, i.e., within Zermelo-Fraenkel set theory plus the axiom of choice.) Bridges, who offers this standard example drawn from intuitionistic stock-in-trade, goes on to claim that such an example indicates 'that the standard theory of computation [namely, classical digital] does not exactly match computational practice, in which we would expect to pin down the algorithms that we use' (Bridges 1994, pp.viii-ix).

A much larger issue intrudes here, one side of which has been exposed by a series of constructivist critics who lodged objections to Church's thesis and classical orthodoxy in the 1950s and 1960s.[112] One important matter, among others, is the extent to which computable functions have to be *really* computable, or in other words, the extent of the idealisation admitted in (the *-ability* of) computability. Another side of this larger issue appears in one of the standard classical objections to the analogue procedures discussed in Copeland and Sylvan 1999: that they are unrealistic, involve excess idealization, and so forth. This question of the extent of idealization countenanced by one or another computational framework is plainly crucial, and plainly a source of relativism. If computability *were* tied to constructability, then absolutism would dissolve. For what is constructable

---

112 For example, Péter questioned whether every general recursive function can be regarded as effectively calculable. Kalmár disputed the converse of this, among other things. For both, see Heyting 1959.

varies at least over time, and perhaps also from place to place, as different notions of constructivity are insisted upon or deployed.

More audacious than constructivist deviations are variations of dialethic or paraconsistent character, which admit contradiction to enter. Here the possibility arises that *more*, rather than less, may be computable than classical theory allows. Take the classical limitative theorems: Gödel's incompleteness results and the Turing-Church uncomputability results. These results exploit the diagonal method, a classically acceptable variation on paradoxical arguments. As incompleteness theorems are straightforwardly avoided in rich systems that admit some inconsistency, so, it might be expected, limitations on computability can be similarly escaped. That is a main question we wish to begin exploring in the present exercise.

8.3 Inconsistency: exit from another Grim tale

Another problematic totality, along with all truths, all sets, all propositions, and so on, is that of all algorithmic functions. Concerning it too a grim tale can be told—were there to be such a totality it would be inconsistent; therefore, classically, there can be no such totality. (This section presupposes some background from Sylvan 1992, where the classical tales Grim tells are retold (see Grim 1984, 1991).) And in turn that grim tale can be retold dialethically: there are inconsistent totalities whose inconsistency is benign. What happens to the totality of algorithmic functions precisely parallels what happens to several totalities that Grim investigates: a 'paradox' is produced by use of a 'diagonal' argument. The *paradox of algorithmic functions* which should emerge under orthodoxy is this: *there are more algorithmic fuctions than all algorithmic functions.* (Compare one of Grim's conclusions, achicvcd by a similar diagonalization: there are more truths than all truths.) Such a paradox is almost reached in Roger's revealing presentation of the theory of recursive functions in his classical textbook, Roger 1967, but then avoided by a classical incompleteness stratagem.[113] In what follows Roger's presentation is adapted, first to yield the paradox.

---

[113] See especially p.11. The textbook is classical in more than one respect. It is not only a standard, widely used text. It assumes classical two-valued logic, which it claims is 'the common practice in other parts of mathematics' and in standard set theory, ZF. It exhibits that curious mix that Ryle discerned in Carnap, of great technical virtuosity combined with a certain philosophical naiveté (consider for example the remarks concerning, and emphasis on, extensionality, p.10).

As usual, an *algorithm* is a routine book-keeping procedure, finite in nature, which can be carried out by an agent prepared to follow instructions obediently but exhibiting no insight or ingenuity; the procedure yields, from given symbolic inputs, corresponding symbolic outputs. All these procedures can be construed, in the standard way, as *derivations* leading from inputs to outputs. A precise symbolism can be set up for these derivations, which uses only a finite number of basic symbols of augmented English. These symbols include function symbols, variables, arithmetic symbols, digits of a range of sorts, and several sorts of bracketing and punctuation, among them a special symbol to indicate the end of a line. As a result any (finite) derivation can be represented as a single finite string of basic symbols. Furthermore, there is an algorithmic test to determine whether or not any given finite string of basic symbols constitutes a genuine algorithmic derivation.

Accordingly, an agent could list, in sequence, all algorithmic derivations, by first examining all strings of length 1, then examining all strings of length 2, and so on. Indeed, one could give a definite, if still informal, algorithmic procedure for developing this list. While the list itself is infinite, each derivation is reached at some finite stage.

A further algorithmic procedure would list just the derivations for algorithmic functions of one variable. Let P_x be the $(x + 1)^{st}$ derivation in this pruned list. Let p_x be the function specified by P_x. Now define a 'diagonal' function d by

$$d(x) = p_x(x) + 1.$$

Evidently the agent has an algorithm for computing d, as follows:- To produce $d(x)$, for given x, the agent generates the list of derivations as far as P_x, then uses P_x to compute $p_x(x)$, then adds 1.

It follows that there are more algorithmic functions than all algorithmic functions. For d is algorithmic, as shown; but it is not on the list of all such functions. For suppose it were included on the list of all algorithmic functions. Then, for some z, d has a derivation P_z and $d = p_z$. Therefore, for such a z, $p_z(z) = d(z) = p_z(z) + 1$. Thus d *differs* by 1 at z from what purports to represent it. By standard arithmetic subtraction, $0 = 1$, a contradiction. So, by reductio, there are more algorithmic functions of one variable than all such functions. d is both algorithmic and not.

The orthodox resolution of the algorithmic paradox proceeds, as might be anticipated, through incompleteness. Consistency is restored by supposing that some functions sometimes are incomplete, lack values. p_z is not defined for all values; it is undefined or takes Kleene's value u at z. Whence arose the theory of *partial* functions, partial recursive functions in particular, from which many famous incompleteness results flowed. There you have, in essence, the Gentlemen's Agreement of the 1930s. Notice that there is an interesting tension here. The introduction of the value u is a step on the road to three-valued, four-valued, and other nonclassical logics, and ultimately to the logic-relative conception of computability that is presently under discussion. But this is not the road that the signatories to the Agreement were intending to follow, and partialisation was regarded as a crucial beam in the absolutist edifice.

Yet other approaches than the orthodoxy that emerged from the tacit agreement were feasible. A range of resolutions had, after all, been proposed for analogous semantical paradoxes, such as those of Richard and Berry; and other alternatives now beckon. One obvious alternative, when confronted with the algorithmic problem, is to try to introduce levels. The diagonal is, so to say, *super*-algorithmic, algorithmic at a *level up*. Under this alternative, there is no one totality of all algorithmic functions. The 'totality' breaks up into a hierarchy of levels. Hierarchization is after all a prime technology for restoring consistency, which finds application not only in logic, with levels theories, but in law, with inconsistent sets of principles reordered hierarchically.

Another relatively simple alternative would be to replace the three-valued Kleene logic by Priest's almost complementary three-valued 'logic of paradox' LP (see Priest 1979). Other related alternatives would start from some other logic allowing properly for overcompleteness (for example a four-valued logic FDE of first degree entailment, or some relevant extension thereof).[114] Exploration of such alternatives constitutes the emerging field of *paraconsistent computability theory*.

8.4 Dialethic machines

It is not difficult to describe how a machine might encounter a contradiction: for some statement A, both A and $\sim A$ appear in its output or among its inputs. Classically, of course, explosion then occurs, any statement whatsoever being derivable, but the machine itself may well not literally explode (as it might in some science fiction scenario). More likely, the machine simply muddles on, unapprised of its predicament; or perhaps it has been instructed—most unclassically—to halt. By contrast, a machine programmed with a dialethic

[114] For details of all these logics, and of the basic paraconsistent setting, see PL.

logic can proceed with its computation satisfactorily. Let us call such machines D-machines of type 1 (D for *Dialethic*). D-machines of type 2 are machines whose *meta* logic is dialethic: for such a machine, M, one of whose states is x, 'M is in x' and ~(M is in x)' may both be the case. D-machines of type 1 are nothing more than Turing machines, albeit a conspicuously useful sort of Turing machine if inconsistent data is in the offing. In the next section we sketch a central idea of paraconsistent computability theory: such Turing machines may be employed to compute diagonal functions that are classically regarded as uncomputable. It is an open question whether D-machines of type 2 compute classically uncomputable functions and, if so, which. We recommend the question to the paraconsistent community. (Another major open question in paraconsistent computability theory is raised at the end of the next section.)

8.5 Diagonalization, undecidability, and the halting problem

At least since Poincaré's criticisms of impredicative definitions at the turn of the century, there has been unease concerning diagonal procedures and definitions, both as to whether they are satisfactory, and as to what, if anything, they show. The issues concerned are bound to persist, because diagonal results are out of the same box, so as to loosely say, as the semantical paradoxes. In principle, virtually every position that has been tried with respect to these paradoxes can be replicated with respect to diagonal arguments. If the orthodox resolution should loose its heavy grip, perhaps that diversity will come to pass. There is already a little evidence that this may happen. Recently Gumanski, for instance, has argued that diagonal predicates (and functions) are not well-defined—or do not exist, in his different ontological terms (Gumanski 1988). They fail to exist because they lead, so it is alleged, directly to contradictions. But unless the notion of directness is tightly curtailed, the objection is very sweeping, and would, like intuitionism, destroy a great deal of appealing and apparently satisfactory theory, including all classical undecidability results.

Orthodoxy is less destructive, but more exploitative, aiming to convert near-paradoxical reasoning into dramatic results. It achieves this dazzling outcome by a division of semantical and logical paradoxes, with logical paradoxes supposedly excised through a size-truncated set theory based on ZF, and with semantical paradoxes emasculated by a levels-of-language theory. In this way it can retain some impredicativity and diagonalization, namely away from grand totalities. Effectively the claim is that diagonal functions cannot be computed (are not total computable functions), because the contrary assumption leads to contradiction. But now that we have more experience in working with inconsistent totalities, we may well be inclined to push on, rather than retreating (for example to partial functions) or adopting evasion tactics. For in fact we may know perfectly well how to compute certain diagonal functions. For example, if the instruction is to increase the diagonal by 1, well then we add 1! Pushing on, a diagonal element is encountered such that $d = d + 1$: $d = d \,\&\, d \neq d$. This gives pause neither to the contradiction-tolerating theorist nor to a D-machine of type 1 that is computing the function. The computation continues, there being no danger that the machine will infer absurdities from this esoteric contradiction.

What of undecidability results? Virtually all these results are established, either directly or indirectly (which is to say, through mappings), by diagonal arguments. Typically, a diagonal function Ψ_m is defined for which it is shown (as, for instance, in Mendelson 1964, p.255) that for some k:

(D) $$\Psi_m(k) = \Psi_m(k) + 1.$$
$$\Psi_m(n) = U(\mu y T_1(n, x, y)).$$

Often that is as far as the argument proceeds. A contradiction has been reached, it is presumed, whence reductio is applied to reject the supposition of decidability.

If simply an esoteric contradiction has been reached, the dialethic theorist can say: So what! Getting decisions sometimes—in diagonal cases—involves computing inconsistently. That can be accomplished, as we have already seen. What is classically undecidable may still be decidable.

But, of course, it is not quite as simple as this. For Ψ_m is a number-theoretic function. *If* the bounds of classical arithmetic have not been exceeded, then by subtraction within (D),

$$\Psi_m(k) - \Psi_m(k) = \Psi_m(k) + 1 = \Psi_m(k),$$

whence $0 = 1$, a classically unwelcome outcome, since, classically, arithmetic then collapses! But plainly there are alternatives here to being clean bowled.

One apparently extravagant option is to allow that

- the functional extension that has been made is not conservative.

(It may be that all recursive functions can be 'represented'; they may still disturb the system). As reflection reveals, this option is not really so extravagant. In particular, what impredicative definitions achieve, as when they facilitate expressionary diagonalization, is sometimes precisely an inconsistent extension. What is more, the definition of the diagonal falls squarely under Russell's intended (but much criticized) characterisation of *impredicative*. For the diagonal given is defined in terms of all of a totality while purporting to be one of that totality. Impredicative definitions are liable to lead to inconsistency: they can render consistent theories inconsistent. Thus what impredicative definitions achieve (likewise axioms or postulates that facilitate them) is *nonconservative* extension. One moral appears to be: do *not* expect diagonal definitions to leave arithmetic classical.[115]

A connected possibility is accordingly that

- the additional functional apparatus has converted arithmetic into an inconsistent arithmetic.

But then subtraction may not be admissible. So arithmetic does not collapse. Mathematics can go on, albeit inconsistently (see, for example, Priest 1994).

There are further options to pursue. Most promising is

- to shed the conditions for consistency adopted by Kleene (as discussed in, for example, Rogers 1967).

Connected therewith is the option of

- allowing multiple computations, which may sometimes yield different results.

So it is with the diagonal. There one computes $d + 1$ by adding 1 to d, and obtains a *different* result from computing d.

The proof of the halting theorem, in all its known variations, likewise proceeds by *reductio*. But, again, in a paraconsistent setting, the derivation of a contradiction is in general insufficient ground for rejecting the assumption that leads to it, which in this case is the assumption that there exists a Turing machine capable of computing the halting function. A leading open question of paraconsistent computability theory is whether or not the halting theorem holds within a paraconsistent framework.

---

[115] An entertaining alternative to classical real number theory also emerges under diagonalization: *inconsistent rational* theory. That is, take difficult diagonals as giving further inconsistent rationals.

9 Part II Fragments

As Chapter Three of the proposed monograph, Sylvan intended to include material on the use of pararelevant logics in the cognitive and computing sciences. The extant draft fragment of this chapter is reproduced below. It was last worked on in July 1995. [Editors]

9.1 On the relevance of pararelevant logics to cognitive and computing science

Many of the applications outlined for logic and philosophy naturally extend to cognitive and computing science, to CS. Such an extension is hardly remarkable, as cognitive science is but an elaboration of epistemology (little more than jumped-up epistemology) and cognitive science is grounded in logic, the hardware of present machinery consisting of two-valued logic circuits, and much of the recent software involving logic programming.

9.1.1 Classical logic versus alternatives: relevant options and relevant themes

It has usually been assumed that where logic *does* enter as regards cognition and intelligence, artificial or perhaps not, *that* logic will be classical logic. But that, so one main theme will be, is very far from the case. Rather classical logic has often been imposed or infiltrated, where it should not have been, often with very unsatisfactory results.[116]

There are several intertwined themes that will be advanced.

- Mainstream logic, particularly classical theory, is inadequate for a range of logical purposes in cognitive science (and mental science).
- Superior analysis and results can be achieved by use instead of relevant logics.

Put differently, wherever logic has applications and roles in cognitive and computing science, relevant logic does also, and relevant logic can often perform these tasks *better*. But there is more to it than that.

For so far the theoretical framework could be one of a generous logical pluralism, where perhaps classical logic still enjoyed pride of place, but was

[116] The theme has been argued in several places, most notably UU.

displaced in certain logical applications, such as where inconsistency entered seriously. Under classical pluralism, classical logic is always preferred only to be displaced by a next ranked logic (for a given purpose) where it breaks down irretrievably. Similarly for the next ranked logic and so on down. While pluralism itself is fine, that sort of classical pluralism is less than satisfactory, yielding unfortunate outcomes. Under relevant pluralism, here sponsored, classical logic is superseded in its prime placement by some batch of relevant logics. Relevant logic is not simply something pulled out for special isolated occasions where classical and mainstream breakdown occurs. Relevant logic offers an all-purpose working logical systematisation. In fact it can be adopted as a uniform general working logic, so that there is no further need to endlessly change logics to suit conditions. Although inessential for present purposes, further theses loom in the background, one of which exiles classical logic from any central or preferred place to the periphery: namely that classical logic (and similarly mainstream logic) is wrong, wrong anyway, except in very restricted circumstances. Accordingly it is not a smart idea to proceed to use it, though of course agents may—as they proceed to use defective aircraft, rust-bucket tankers, and so on. The broad relevant program, of classical displacement, will emerge, in increasing detail, as things proceed.

Classical logic is superseded by a batch of interrelated relevant logics because there is a batch of logical tasks to perform. As there are many logical *connections* to explicate—implication, entailment, conditionality, reason-conferment just some among them—with various different explications, better or worse, on offer, so there are many different relevant logics. A main goal has been and remains correct, paradox-free systems. Plainly that goal is not sharply defined. As, moreover, there is now a staggering array of logical systems to choose from, for these and various other essential purposes, so, within that welcome plurality, there are *many* relevant logics, of different kinds. Along with many sorts of relevant logics, there are many sorts of paraconsistent logics. A main reason being that, as we have seen, there are intimate connections between these sorts of logics, especially in their treatment of negative paradox arguments, those that spread inconsistency, or falsehoods, profligately. Problems of choice of system can in fact be substantially postponed, for the following reasons:- At the first degree where many of the applications can be confined, there is essentially only one relevant logic, main choices arising at higher degrees. Further, apparent competitors, other sorts of first degree sociative logics, can be seen as arising out of relevant logic through further requirement, as for instance containment logic arises through adding a containment requirement.

Relevant logics have *wide*-ranging applications in CS and AI in large part *because* of their paradox removing capabilities—many, many other paradoxes are generated by paradoxes of implication which relevant logics were designed to eradicate—and *because* of the capacity to handle incompleteness and inconsistency, incomplete or inconsistent data especially. Significant corollaries for CS and AI are virtually immediate. In exhibiting distinctive contributions from relevant logics we begin with *one-place* functors.

9.1.2 Information, knowledge and data; action and process theory

Relevant theory can unproblematically include data bases and knowledge bases which are inconsistent or incomplete; the theory can allow for direct processing of inconsistent or fuzzy information. Some of the ground for this was prepared some time ago, with *relevant* analyses of *content* and *information* in place of modal accounts such as Carnap's, and *relevant epistemic* logics in place of the modal systems of Hintikka and others. For example, as regards content, all the bizarre modal results about necessary statements all having the same content, namely none, and a single isolated contradiction delivering total information, are eliminated. Inconsistent information ceases to be an insuperable obstacle, or indeed an obstacle at all. Similarly in epistemic logic, obstacles concerning inconsistent beliefs vanish. And all those requirements as to ideal agents, consistency, perfect rationality, and so on, that render modal representations effectively inapplicable to more *ordinary* agents and regarding more *normal* intelligence,[117] are likewise removed.

Similar advantageous features obtain wherever data or knowledge enter into computing science: with knowledge bases and data bases, in expert systems and other systems which operate with such bases. They also occur as regards dynamic development such as revision of bases, theory change and updating, and the like. For here too the prevailing theory, but recently elaborated, is modal in character, so that inhibiting and damaging consistency and other restrictions can once again be relevantly removed.[118] A system can work with and through inconsistency.

---

[117] Also to limited robotic and agent systems.

[118] For a pioneering investigation see Fuhrmann 1988 and for a detailed treatment see Restall and Slaney 1995.

Treatment of dynamic systems leads easily and naturally to, what is really a special case, action theory. Cognitive science is bound to be much concerned with agents, artificial (such as robots) and biological (such as humans), and their actions. A theory of such matters has become known as action theory, a subtheory in turn of process theory. Naturally in action theory actors include not only some humans and various animals but also actors that are other than biologically natural individuals, for example robots, artifices, conglomerates, clubs, etc.

One-place functors that play a large role in action theory are those in the vicinity of intention: not merely intending itself, but wanting and preferring, desiring and deliberating. So let us consider, as a more sustained example, functors from action theory such as *intention* and like functors—*doing, striving, intending* and so on. Indeed these functors tend to play a grander role than action theory really warrants, because of the mistaken assumption that all (normal) action is intentional, along with the mistaken conflation of intending with wanting or (when that is removed) the mistaken location of intention within the want-belief nexus. Defective accounts of intentionality abound, partly because there is heavy pressure to reduce it to something else. Though it has a logic rather like belief, the notion is very different, with a quite different range of application. Nor is an agent's intention tied to what it wants, or wanted, to do, because it may have to perform (intentionally) other actions it does not want to do, or have done, in order to achieve what it does want or values. Nor is an action intentional when it realises approximately, or indirectly, what is wanted or sought, because, even though directed, things may still go wildly astray: successful direction is something more. And so on, through subtler and more oblique combinations of wants and beliefs. Want-belief reduction is, once again, not what is wanted: intention needs independent investigation.

Let us introduce the functor $\mathbb{I}$, construed awkwardly, in order stay within the propositional setting, as: the given agent intends that, or: that ... is intended by the agent. Thus $\mathbb{I}_x A$ reads: *(agent) x intends that A.* A certain familiar amount of twisting is needed in order to push variant forms into *that*-forms, for instance "x intends to do Ψ" becomes "x intends that x do Ψ". Then a preliminary analysis of "x does A intentionally", and likewise, it is sometime imagined, of "x does A deliberately", takes the conjunctive form: $\mathbb{D}_x A \,\&\, \mathbb{I}_x A$, where $\mathbb{D}_x$ reads: *(agent) x does A.* (But as is well known, from Montague and Gettier, conjunctive analysis of modifiers and intensional conjuncts is liable to produce paradoxical or bizarre outcomes. For instance, it is a joke that large mice are large, that motivated doings are done.) In fact deliberate action is separate from intentional action, for all that their equation is sometimes suggested. Plainly an agent may deliberate— consider alternatives and weigh up considerations mentally—without reaching a result or forming an intention. If an agent takes action deliberately, for instance because some time-line expires, then, though intentionality in the weak sense of mental activity may be met, the agent may nonetheless have formed no clear intention, no design, not be set or bent on an object. The breakdown of the converse is clearer still. An agent may form an intention, immediately, without deliberation; no consideration, no pondering, no requisite processing. Deliberation demands, then, further investigation, beyond that we intend to direct towards intentionality.

While a logic for $\mathbb{I}$ is meagre, it is not null. Thus the idea of disposing of intentions as a bundle of more or less arbitrary propositions (mere *that* p's, on Segerberg's more recent 1982 account) can be dismissed. But certainly some of the logic that has been laid on intention is a bit thick. For instance, Segerberg 1982 has variously supposed that an agent always has some intention; indeed just one intention, i.e., in effect $\mathbb{I}_\alpha A \,\&\, \mathbb{I}_\alpha B \rightarrow .\ A = B$. But an agent can have a bundle of intentions, more than one of which may be operational in propositional settings. Further, an agent can act, for example on the spur of the moment, or in response to an immediate issue, without always drawing upon a ready-made intention set or forming a further intention. But logic there is. The propositional logic of $\mathbb{I}(A \,\&\, B) \leftrightarrow \mathbb{I}(B \,\&\, A)$, $\mathbb{I}(A \,\&\, A) \leftrightarrow \mathbb{I}(A)$, $\mathbb{I}(A \,\&\, (B \,\&\, C)) \leftrightarrow \mathbb{I}((A \,\&\, B) \,\&\, C)$, $\mathbb{I}(A \,\&\, (B \vee C)) \leftrightarrow \mathbb{I}(A \,\&\, B \vee\!\!.\ A \,\&\, C)$, etc. What ensures all these? Even if $\leftrightarrow$-replacement fails, as it may for stronger implication, $=$-replacement does not. That is, the rule, $A = B, D(A) \,/\, D(B)$, with $\mathbb{I}$ in its scope, will provide such combinatorial results.

As observed the logic of intention resembles, at a propositional level, that of belief, and similar trouble spots are encountered. But of course the logics are different, evidently when other functors are brought in; for instance knowledge implies belief, while knowledge does not imply intention. Trouble spots include principles yielded by first degree $\rightarrow$ transmission, such as $\mathbb{I}(A \,\&\, B) \rightarrow .\ \mathbb{I}A \,\&\, \mathbb{I}B$ and $\mathbb{I}A \rightarrow \mathbb{I}(A \vee B)$ and its mate, and also the converse of the first, $\mathbb{I}A \,\&\, \mathbb{I}B \rightarrow \mathbb{I}(A \,\&\, B)$, adjunction. But the supposed trouble with the first, intentional simplification comes from rendering the conjunction intensionally, as if the components were related. Thus it is claimed, correctly, that an agent may intend A and B in concert without intending each separately. Only then the claim should be symbolised differently, for instance through such intensional composition propositions as $\mathbb{I}(A \circ B) \rightarrow \mathbb{I}A$, which get duly rejected. The objections to the second principle and its mate are different. They include such observations as that an agent who intends A may find B irrelevant or may even lack the information or conceptual apparatus B presupposes. As such observations themselves suggest, additional baggage like that B brings with it can be shed by using an analytic implication or, in this content, relevant containment logic. Closure of intention under the first-degree implication of such a containment logic will deliver intentional principles like $\mathbb{I}(A \,\&\, B) \rightarrow \mathbb{I}B$, $\mathbb{I}B \rightarrow \mathbb{I} \sim\sim B$ and $\mathbb{I}B \rightarrow \mathbb{I}(B \vee B)$, but not $\mathbb{I}A \rightarrow \mathbb{I}(A \vee B)$. Such a rule we accordingly adopt (semantical details are readily pieced together from previous work).

Adjunction is open to objection also, though not on all the grounds that afflict addition, since no new propositional content (in the shape of irrelevant B) is introduced. Both A and B are given, and virtually all agents will have a suitable concept of conjunction. Even so an agent may not put intentions together, even though it appears reasonable to do so. If intending involves 'having in mind' then adjunction will indeed fail for some agents. This is not so remarkable, and not destructive of logical theory. For certainly adjunction does break down functors, such as assertion and explicit belief, not so remote from intention. A strategy, which handles the business takes adjunction as an optional extra, which holds for certain agents, namely *fully adjunctive* agents. But adjunction is not taken as universally valid.[119] The strategy noted is part also of a grander problem-resolving strategy, which envisages different logics for different sorts of agents.

We are now likely to encounter, but are well prepared for, several puzzles Segerberg 1982 has assembled, puzzles which provide (as Russell might have said) a fine test for any logic of intention.[120] Most of these puzzles are in fact familiar from other reaches of intensional logic, and indeed met in much the same way through relevant logic in the case of intention as elsewhere.

The first puzzle turns on the impact of incompatible intentions through adjunction and replacements. Oedipus intends A (for example, that he marry this women, viz Jocasta) and intends B (for example, that he avoids marrying his mother), so, as an adjunctive agent, he intends $A \,\&\, B$. But $A \,\&\, B$ is *de facto* impossible, Oedipus intends to do what is impossible. But then, classically but not relevantly, $A \,\&\, B = F$, the false proposition (in Segerberg's terms $A \cap B = \emptyset$, the null set). But Oedipus never intended F ('$\emptyset$

[119] A feature easily accommodated in nonsystematic semantics.

[120] Segerberg works with a functor Int which applies to events, but is applied in examples to infinitival clauses (a straightforward transformation, mostly of *that-* clauses). In any case, the event structure, a Boolean algebra, is homomorphic to what is assumed to be a propositional structure.

was surely never an intention of Oedipus', p.250). Relevant logic almostly automatically resolves such puzzles, since the classical equation of distinct impossible propositions with one another and with the False fails.

While Segerberg rightly observes that the puzzle is not so much an objection to adjunction as to replacement, he goes astray in supposing that it can be removed simply by strengthening replacement to a strict (modal) form (see pp.251-2). For intentions may be directed towards logical impossibilities. For example, while Hobbes intended to square the circle, some of his contemporaries intended to design a perfect perpetual motion machine; but they did not thereby have the same intentions because both types of intention were impossible. Here, as elsewhere, the modal framework can and must be transcended: possible worlds give way to wider classes of situations.

However Segerberg does not see how to escape the possible worlds strait-jacket, as he effectively concedes in response to 'the second objection' (pp.252-3), which offers almost a paradigmatic example of what relevant theory was intended to accomplish. 'It does not seem plausible that if I intend to close the door, then I intend to close the door and to visit Japan or not.' So objected a perceptive referee to the following sort of exhibition of irrelevant tack-on that modal-based theories inevitably validate:

Junktion $\mathbb{I}A \leftrightarrow \mathbb{I}(A \,\&\, (B \vee {\sim}B))$.

The result is inescapable in modal theory so long as $\mathbb{I}$ is what it appears to be, a propositional function, because modally that-A and that-$(A \,\&\, (B \vee {\sim}B))$ amount to the same proposition; they take the same values in every possible world (cf. Segerberg 1982, pp.258-3, who remarks that 'the objectionable feature is deeply grounded in the semantics presented ... How to develop an interesting system without this property is not clear to the author'). As $A = A \,\&\, (B \vee {\sim}B)$, Junktion results at once by substitution. But in relevant logic $A \not\models A \,\&\, (B \vee {\sim}B)$ (as Fitch in effect observed). For there are incomplete (possible) situations where $(B \vee {\sim}B)$ does not hold though A does. So Junktion does not follow, and indeed is easily counter-modelled. For 'intention situations' are usually far from complete.

The third puzzle is an elaboration of the first and second, that any modal theory 'will be too strong if one has the logic of intentions of real people in mind' (p.253, amending the context). Prior had made a similar complaint concerning (modal) logics of belief and assertion. And Segerberg himself leads into the 'third objection' from a discussion of 'the paradox of the logically omniscient subject: in most epistemic logics, an agent who knows one logical truth knows them all' (p.253). Such a paradox disappears in relevant epistemic logic, along with epistemic analogues of Junktion (for example Segerberg's $KA \leftrightarrow K(A \,\&\, (B \vee {\sim}B))$, because different logical truths are differentiated. In relevant theory there is no need to be so far removed from 'real people' (even if *some* real people may severely test the theory through their unreasonableness and logical perversity) or from 'real robots'.[121]

The puzzle, as duly elaborated, appears to come to this: 'there are real life situations that cannot be modelled in even the weakest of the' modal systems considered (p.253). But it is, so it is assumed, modal theory or nothing, no logic of substance. If the demand to model real life situations and intentions 'is accepted, trying to meet it would lead to a logic of intention so weak as to be void of content' (p.254). Segerberg proposes to bulldoze through this dilemma—modal logic, paradoxes and all, or not logic—in the received destructive fashion: suppression of natural people and the artifice of highly idealised agents. Whereupon Segerberg feels able to stand modal theory up again (alongside modal game theory and decision theory, which are presumed not to be damaged by analogous dilemmas[122]), and to display his favoured modal system as 'a logic of *rational action*' (p.254). The verdict should be that this modal industry does not succeed.

For there are rational actors who make relevant discriminations modal systems cannot accommodate, in violation of Junktion for example. The availability of relevant logical alternatives, to which such agents can certainly appeal, shows that the dilemma modal theorists try to erect is a false one. In between the false choices offered are relevant alternatives. A respectable logic can remain after modal idealisations are shed, and with Adjunction, the example Segerberg focusses upon, abandoned as well. Logics of assertion, which should be of this very sort, illustrate the matter nicely. For an assertor may well assert B and also assert C, without thereby asserting their conjunction. But such logics of assertion, though no doubt subtle by crude modal standards, are not difficult, nor devoid of content or interest. So it is also with relevant logic of intention; they are not nothing, they can serve real people. They may also serve viable artificial agents. Established rational decision theory is a thoroughly modal enterprise, which brings in its train new

---

121 Real machines have limited capacities, confined search spaces, are liable to breakdown, etc.

122 But of course they are. A further enterprise, analogous to the present one, is relevant decision theory, which enables many puzzles of decision theory to be straightforwardly resolved.

expected puzzles and paradoxes. It likewise stands in need of pararelevant upgrade, a beginning upon which is made below (in part III).[123]

Similar kinds of considerations operate right across the range of functors that matter in AI and its applications. They effect not merely information storage and processing, and action theory, but dynamic and process logics, theory-change theory, relevance in discourse, expert legal systems, and so on. Some of these examples intricate two- or more place functors.

9.1.3 Illustrations involving two-place functors

So far the focus has been on *one-place* functors added to a working relevant basis. *Two-place* functors are bound also to be an important part of a relevant logical package for AI. Among these functors are the following:

- Conditionality, symbolised $>$, read "if ... then".

While in some places it is now well appreciated that classical logic is hopeless in affording a satisfactory representation for conditionality, in other places this obvious information has not yet penetrated an amazing tangle of classical defences. Consider a couple of examples from a long list of classical ridiculousness.

It is not true that if she is over forty she is still young. Therefore if she is still young she is over forty (instantiating $\sim(A \supset B) \therefore B \supset A$).

If Saul comes from Georgia he is a southerner. Therefore if Saul rides a bicycle to work he is a southerner or if he comes from Georgia he owns a bicycle (instantiating $A \supset B \therefore (C \supset B) \vee (A \supset D)$).[124]

A modal theory of conditionals can offer some relief, but insufficient, as ridiculous results still emerge, ridiculousness a relevant theory can shed. Nonetheless the relevant semantical theory resembles semantics for the modal theory, except that once again inconsistent and incomplete situations feature. The situations are what alter the impact of the look-alike semantical rule

$$I(A > B, a) = 1 \quad \text{iff} \quad s(a, [A]) \subseteq [B].$$

As a result the logical and syntactical theory for relevant conditions is tighter than the modal theory, and free of the obvious paradoxes that detract from the modal theory, and distance it from natural language.

---

[123] *This reference is to part III of the proposed monograph, which was to include a chapter entitled "Rational decision theory without paradoxes". Unfortunately, no drafts of this chapter have been found.*

[124] For many other similar examples, mostly drawn from others, see RLR I, p.11.

Analogous points transfer to CS. In the first place, we want rules in CS operations, regularly formulated through "if ... then" (variously formulated $\rightarrow$ or $\leftarrow$) to be useful, and not lead us astray, as classical and modal rules may do (especially when noncontingency enters). For instance if a rule $A \rightarrow B$ says roughly "in order to proceed to B, check out A", we should want A to be relevant to B (otherwise what is the point?) In particular, we do not want to be told, in the presence of contradiction, to proceed anywhere at all (or else to stop, as things are haywire).

Secondly, a growing segment of AI concerns discourse representation, paraphrase of natural language into a suitable formal system, typically into modified quantification logic. The importance of such exercies is evident: in practice for facilitating communication of humans with computers and, especially where they speak different languages, with one another; and theoretically for enabling improved understanding of discourse, its various roles and how it achieves these. For reasons already outlined, notably, in the case of conditionals—one using very many cases—classical logical theory is utterly inadequate for such representational purposes; what *it* offers is discourse distortion.

Discourse representation leads through to the representation of dialogue, and issues of interchange and communication between intelligent systems and machinery. Obviously too dialogue has interconnections where parallel processing, where, too, relevant logic has direct application (as linear logic has shown; also the rules of connections fission and fusion).

A satisfactory theory of conditionals, along with corresponding theories of implication and entailment, gives pararelevant systematization a huge advantage over mainstream theories in discourse representation, another important field for computing science.

- Containment, symbolised $\subseteq$, read "contains".

The intuitive notion is that where $B \subseteq A$ the content of B is contained in that of A. In critical developments of relevant containment logic, content was assessed as in modal analogues, namely Parry logics, through the set of variables concerned. Relevant containment $\twoheadrightarrow$ itself is defined through the connection

$$A \twoheadrightarrow B \leftrightarrow . A \rightarrow B \;\&\; B \subseteq A,$$

and the characteristic features of its logic can be developed from there.[125] A most important feature for what follows is that no new concepts are added in

[125] *For further elaboration see Chapter Seven. [Editors]*

through addition principles: relevantly valid principles such as $A \rightarrow. A \vee B$ are not correct for containment.

Adapting the *style* of containment logic (but perhaps in combination with conditionality) affords a way of analysing other two-place functors of considerable AI significance, notably *reasoning*, which is examined shortly. Parts of this developing theory can be put to work almost immediately, in showing how to deal with one facet of the general frame problem, really a bundle of problems regarding the frame within which an automaton or problem-solver operates.

9.1.4 A control problem within that problem complex known as the frame problem

A problem solver may be swamped by a sea of axioms and other information, even if not by those postulates updating the frame (which is now automatically updating). How does it stay focussed on what is relevant and not fritter valuable time away on irrelevant deductions or other processing. This is the processing control problem (Dennett's 'frame problem'), for a solution to which appeal will be made to relevant logic, and more generally to *relevantly focussed* processing.

The idea is, in brief, that the automaton selects information or axioms assessed as relevant to the task or problem at hand and processes that information through a suitable relevant logic, which introduces no new extraneous information. It works directly with what is inferentially relevant. Let us consider deductive processing; similar points will arguably extend to other types of logical and heuristic processing and reasoning.

The logical apparatus of an automaton typically enables it to make (potentially) infinitely many inferences from the information it has stored. But much of that information and many of the inferences it might make, even *if* classically valid, will be irrelevant to solving specific issues confronting it. A mass of irrelevant information will be needlessly and wastefully piled up and will block processing channels. A significant part of the problem reduces to that of limiting inferences to those relevant to the issues at hand (similarly for perception).

A relevant containment implication, an implication carrying relevance and guaranteeing containment, which does not allow the iterated introduction of extraneous material, can accomplish much of what is sought. For such problem-solving situations the automaton's logic is a relevant containment one. Thus it does not derive extraneous material, or waste important time piling up irrelevant consequences of the problem data; it goes direct to what is relevant. This *internal* resolution of the logical problem can be contrasted with Dennett's external tagging suggestion. Dennett's designers tried to teach their newest and best model '*the difference between relevant implications and irrelevant implications* ... and to ignore the irrelevant ones'. So they developed a method of tagging implications as either relevant or irrelevant to the project at hand. 'But this robot-relevant-deducer spent its time listing implications and tagging as irrelevant those it should ignore, when urgent evasive action was required. An automaton programmed with a relevant containment logic would spend no time on such irrelevant details (or on gruish properties such as Fodor's *being a pigeon*, which relevant procedures can straightforwardly exclude, as not genuinely depending upon their arguments). Given relevant inputs (a significant control matter), it would arrive, and arrive directly, at only tightly relevant consequences.[126]

The appropriate strategy is glimpsed by Glymour and also by Haugeland. Thus Haugeland: 'in this respect, the frame problem is analogous to the knowledge access problem ...; in both cases, the issue is how to "home in on" relevant considerations, without wasting time on everything else the system knows. Then the challenge is not to decide for each fact whether it matters, but rather how to *avoid* that decision for almost every bit of knowledge' (RD, p.82). Rather more explicitly, Glymour: 'the way to compute relevant consequences is generally *not* to compute consequences and to test them for relevance. ... The reason is that such strategies require a lot of things be generated that will in the end be rejected as irrelevant, and thus waste computational resources. The best solutions to frame problems are algorithms that generate the relevant without ever looking at the irrelevant' (RD, p.70).

But they both pass on. Haugeland slides to the standard, cheap test and sleeping dog, computing approaches to the problems and their inadequate ways of trying to focus upon what is relevant and bypass what is irrelevant. Glymour jumps to a generally unsolvable halting problem, *but* in relevant satisizing form: when to stop relevant computing, when enough information is enough? Both pass on without indicating relevant technology, such as relevant containment logics.

---

[126] Pylyshyn wrongly contends that 'general reasoning' allows the 'connect[ion] of any item of knowledge or belief with any other' (RD, p.viii). The reason he offers is that inference, even relevantly restricted can in principle go anywhere. For connections like relevant containment this is not so. Such relations also undercut his speculation 'that there appears no way *in general* to index beliefs so as to exclude certain obviously irrelevant paths of inference in advance, without at the same time excluding some obviously relevant ones' (RD, p.viii). It is not a matter of indexing statements; it is a matter of controlling premiss selection and inferential procedures. And this can be accomplished in several allied ways (see further DRL).

Again such relevant processing and theorem proving has yet to be implemented. But in this case—as contrasted with vision and sensory implementation—though the difficulties to be surmounted are formidable enough, the general route to relevant theorem-proving packages is moderately clear in broad outline. As with vision implementation, the idea is again a satisizing one, to shed excess information and loadings, to avoid too much. Unnecessary or irrelevant information is shed by tighter relations, such as those of restricted sensitivity and relevant consequence.

An Aside: Applying relevant logic in such a direct search fashion has the effect of significantly reducing the search space. It has been speculatively suggested that use of relevant logic might have this effect more generally, beyond the confines of the control problem considered. It is already appreciated that there is technical divergence between classical and quantum computability. It appears moreover that this relativity extends more widely than physical systems and analogue systems, to logical systems (of which physical systems are special postulatized cases). So there will in principle be relevant computability, dialethic computability, and so on, depending on the general logical systematization adopted. My conjecture is that some very interesting results will emerge from investigation of these matters.

Reduction of the search space is not the only advantage that may accrue. Other related advantages may be reduction in complexity and higher order simplicity.

9.1.5 Reasoning, its types, its automation

As treatment of 'the' frame problem indicates, a cognitive subject proceeds to try to solve problems, take action, and so on, by reasoning. Furthermore, especially with artificial intelligences, this reasoning typically proceeds through generation of theorems in an automated reasoning program or the like. Indeed such an automated reasoning approach affords the underlying pattern of many efforts in artificial intelligence, for instance as regards expert systems.

There is no doubt that reasoning is critical. Reasoning does constitute a major part of problem solving and of components of intelligence, and other essential normative notions such as *rationality* turn upon it. However reasoning, which has been conflated with implication, is a notion which has not been satisfactorily explicated in standard logic. Certainly, the assumption is now commonplace that logic, classical logic even, takes care of reasoning. But that assumption is very far astray. A corollary, of AI relevance, is that the fashionable idea of automating reasoning by mechanising proof procedures in classical logic, is misguided.

The case against the assumption has several parts. In the first place, the logic of *reasoning for* and *reasoning against* is very far from classical; it is progressive, non-monotonic, and relevant. While it appears to admit of relevant explication, 'classical theory' would have to be much changed, revolutionized, to include it. The basic positive notion *reason for* can be represented $A \overset{+}{\supset} B$ read "A is a reason for B" (analogously *reason against,* $\overset{-}{\supset}$). In the special case of ponible deductive reasoning, this connection can be satisfactorily analysed, in a way parallelling relevant containment, as

$$A \rightarrow B \,\&\, A \multimap B.$$

i.e., A implies B and A is prior information to B. Much as ordered information may be re-expressed in terms of content, so $A \multimap B$ can be spelt out as $s(A) < s(B)$.[127] More generally, where reasoning may not be deductive, implication is expanded to plication, $\circ\!\!\rightarrow$. So the full recipe is

$$A \overset{+}{\supset} B \quad \text{iff} \quad A \circ\!\!\rightarrow B \,\&\, s(A) < s(B).$$

It is tempting to equate plication with conditionality, $\circ\!\!\rightarrow$ with $>$. Then nice semantics are forthcoming for *reason for*.[128]

Perhaps not all reasoning is the giving of reasons, for or against (this or that, justifying this opinion or conclusion, that action or course). It depends how far 'reason' and 'reasoning' are stretched, whether they include making comparisons, drawing analogies, and so on. But certainly such giving-of-reasons reasoning for or against is a large significant, yet logically neglected, part or reasoning. Such reasoning not only comprehends a variety of argumentation procedures, as in dialogue, discussions, cross-examinations, material at the margin of contemporary logic and still scorned by many logicians; it also includes the myopically-focussed part looked at by contemporary schools, the substantially context-independent drawing or deriving of conclusions and consequences from premisses or assumptions—what might be called *standardized* or school reasoning.

Standardized reasoning is subjected to various cross-classifications, which appear to be however largely outgrowing their former usefulness. One long important type of cross-classification is that, hardly well-characterised or exhaustive—into deductive, inductive, and perhaps abductive, eductive, subductive and so on. It was a classification already bothering some (Joseph

127 *See Chapter Six, section 3 for more detail. [Editors]*

128 *For details and arguments see Chapter Six. [Editors]*

among others) at the turn of the twentieth century, which has been variously subverted in recent cognitive science, with induction and abduction coming to play roles substantially severed from their received meaning. However an improved classification shakes down.[129] Nondeductive processes included have significant applications as regards intelligent automata, which need to update and generalize information held, to learn from environmental inputs, and so on. Indeed such issues provide other aspects of general frame problems, broadly maintaining adequacy of operating frame-work.

A different cross-classification in increasing use, gets made not in term of *content* or style of the argumentation connections at all, but rather through the *means* by which such argumentation is accomplished. Such means divide roughly into natural reasoning (often erroneously equated with that carried out by humans) and artificial, with automated or machine reasoning the main division of the artificial (the division sometimes made into machine-assisted or not, though fuzzy, is certainly exhaustive).

From this angle the fashionable terms "natural-deduction" and "natural logic" are entirely anomalous. On the one hand, natural-deduction may perfectly well be automated, and so not be natural. On the other hand, natural-deduction with its usual highly artificial (not to say precious and precarious) rules and restrictions is quite far removed from most of the reasoning carried out by natural creatures, bearing only a casual resemblance to some of it.

Theorising humans have a special propensity for claiming as the whole, or essence, of some enterprise what is only a part, often an atypical part, of that enterprise. Reasoning and logic have been frequent casualties. It is now a joke that Kant, Whately and others contended that logic was a complete, closed and largely exhausted subject. Unfortunately natural-deduction and natural-logic are not yet jokes. Nor are the competing claims of set theory and of category theory to wrap up all of mathematics treated as simply amusing (you have to get serious to take them as ludicrous). Still less is what goes by the name of 'automated reasoning' on the way to becoming a joke; on the contrary, it is treated with deadly seriousness and solemnity, and frequently at such a pace that it is difficult to observe how *little* of reasoning gets into it.

Automated Reasoning (AR) is thus far not the *automation* of *reasoning*, but (isolated exceptions apart perhaps) the automation of fragments of classical logic, of certain (first order) classical theories. The whole great sweep of *reasoning* is passed by—as for instance in many standard texts,

---

[129] *Certain elements of such an improved classification already figure in Chapter Six. [Editors]*

where AR just is the automation of machine-confinable parts of classical logic.

True, the techniques for automating classical logic—a logic which lends itself especially well to machine transcription, since the machines are presently made in its image—can *sometimes*, usually with considerable difficulty, be adapted to nonclassical frameworks. A main reason for obstruction is that many classical tricks and shortcuts, such as normal forming and resolution techniques, do not adapt at all well, or at all, to intensional logics, and still less to the largely unformalised expanse of reasoning beyond contemporary, little, deductive systems.

9.1.6 Elaborating relevant themes and a deep relevant program

The appearance perhaps emerging of a messy, or even opportunistic plurallism—with a different logic tailored for each purpose—is deceptive. For the whole logical structure involved is fabricated from a select single logic (or a simple narrow band surrounding that selection). At bottom there is a single structure, universal pararelevant logic, an older *uniformity theme*.[130] The idea of construction from a single band of sources, *D* systems (of depth pure relevant logic), takes the following lines:- Firstly enthymematic representations. Out of these logics other logics, relevant and irrelevant, are generated, in terms of the types of suppression that they admit. For instance, relevance logic *E* allows the suppression of relevant identities, intuitionistic logic of all truths, classical logic positive suppression of truth and negative suppression of falsehoods. Further, other relevant logics that matter for applications, like relevant containment and reasoning, are generated by obvious additions, as already indicated. Modal logic upgrading is achieved by addition of intensional functors, as also shown.

Out of this universal pararelevant approach come not merely applications and criticism of rival irrelevant approaches, especially mainstream, ones. There also emerges an alternative relevant program, which would change ways of thinking in CS and about AI, for instance about reasoning and intelligence in general. In this sense, relevant logic is not some adjunct, like modal logics, to classical, useful for some specific purposes; rather it represents a rival paradigm. It is a deep, and strong, alternative program—though not exclusivist, because it makes some room for classical theory.

---

130 Advanced in UU. Of course, pararelevant logic is not the only logic which pretends such universality; traditional logic used to, classical logic now does also.

This strong program, doing CS and AI things relevantly, would be pursued in *softer* and *harder* ways, as such examples as the design of a *relevant intelligence* will reveal. The soft way adopts relevant logic in the software, in the programming languages, and so on. The hard way would also alter the hardware (presumably no great feat now that many-valued chips are available). It would rewire computing machinery using a four-valued basic logic—with values true, false, both and neither—congenial to relevant systematization. Machines would change, for the four-valued computer would offer a great improvement.

Dominant logic is classical two-valued. Computing hardware is presently shaped in its two-valued image. There is nothing essential, or god-given, about this. Machines could be constructed, just as readily often enough, to match any finitary system of valuations, not just *two*. It may be tempting—many succumb to such temptation—to suppose that two-valued is simplest. But of course numerically it is not: one is simpler (and avoids disadvantages of evenness) and zero is simpler still. What could be simpler than no values. No doubt it will be justly claimed that, with fewer than two values, *adequacy* is lost, adequacy in characterising negation in particular. But if adequacy is now required, then it can be cogently argued that two values does not deliver it. Two values does not permit a satisfactory negation duly allowing for incompleteness and inconsistency; for that it is advantageous to proceed directly to four values (space-time dimensionality). Two values certainly do not permit a satisfactory implication; again four does better, at least within a limited setting (of 'first-degree implications'). Four values are needed furthermore, so that machines can be in inconsistent and incomplete states, not merely transit through them between states.

It is a present working hypothesis that four values (or the effect of four values) in combination semantically with situations (or worlds), i.e., an effective four-valued worlds semantics, *is enough* for adequacy.[131] The hypothesis can be proved, after a fashion, by showing that all discourse however postulationally governed, can be fitted into such a framework. But the proof is only 'after a fashion' because it cannot guarantee that the representation is pre-analytically adequate. Also the hypothesis can be argued for, in particular by explaining how it does not leave out what may seem to be left out. For example, major pressure for advancing to infinitely many values derives from matters concerning vagueness and fuzziness. However such values admit of simulation through many finite values plus worlds.

---

[131] No doubt though both significance and context have also to be taken into due account, as can be done. As regards context, see research on logic and accounting.

Moreover functors concerned such as "It is indeterminate that" work like modal functors, which are better evaluated through finite values over worlds then by infinitely many values (which leads to bizarre determinacies).[132]

What is the point of all this further ado, so deepening the relevant program? Briefly, there are many deficiencies, anomalies and puzzles in the standard classical program for AI, that can be surmounted. Furthermore, the relevant program affords a way of explaining intelligence seemingly closer to natural intelligence, of offering a less artificial artificial intelligence than the standard program. Some of this we have already seen in the ideal agents with perfect capacities, omniscience, and total information, required for the standard program. More we have glimpsed through frame problems. Greater generality can also be gained.

9.1.7 On broad claims of logic, pararelevant logic particularly, in issues concerning intelligence, natural and artificial

There is a sense in which *all* of what concerns intelligence theoretically falls within broadly logical reaches. For it is all a matter of processes of certain sorts, treatable in process theory, the general theory of processes. (Process itself *broadens* and *comprehends* argument, inference, inferential steps, and the like, all of which are types of process.) But there is another more familiar sense of logic, in which intelligence, especially natural intelligence, has much less to do with logic than has been supposed. Moreover, as we have begun to see, classical logic, its variants and its adjuncts, are quite inadequate to *ordinary* intelligence. Many *philosophical issues* are quickly stirred up: what makes for intelligence? where does rationality, reasoning, etc., fit into scheme? Above all, what has *logic* got to do with it?

Although of course interconnected, there is a surprising extent of independence between these different items, rationality, intelligence, argument and so on, that CS and AI usually try to combine, and certainly confuse. With *rationality*, for instance, the fit is only loose. Rationality is a matter of certain sorts of systematic procedures, routines, regularly applied? Some intelligence is necessary (to apply procedures?), but far from sufficient. An intelligent item may be significantly irrational (as with psychotics). Accordingly, there are outstanding issues. There are several items and categories in the generous intelligence basket that do not fit together as neatly as has been supposed, such as the fit of reasoning with operations in modern formal systems.

---

132 *See Chapter Sixteen for more detail. [Editors]*

Intelligence manifests itself in certain processing (or manipulation) of information to given ends, a central one of which is *problem solving*, others are drawing inferences and making decisions. Only certain sorts of data processing by subjects (individuals or groups) exhibits intelligence. A super-camera, for example, runs through a sequence of sophisticated processing steps, but it does not display, or get assigned, intelligence; it exhibits no appropriate problem solving capacity. In problem solving intelligence is typically displayed in action, whence its exaggerated place in psychology textbooks on intelligence. In each case problem solving takes the form of a sequence of processes leading towards a proposed solutions.[133] Not any old sequence will serve, but only those governed by adequacy. For instance, steps need to be able to be suitable *varied*, and to be *reasonable* under retrojustification. But, they are not governed by consistency and maximality requirements, as the classical paradigm presumes.

It is in part because of distortion by these sorts of assumptions, assumptions such as that rational agents are consistent and even consistent maximizers, that we are still far from a satisfactory account of natural reasoning and intelligence and of empirical logic, as practiced for instance by intelligent humans. (Examples of such distortion are illustrated by preference logic.) The situation is even worse as regards intelligent animals lacking advanced symbolic languages, because of sustained attempts to deprive them of the conceptual apparatus in terms of which some sort of combinatorial logical structure could be developed.

There are, without doubt, difficulties in setting down what any of these critical things—intelligence, rationality, reason and so on—involve, difficulties compounded by essentialism, reductionisms (such as behaviourism), and other quagmires. One reaction has been evasion behaviour. For example, a 'black box' picture of intelligence is taken; sloganised in the rather too behavioural: 'intelligence is as intelligence does'. Yet, when combined with other inputs concerning intelligence as comprising capacities and abilities to engage in intelligent activities and practices—what intelligence does—that comes almost close to a satisfactory account of intelligence as most of us are likely to encounter (there are certainly worse about). Refined somewhat, it consists of these parts: Intelligence is an attribute, that attribute exhibited by intelligent subjects. In turn, a subject is intelligent if it has a capacity to undertake and ability to perform enough of a certain range of activities and practices, namely solving or reducing problems, drawing inferences, comprehending directives, making decisions,

---

133 A fuller account would spell out *problem*, *solution*, etc., for elements of which see TM.

recognizing items, and so on. The list is open-ended, which does not matter, but closing it would induce a certain circularity. The list should not be too demanding (for example, including ability to solve complex mathematical problems, or capacity for advanced linguistic behaviour), else many subjects ordinarily accounted intelligent are inadmissibly excluded—including rather too many humans. However, without further detailing of what gets into the list (and preferably what does not), this sort of account remains inadequate, apparently like all others so far.[134]

Among defective accounts of intelligence is that Hume is said to have presupposed as unproblematic, namely a 'capacity for appropriate and flexible responses to a variety of environmental changes or events'.[135] It seems that such unintelligent items as computerized weathercocks would satisfy this account, as would flexible 'low-level' lifeforms. But high redefinitions appear more abundant than such low accounts. Among the excessive definitions is that of the *Concise Oxford Dictionary* which offers the following: having or showing (us u., a high degree of) understanding. Unfortunately for the definition, a subject can be intelligent without exhibiting much in the way of understanding (for example, it regularly manages to solve problems by empirical and heuristic methods).

Under what they call an empirical approach (resembling the list of capacities and abilities approach above)—as opposed to inadequate behavioural approaches—McCarthy and Hayes contend that a system is intelligent if it has a model of the world that is adequate to its goals. This definition is said to have two parts:

1. epistemological, which 'requires that the world be represented in such a way that solutions to problems can be shown to follow from the facts stored therein'.

2. heuristic: it is 'the mechanism that uses the representation of knowledge to draw inferences, make decisions and solve problems'.[136]

Appealing as features of this account are, it too is both underspecified (in list respects) and too demanding. Even very highly intelligent creatures often do not have a model of the world, still less one that is adequate to their goals. Lesser ones mostly do not (I do not). Moreover the account makes

---

134 Expanded in an obvious way it does guarantee a basis for the following still contentious claim: '... it is common place knowledge that computers behave intelligently' (Graham p.82) *[reference unknown—Editors]*. The basis given is that they can win chess contests, perform complex calculations, guide missiles, ...

135 Thus Graham, p.65. *[Reference unknown—Editors.]*

136 Quotes from Jackson's report upon and apparent endorsement of McCarthy and Hayes, p.65. *[Reference unknown—Editors.]*

philosophical presuppositions (for example, as to 'the world') which non-realists do not accept (I do not).

We are already becoming enveloped in the rich mix of notions that enter into cognitive sciences, that first desirable enlargement of AI:[137] not just intelligence, artificial *and* natural, between which there is no sharp separation, but also rationality, reasoning, logic, etc. Logic and also (also differently) rationality have *lesser* roles than has often between assumed. Intelligence can be exhibited in processing that appears irrational, for example where such practices as divining, magic or whatever enter essentially, where holistic as opposed to analytic methods are used, where heavy use is made of chance (or analogy or whatever), and so on. No doubt some such processing can be encompassed by broadening *logic*, away from its modern textbook usage back towards traditional expositions, but even so a range of intelligent action and practice appears to escape capture. Many philosophical issues concerning intelligence remain open here. As it happens, intelligence drops out in much of what now passes for *AI*, which is concerned with 'smart' (i.e., computer involved) machine processing: *intelligent* here suffers the fate of *smart*.

Observe that AI comprises but a very proper part of mental science. Cognitive science comprehends more, but falls significantly short of all that is involved in mental science. Serious defects become apparent when attempts are made to account for what involves emotional phenomena, for example consciousness.

9.1.8 Further relevant ado, including meeting criticisms

While pararelevant systematization may enjoy great theoretical advantages over classical, above all in shedding noxious bundles of paradoxes and puzzles, it still has—as well as meeting criticisms, such as that it has left out crucial things—to show its capabilities on the computing floor, in the laboratory, workshop and factory. Otherwise, like other esoteric theories, it will be little or never used; at best like better accredited general relativity theory it will be used but rarely when classical systematization breaks down badly. As the exalted comparison suggests and previous argument brings out, one regular objection, that pararelevant systematisation is too hard, would carry little weight. More telling are the objections that there is little implementation, in particular little or no effective pararelevant theorem

---

[137] A further separately desirable englargement expands cognitive science into mental science.

proving on computing machinery. While not entirely correct, it is close enough to true to be damaging. Implementation is needed. But there is no need to fear results; rather implementations should help confirm various virtues of relevant applications.

What has been sketchily considered is far from exhausting relevant applications. For example, relevant dynamic and process logics have applications to program verification if classical ones do. And so on. Virtually *whatever* applications modal-type logics, tense logics, epistemic logics have, relevant logics can imitate them and often can *improve* upon them. Showing as much in requisite detail is one part of developing, sharpening and confirming the strong relevant program. In sum, as in many places intellectual, there is much to be done.

Indeed more. For relevant logics and their enlargements are only one part of the larger logical story concerning intelligence and cognition and, more comprehensively mind. Another part of that story involves item theory, i.e., generalised object theory. But that is a story for another occasion.

9.1.9 Appendix: other issues

Apart for several issues on which there needs to be more information (for example, theory change and data base updating) there are several pertinent topics that have scarcely been addressed at all. These include such diverse matters as logical programming and program verification, semantic and inheritance and other networks, automata theory.

Part III

Philosophy of Science, Probability and Non-Deductive Logic

10 Confirmation Without Paradoxes

Co-authored with Robert Nola, this essay originally appeared as Sylvan and Nola 1991, in an edited festschrift in honour of Paul Weingartner. [Editors]

Our modest aim is to show how, within a suitable relevant logical framework, *all* the paradoxes of confirmation vanish—not merely those Hempel and Carnap adduced, but those uncovered by Good, Goodman, Glymour, and others. But, naturally, even removing these paradoxes wholesale still leaves much logical work to be done, work we shall not try to complete here. What we shall indicate however, in the final section, is convergent work undertaken independently by Paul Weingartner and collaborators, which resolves paradoxes in an analogous fashion.

Let us adopt the setting for a logic of confirmation made standard by Hempel, with one important exception. That exception will make all the difference. The logic will contain statemental connectives, & ("and"), $\vee$ ("or"), $\sim$ ("not"), and a conditional > ("if ... then ..."), and also quantifiers, $\forall$ (universal, "for every") and $\exists$ (particular, "for some"), and as well two further connectives or statemental relations: $\rightarrowtail$ of confirmation (read "that ... confirms that ...") and $\rightarrow$ of entailment.

It can be left open for the present whether $\rightarrowtail$ permits of nesting, that is, has a higher degree logic; it will be supposed that $\rightarrow$ does (though this is inessential, and could easily be revoked and first degree system FD adopted).

The assumption we will *not* make is that, taken for granted by Hempel and most who followed, drawn from the dominant logical paradigm, that of (narrow) classicalness: that the conditional is the material conditional, and entailment some sort of strict implication (or a metalinguistic equivalent). As a result, confirmation too, which is intertwined logically with conditionality and entailment, will not be (narrowly) classical. Nor will we assume, along with the liberalised classical paradigm (i.e., essentially the modal extension of narrow classicalness) that the conditional is a modal functor, evaluated over possible worlds. For while the liberalisation would enable significant improvements in confirmation theory (as modal analogues of the theory which follows will show), it would not take us out of damaging classical precincts or to a paradox-free theory.

The *paradoxes of confirmation are, in the main, of a piece with the paradoxes of implication and conditionality, and removed along with them.* The uniform strategy (discussed in UU) consists in transfer to a relevant logical framework.[138] The trouble with classical theories, whether liberalised or not, is their classicality, that they are intrinsically paradoxical and *much* too strong. Adopting instead a relevant strategy, we shall suppose that the logic of connectives &, $\vee$, $\sim$, $\rightarrow$ is the logic of first degree entailment, FD, or else a conservative extension of that, that is, that it is a relevant logic (for which generally see RLR I).

Similarly we suppose that addition of the quantifiers gives a relevant quantified logic, and that the conditional is a relevant conditional; so the total logic against which confirmation is evaluated is a quantified relevant logic with conditionality. Observe that adherents of the classical paradigm can hardly object to what is included, since it is included in their story also (whether liberalised or not). What they can object to is what is excluded: but that is less than convincing as it includes, among other things, a big bundle of paradoxes![139]

To arrive at a relevant logic and accompanying semantics of confirmation, we shall try to proceed from what is more secure (such as intersubstitutivity principles) to what is less (such as converse consequence conditions discussed by Hempel). In brief, we build up from weakness (as was the main Taoist-Cartesian methodological stratagem of RLR I) rather than adopting the more difficult, if theoremwise intertwining, classical procedure of cutting down from damaging strength. The goal is, of course, a relevant logical theory free from paradoxes. Nonetheless, a subsidiary aim is to achieve a substantially *logical* theory, to accomplish as much as we reasonably can with logic alone, without undue appeal to outside distinctions or information (which may however enter in applications): in particular, without undue appeal to types of samples, or classes of conditionals or hypotheses, or 'theoretically barren contexts' with no illicit background information, or 'total available information'.

---

138 There is an objection, we can afford to ignore for the present, that this sort of transformation of framework makes philosophy, and removal of paradox, too easy—as if this were a defect!

139 The bundle of paradoxes also serve to render unpersuasive Goodman's 1965 soothing discourse, in his "dissolution of the old problem" of induction (pp.63-6), concerning the comparative success of the recent codification of deductive logic, as to opposed to confirmatory logic.

10.1 How to fail prominent paradoxes of (instantial) confirmation, without really trying logically

A striking feature of more prominent paradoxes of confirmation is that they depend hardly at all on the specific shape of 'the' logic of confirmation; it is almost enough that the logic supply substitutivity of coimplication principles. But these paradoxes do rely heavily on the formulation and logical principles governing traditional A -propositions, propositions of the form "all A's are B's" or "every A is a B", or (making contemporary assumptions as to their logical analysis) universal conditionals, i.e., propositions of the form "whatever x, if Ax then Bx".

Consider, to begin, the confirmational paradoxes, linked to the Nicod criterion,[140] that Hempel 1965, chapter 1 states. It is well-known in relevance circles that the most outrageous of these paradoxes are automatically removed upon replacement of the material conditional, ⊃, by a satisfactory implication or conditional. Thus in the formulation of universal 'lawlike' statements such as Hempel's working example "all ravens are black" (*ARB* in traditional notation), which he symbolises

S_1: $\forall x(Raven(x) \supset Black(x))$,

simply replace the material-conditional ⊃ by a decent conditional, > say, giving

S_1': $\forall x(Raven(x) > Black(x))$,

140 According to this criterion (for example, as presented by Hempel 1965, pp.10-11), instances of A's which are B's (AB's) confirm "all A's are B's", while instances of A's that are not B's disconfirm "all A's are B 's". Under standard assumptions the first part of Nicod's criterion is tantamount to what is variously called instantial confirmation, simple induction, the straight rule. An emerging goal in the ensuing text is the formalisation of a logic of confirmation that can work satisfactorily with instantial confirmation.

There is however a significant ambiguity in the literature as to the strength of the Nicod criterion; noticeably, as to whether instances are necessary as well as sufficient for confirmation. Carnap 1962, for instance, burdens Nicod's criterion with such necessity, presenting the criterion thus: 'the law "all swans are white" is confirmed by white swans and only by these' (p.468). But, in fact, such a presentation considerably exceeds what Nicod stated, specifically: 'It is *conceivable* that we have here [with confirmation and invalidation] the only two *direct* modes in which a *fact* can influence the probability of a law ...' (see Carnap 1962, p.469; italics added). Because there may be indirect modes, such as through argument, necessity fails in general. Here we consider only the basic weak form of the Nicod criterion, which is very weak. As will become plain, we do not want to invest too much in an *unsupplemented* Nicod criterion.

which reads: "for every x, if x is a raven then x is black".[141]

Then the 'remarkable consequence' to which Hempel 1965, chapter 1 draws attention (p.12, repeated p.15, n.23): that 'every universal conditional can be stated in a form for which there cannot possibly exist any confirming instances' is automatically voided. For it depends, in decent conditional terms, on the blatantly paradoxical transformation of $R > B$ into $(R \& {\sim}B) > (R \& {\sim}R)$, and so in effect takes it for granted that $R > B$ iff ${\sim}(R \& {\sim}B)$, that the conditional is material. Similar remarks apply to Hempel's contention that 'we have therefore to consider as confirming for S_1 any object which is either no raven or also black (in other words: any object which is no raven at all, or a black raven)', which helps him to his paradoxical conclusion 'that any nonraven represents confirming evidence for the hypothesis that all ravens are black' (p.15). For the contention turns on the equivalence of S_1 with

$$S_3: \quad \forall x(Raven(x) \vee {\sim}Raven(x))) \supset ({\sim}Raven(x) \vee Black(x)));$$

and thus, in decent conditional terms, it relies in the paradoxical transformation of $R > B$ into $(R \vee {\sim}R) > ({\sim}R \vee B)$.

What has not been so widely noticed, in relevance circles or without, is that *all* the paradoxes of confirmation Hempel adduces depend upon how universal conditionals are formalized. One reason for this blind spot within relevant theory has been that relevance logic is not equipped with a satisfactory theory of conditionals, and that universal conditionals have been symbolised using an implication $\rightarrow$ (usually that of relevance logic R), which ensures contraposition. Thus, just as S_1 is logically equivalent to

$$S_2: \quad \forall x({\sim}Black(x) \supset {\sim}Raven(x)),$$

so $S_1(\rightarrow)$, i.e., $\forall x(Raven(x) \rightarrow Black(x))$, is logically equivalent to (i.e., here coentails) $S_2(\rightarrow)$, i.e., $\forall x({\sim}Black(x) \rightarrow {\sim}Raven(x))$.

Thence all the jibes and jokes, drawn out of very generous applications of the Nicod criterion—about how objects which are nonblack nonravens, such as (it is said) red pencils, green leaves and yellow cows, one and all confirm "all ravens are black"—appear to transfer intact to relevance logic.

---

141 For the present connective $>$ is an undifferentiated conditional, such as is studied formally in Chellas 1975 and in PLI. Only later do we begin to address the issue of whether it should be restricted to some subclass of conditionals, such as lawlike, subjunctive, etc. But that additional information is, as we have found from discussion, extremely important in trying to budge hard-line empiricists from their single-extensional-world fixation as regards evidence.

Fortunately for relevant theory, it is not a difficult feat to argue that the formalisation $S_1(\rightarrow)$ of *ARB* using a contraposible law-like implication is a *mis*formalisaton. To begin, universal generalisations like *ARB*, even if universally true, do not represent *law-like* connections. Colouration in birds is particularly liable to variation, not only between species as with swans, but within them. Unusually coloured members of many species occur in the wild, while in captivity such features such as albinism are more likely to survive and be noticed. The upshot is that colour is not connected in a universal law-like way to species; the connection is of a less rigorous kind. As it happens, some species popularly called ravens are not black but have grey or white in the plumage, for example *Corvus spendens* from the Indian subcontinent. Perhaps Hempel intended to refer to the large Northern Raven, *Corvus corax*, which is almost always naturally glossy black. Since the connection may well not be lawlike (even with "raven" duly restricted), a better modern representation of *ARB* is by way of a (qualified) universal conditional. But why a material, and contraposible, conditional?

Should the universal A-form, regularly equated with a universal conditional, be differently represented through a conditional connective which does not sustain contraposition, the nastiest paradox of classificatory confirmation is stopped in its tracks. In fact a range of connectives which are candidates for expressing universal A-statements, do not guarantee full contraposition: not only $\rightarrow$ connectives of weak relevant logics such as system B, and connectives such as *such that* and conditional assertion, but more important, conditionals themselves. Since the recent development of possible world semantics for conditionals, which fortuitously flunked contraposition, contraposition has become under increasing suspicion. Acclaimed counterexamples to contraposition have begun to appear; supporters have even begun to speak of 'the fallacy of contraposition'. Indeed given the standard Stalnaker-Thomason-Lewis *et al.* type of theory, contraposition *has* to be rejected. Were contraposition admitted the whole new modal structure would collapse; the conditional would fall back into the material conditional, conditional logic into classical logic.[142] With hardening American opposition to contraposition of conditionals, it is a trifle surprising that the new theory was not applied to confirmation theory. For the remaining Ravens paradox is then easily broken (perhaps too easily broken some will again say: philosophical problems must be conserved!). For the argument to

142 Briefly, contraposition enables derivation of a (forbidden) prefixing rule $B \rightarrow C \,/\, C > D \rightarrow B > D$ from a corresponding (admissible) suffixing rule. But by prefixing $B \,\&\, C \rightarrow C$, the dreaded augmentation $(C > D) \rightarrow (B \,\&\, C > D)$ results, which leads to collapse into classical logic (as several expositions show).

the paradox depends upon the logical equivalence of $R > B$ with $\sim B > \sim R$, and their instantial confirmation thus being the same.

A minor hitch is that there are evident expository weaknesses in the modal case against contraposition. One serious weakness of modal conditional theory is that no particularly striking counterexamples to the ancient received principle of contraposition are developed. Indeed often no counterexamples are offered at all (Nute in a whole book), or but formal countermodels are presented. While formal countermodels are fine (it is quite evident that negating propositions may detrimentally affect their similarity features and the distances of their worlds from others), they are, when applied to natural language examples, only as good as the more intuitive data to which they must appeal. The sorts of counterexamples that are sometimes preferred, such as 'weather counterexamples', suffer from evident limitations. Examples like "if it should rain it will not rain heavily", while they certainly do not ensure contraposition (to "if it will rain heavily it should not rain"), are open to the objection that the "if" involved is a condensed "even if"; but, the objection continues, the logic of *even if*, which is different, is not what is at issue. Examples like "if it rains, we'll get wet / we'll have an umbrella", while they do not on the surface contrapose, do admit of twisting or transforming into apparently contraposible forms, such as "if we don't get wet, it won't have rained". Nor do such soft counterexamples transfer what plausibility they have to 'scientific' conditionals like those linking blackness with ravenhood.

In the latter regard, as elsewhere, relevant theory (which can make full use of the soft counterexamples) fares rather better. The reason is that such natural history conditionals as "if that is a raven then that is black" exhibit enthymematic features, amounting, when background is taken up, to "if, *ceteris paribus* (things being normal), that is a raven, then that is black", and thus ruling out cases like "if that is a raven which just flew through an operating spray paint unit, then that is black"! With background so taken up, contraposition involves an application of antilogism, in some such form as:

$$A \,\&\, C \rightarrow B \,/\, {\sim}B \,\&\, C \rightarrow {\sim}A.$$

But antilogism is relevantly invalid (for example, it leads from Simplification directly to irrelevance: $A \,\&\, B \rightarrow B \,/\, {\sim}B \,\&\, B \rightarrow {\sim}A$). Thus the validity of contraposition for tight suppression-free implication does not extend to derivative or enthymematic forms, because the extension depends upon antilogism. For analogous reasons it does not extend to relevant conditionals. Finally, and also bound up with the issue of background information, there are some long-standing counterexamples to contraposition, such as that based on or around the hoary "if the match is scratched, the match will light" (cf. Goodman 1965, pp.5-6; for other examples see RLR I, p.43).

To summarise progress thus far: with a relevant conditional connective, $>$, *all* the paradoxes of confirmation linked by Hempel to the Nicod criterion vanish, the hardest one because contraposition for the underlying conditional, i.e.,

$$(R > B) \rightarrow (\sim B > \sim R),$$

fails. Now Hempel was aware that derivation of the paradoxes rested upon the controversial assumption that the universal material conditional adequately expressed relevant universal A-forms. 'Perhaps', he says, 'this customary mode of presentation has to be modified; and perhaps such a modification would automatically remove the paradoxes of confirmation?' (p.16). But then he completely dodges his question—to which we are simply responding affirmatively.[143] The *only* variations on the $S_1(\supset)$ expression of "all ravens are black" that he looks at are these: firstly a rather unconvincing alteration which merely adds existential loading, not varying the main connective; and secondly, more plausibly, a field restriction. Yet, universal material expressions of A-forms were heavily challenged from their very inception; it was, and still is, widely maintained that "all ravens are black" expresses something very different from "everything is either not a raven or is black", something about ravens and not about everything, at a minimum they were said to be about different things. No doubt the field restriction suggestion was intended to head off such criticism (as recent appeals to assertability are supposed to head off criticism of the material conditional as capturing conditionality, at least in respect of truth conditions). But the move deservedly failed, as more recent rejections (by Bhaskar 1978, and others) of contraposition as the defective step in the generation of Hempel's main paradox serve to show.[144]

---

[143] As did Goddard 1977 before us, whose route we are so far renegotiating. Goddard also supplies significant additional information, needed in making reformulation empirically convincing, which we shall not repeat.

[144] Thus, as Bhaskar 1978, (p.224) remarks of this paradox, unearthed by Hosiasson-Lindenbaum: 'Hempel's paradox may be resolved quite simply once the significance of his intuition, viz. that propositions about shoes and herrings are irrelevant to the truth of propositions about ravens, is grasped. If laws are statements about things and there must be some reason other than instances for accepting them, then Hempel's paradox may be resolved as follows: If "all ravens are black" is law-like there must be a reason, located in the nature of ravens (not in the nature of black), why ravens are black. "All ravens are black" is a truth about ravens, not about colour. Hence the contrapositive "all non-black-things are non-ravens" has no bearing on it. The logical subject of a law of nature is a (natural kind of) thing. Hence there

While such a relevance shift is, then, a sure-fire way of knocking out all the prominent paradoxes of confirmation, in a way which fits elegantly into a uniform relevant elimination-of-paradoxes program (see again UU), it may appear open to criticism as unnecessarily deviant from the classical mainstream. For given the conditionalizing strategy adopted, most of the prominent paradoxes actually considered can also be removed within a modal theory of conditionals. Such criticism is however readily seen to lack merit. The reason is that analogues of damaging paradoxes are easily designed within modal conditional theory, by adapting paradoxes of strict implication. Merely most conspicuous are those that apply the modal (dis-) connection, $B \,/\, A > B$, not of course a principle of relevant conditional theory. For then anything that is black, such as a black dot or black boat, modally confirms "all ravens are black". Moreover there are other groups of paradoxes, where a relevance approach seems mandatory; most notably Glymour's 1980 paradox-driven criticisms of the hypothetico-deductive (H-D) methodology, as for instance hopeless. As Waters 1987 in effect explains, while classically-based logics, such as modal theories of conditionals, do not remove Glymour's paradoxes, relevant logics can (since his paradoxes turn on such propositions as that $H > T$ follows from T; see Waters 1987, p.461) But before we attempt, varying Waters 1987, to salvage H-D methodology, there are further weaknesses of instantial confirmation we should defuse.

---

is a logical asymmetry built into its structure, reflecting the site of the mechanism designated, in virtue of which its terms are not equivalent and contraposition is prohibited.'

The failure of contraposition is indeed an almost direct inference even from fashionable Bayesianism. Consider the work of some who proceed to immerse themselves in the problematic intricacies and complexities (bound up with cardinality speculations) of Bayesian analyses of the confirmational paradoxes such as the Raven. A striking example is Chihara 1981 who first sets out as follows:

'the salient intuitions that give rise to this puzzle:

(1) Why do we judge that our observations of white shoes and the like that seem to constitute *positive instances of* the contrapositive of h (i.e., "contrapositive instances" of h) fail to confirm h, when the observations of black ravens do seem to confirm h?

(2) Given that many philosophers regard the observation of white shoes as confirming the contrapositive of h, how can this judgement be reconciled with the intuitive opinion that such judgements do not confirm the logically equivalent h?' Chihara 1981, (p.440).

Evidently *that* notion of logical equivalence (strict equivalence of certain universal material conditionals in standard and Chihara 1981 representation) does not preserve confirmation; nor similarly does taking contrapositives.

10.2 Variations upon the prominent paradoxes, and the weakness of instantial confirmation

To deal with some of the many variants upon the more prominent confirmation paradoxes, further work has to be done. That work begins by reforging some neglected distinctions; it is made necessary in part by the weakness of instantial confirmation. Indeed there is a unifying theme of weakness running through the present exercise, weakness not being confined to use of very economical logics by recent overpowered standards. (Coincidentally, it is by keeping logical matters appropriately economical that we are able to obtain a spill of confirmation logics, where others have despaired of producing any.) A remarkable feature also of instantial confirmation—where, for instance, *one* black thing confirms that all things are black—is its extraordinary weakness. For presentation of a single black thing does not give much support to a universal generalisation as to things (nor will it increase probability more than minutely, or at all in an infinite universe). Certainly it does not increase *credibility* of the generalisation, in part because we have much contrary evidence available. One of the many confusions to get past in reaching a satisfactory logical theory of confirmation is that between confirmation and increase of credibility (likewise guarantee or increase of plausibility, 'subjective' probability, etc.). The effect of the equation Goodman 1965 makes, of confirmation with credibility increase (made implicitly from p.72 on), is instructive. For it is in this way that he proceeds to generate both a demand for a distinction between accidental generalisations and lawlike ones, and a requirement for total evidence. We aim to avoid such requirements as long as we can; for they tend to militate against independent *logics* of confirmation. Issues as to what is credible, plausible, to be expected, do invoke other evidence, running beyond any mere instantial confirmation. Mere confirmation by instances (for example, of some black things) does not 'create expectations' (for example, that everything is black) or assign credibility (*pace* Goodman 1965, p.82), because we have too much other information which we bring to bear in assessing credibility and generating expectations.

Excessively strong notions of classificatory confirmation have become entrenched in the literature, notions which have been routinely explicated by way of classical probability measures. Neither bodes well for instantial confirmation. To get past these difficulties, let us go back to basics, to preanalytic and dictionary senses of *confirm* (of *confirm* in the usual classificatory, qualitative form). It is soon evident, from a perusal of dictionaries, that there are two independent dimensions of variation in the notion. One direction in variation is quantised: between a threshold form (*firm*) and as growth form (*firmer*); and the other direction is in strength: from contributing to firmness all the way to establishing. Thus, expanding on the division of classificatory concepts Carnap 1962 eventually reached (in his second edition pp.xv-xvii), we distinguish:

Level form: A contributes to making B *firm* (A lends some support to B).

Increase form: A contributes to making B *firmer* (A *lends further* support to B).

In each case contribution may be *to some extent*, from a certain minimal level up, perhaps up to making firm itself or fully firm.

Different logics go with the different forms. Whereas the increase form excludes reflexive confirmation, $A \rightarrowtail A$, and so flunks the entailment condition in Hempel 1965, chapter 1 (that where A entails B, A confirms B), the level form can retain both reflexive confirmation and an entailment condition. Evidently, too, the level form is the appropriate form for instantial confirmation; for a further instance may contribute little or nothing to make a universal hypothesis firm*er*. (By contrast, probabilistic relevance, which we shall want to distance from confirmation, goes rather with rendering firmer.) Evidently too, as before, an instance or further instance A' may contribute comparatively little, certainly not enough to make a hypothesis B' firm *on its own*; i.e., A' may not make B' firm though it contributes, perhaps minimally, to making B' firm.

It is because of the partial character of basic confirmation, *contributing* to making firm rather than doing it on its own, that transitivity is rightly in some doubt for confirmation. For instance, 'although the pivot supports the see-saw, and the see-saw the child at one end, it is not in general true that the pivot supports the child' (Hesse 1974, p.142). In particular, much as with *sorites* paradoxes, a long chain of transitivities may remove requisite contributions to making firm.

It ought to be evident that firmness differs from probability, simply from dictionary senses of *firm*—fixed, stable, steady, constant, enduring, secure, etc. "Probable" is not in standard lists, for obvious reasons. For whatever firmness amounts to, it may diverge from probability; what is or was improbable may nonetheless be firm, and what is probable may not be firm. It is accordingly remarkable how many publications proceed to assert or assume, often at the outset, that firmness, as regards hypothesis or propositions, is probability.[145] Thus for instance, Rosenkrantz 1980,

145 Though remarkable to outsiders, the phenomenon is not so hard to explain, for example

parenthetically, in the first dubious statement in a lengthy article of similar character: 'That general laws are confirmed (or made probable) by their positive cases is, perhaps, the clearest formulation of the traditional principle of induction' (p.78; by contrast, he begins his book by saying 'we ordinarily think of induction as "learning from experience"', p.3). Similarly Swinburne 1973 starts a whole book from the unconfirmed assumption that confirmation is nothing but making probable. His first statement, soon repeated in variant forms, runs: 'Confirmation Theory is the theory of when and how much different evidence renders different hypotheses probable' (p.vi, similarly p.1). Similarly Salmon 1975 takes it as axiomatic that all forms of confirmation reduce immediately to relations of probability, and that the normal form is making more probable. 'When we say that a test confirmed a hypothesis, we would normally be taken to mean that the result was positively relevant to the hypothesis' (p.13). Such is the effect of living too long in a narrow community; a wider sampling would quickly shatter such armchair confidence. Insofar as an attempt is made to adduce evidence for, or argue for, a probabilistic analysis of confirmation, typically as positive relevance, it depends upon sliding from confirmation to certainty increase or to expectation (thus for example Rosenkranz 1982, p.80).

We are now placed to consider the sorts of variations upon the confirmational paradoxes 'afficiandos of Bayesian statistics' are fond of urging against instantial confirmation, examples which according to van Fraassen 1983, 'show the idea's real bankruptcy', that it was 'absurd to take instance confirmation as the key to anything. It would seem rather that this was a mistaken paradigm, a false ideal, that misguided philosophy of science for decades' (pp.322-323). The first often repeated or varied example, supposed to underwrite such extravagant claims, begins from

> a story of a storm at sea: a ship, lost in a storm during the night, finds itself near an unknown island at dawn. The captain, consulting his charts and reference books, concludes that this must be either island X or island Y, which have a very similar physical appearance. But there is a difference in the bird populations. On island X, 5% of the birds are black crows, and 95% are non-crows, while on island Y, 95% are

through the following: the reductive drive, the epicycling elegance of Bayesianism in the face of paradoxes, and its number-crunching appeal, the sheep effect, etc. Furthermore, 'probability is the best explored measure of the confirmation relation' (Hesse 1974, p.144). A major mistake in recent approaches to con*firm*ation is indeed the headlong rush into classical probability analyses, thereby entrenching paradox. An obvious alternative is to try to clarify pre-analytic notions first, only *later* to see whether reduction may work.

> black crows, 1% are white crows, and 4% are non-crows. The bosun goes out and returns shortly waving a bird: a black crow. The captain, who took a philosophy of science course with Hempel at Princeton, but was not perhaps one of the brightest students, announces solemnly that they have now increased evidential support for the hypothesis that here all crows are black, hence that they are on island X. The crew suspects they must be on island Y (p.322).

We suspect that the captain should have failed an instantial confirmation course. While a black crow here confirms (hardly much *increases* evidential support) that all crows here are black, nothing justifies the "hence" or the leap to the detached claim that they are on island X. The leap is probably made on the faulty assumption—made by other authors who present versions or variations of this example (for example, Good to whom the original is due, Chihara 1981, p.431)—that positive instances increase confidence in a hypothesis, bestow credibility or plausibility; *further* features that may encourage a leap. But those further features are here undermined by the background information concerning birds on the islands.[146] What the example reveals is not the bankruptcy of instantial confirmation, but the *gaps* between mere confirmation and such (different) matters as increased confidence, credibility and plausibility, gaps afficiandos of Bayesian statistics commonly ignore.

There are further features of this example which are more clearly exposed in van Fraassen's 1983 second example (adopted from Rosenkrantz):

> [...] we go to a philosophy of science conference in a hotel, where a number of philosophers check their hats with the bell boy. This boy is rather confused, and eventually returns the hats at random. An onlooker frames the hypothesis that every man will receive someone else's hat, and here is some evidence he obtains:
>
> E_1, The first philosopher receives the second one's hat.

[146] From our perspective, Hempel 1965 made an unsatisfactory response to these sorts of examples, in claiming that they make illicit use of background information (on the difficulties of such a response see, for example Chihara 1981, pp.431-2; Rosenkrantz 1977, pp.39-40). Of course, Hempel had previously made a similar unsatisfactory response to the original paradoxes. We do not suppose, with Hempel, that the confirmational paradoxes are a sort of *illusion*; rather they are a product of classical reasoning.

> E_2, The second philosopher receives the first one's hat.
>
> Thus the first philosopher, and a moment later, the second one, provide positive instances of his hypothesis. Do they add evidential support for it? I said 'a number of philosophers', but did not say which number. If the number was *two*, the E_1 and E_2 completely verify the hypothesis. But if it was three, they completely refute it. (p.322)

It appears that van Fraassen was in too much of a hurry here. For where the number is three, 'they', namely E_1 and E_2, do *not* completely refute the onlooker's hypothesis. What refutes that hypothesis is the *further* evidence

> E_3, The third philosopher receives the third one's hat,

together with some accompanying details (which van Fraassen shouldn't have neglected) such as that the third philosopher is a man. But that further evidence does *not* prevent E_1 (or E_2 and accompaniments) from confirming the hypothesis of the onlooker (who may not have been chauvinistic). Certainly, they do not supply confirming evidence which make that hypothesis *credible* in the light of further information, namely E_3; but, short of confusion, that is not to the point (van Fraassen himself proposes to take for granted the distinction between credence and evidence, only a couple of pages earlier, p.320). Furthermore, a set of evidential statements may contain component statements which (separately) both confirm and disconfirm a universal hypothesis. There is nothing particularly unusual even about this (consider for example the set of planets providing evidence for Bode's law).[147]

Even though instantial confirmation remains standing under the assault of such examples, additional weaknesses in it are thereby disclosed, well-known weaknesses. In particular, because of its features—including its formally advantageous mechanical applicability and consequent insensitivity to, and inability to take due account of further features of evidential situations—it does not, on its own, afford a reliable basis for inference. There is no reason, however, why instantial confirmation should not be supplemented by a battery of further apparatus, some of it adopted from statistical inference theory, serving to fulfil requisite further functions. For, of course, the *character* of the instances—their number, variety, etc.—matters. Such further information *is* required, in order to lift contributions to making firm

[147] What we may not have, on classical approaches, is a statement which both confirms and disconfirms or refutes a hypothesis. But this is readily blocked by excluding various aggregation principles, as will become apparent.

up to unenthymematic making firm, genuinely making firm on its own. Nor is there any reason, given its nonexclusivity, why it should not be supplemented by other procedures commonly taken to deliver confirmation, such as hypothetico-deductive methodology.

10.3 Hypothetico-deductivism, scientific confirmation, and the verification principle

The hypothetico-deductive (H-D) method is 'hopeless as a method of confirmation because it cannot maintain the condition of relevance between a hypothesis and the evidence which confirms it' (Waters 1987, reporting Glymour's claim, p.453). The criticism of the method strongly suggests that 'the problem is located in the logic we use to reconstruct the H-D method, not in the method itself. ... if we substitute relevance logic for classical logic, the H-D method can no longer be used to confirm hypotheses with irrelevant evidence' (Waters 1987, p.453). We concur, although we see rectifying H-D methodology as but an early stage in a grander theory of scientific testing and confirmation. We shall track along the route Waters has marked out, though with some variation. One reason for variation is that the relevance logic R which Waters 1987 deploys, while it may have something in its favour as formalising law-like connection through its relevance implication, does not really begin to offer a logic of conditionality or confirmation. It would succumb, if it were supposed to do so, to Hempel's Ravens paradox, by virtue of contraposition, to Carnap's confirmation counterexamples, and so on. A different, weaker and deeper logic than R supplies, is required for confirmation—one that resembles much more a relevant theory of conditionality.

The basic content of the H-D method is well conveyed by figure 10.1 below (after van Fraassen 1983, p.232):

Auxiliary Information or Background Theory (T) + Theory or Hypothesis Under Test (H)

↓

Consequence + Data or Evidence (E)

↙ ↓ ↘

Contradiction of Consequence by Data	Neither	Implication of Consequence by Data
REFUTATION		CONFIRMATION

Figure 10.1 The hypothetico-deductive method

In the *critical* case of confirmation of H by E, it is assumed as necessary at least that $T \,\&\, H$ together entail E but that T alone does not entail E (otherwise H would not be tested). 'Glymour's, [hypo]thesis is that these conditions cannot be augmented to characterise a satisfactory method of confirmation' (Waters 1987, p.454). Perhaps surprisingly to outsiders, these are already difficult requirements for mainstream theories of entailment and deducibility, so extra conditions are immediately imposed—which therewith also cut off significant applications—such as that T is not necessary, and E and H are separately and jointly consistent. By adopting a good, relevant, account of entailment, we can fortunately dispense with these complicated and restrictive requirements.

Even so, several supplementary remarks and details apparently should be added to the core account of H-D confirmation, which is simply the following: H is confirmed by E, relative to T, only if $H \,\&\, T \rightarrow E$, where $T \nrightarrow E$. As a first detail, E is assumed true; if E were to be false, then there could be no confirmation, but possibly refutation, of H. As often pointed out, there is a Duhemian objection to the refutation of H since H can, allegedly, never be sufficiently isolated from the context of its auxiliary hypotheses in order to face falsification by itself. A weaker position would be that for falsification of H, T must be shown independently to be true, or at least not a contributor to the falsity of E. Falsification is not a straightforward matter; nor is confirmation despite claims of an asymmetry between confirmation and falsification. Assuming E is true, it does not follow that H is true; to so infer is to commit the fallacy of affirming the consequent. Is H then confirmed?

It is claimed that two principles of confirmation must be at work if we are to infer that H is confirmed. First, we are said to need what Hempel 1965, chapter 1 (misleadingly) calls an 'Entailment Condition': an implied proposition confirms whatever implies it. In the account of the H-D method, we can infer that E confirms $(H \& T)$ on the basis of the Entailment Condition. Second, to confirm H we are said to need an instance of the Rule of Simplification, viz., $H \& T \rightarrow H$, and what Hempel 1965, chapter 1 calls the 'Consequence Condition' for confirmation, viz., in the case above, given $H \& T \rightarrow H$ and given E confirms $(H \& T)$, then E confirms H. In fact successive application of the Consequence Condition yields that E confirms each conjunct of $(H \& T)$. But the H-D method only needs supplementing with these principles of confirmation *if* it is to yield a workable theory of confirmation in terms of an already assumed confirmational setting. *Otherwise*, it can be taken as an account of (relative) confirmation in its own right, saying when (relative to T) E confirms H.[148]

In fact Glymour 1980 has not established his no-satisfactory-augmentation hypothesis,[149] but he has knocked down various proposed augmentations of the basic requirements (most conspicuously those of Merrill 79) and he has listed 'outstanding problems' for the H-D method, which any respectable account will now have to be able to handle. The first of these problems is that on the account given, confirmation of consequences of the background theory T is excluded. As Waters 1987 remarks, 'this might not seem troublesome; for as long as the [an appropriate] subtheory conjoined with the hypothesis is logically equivalent to the original theory we should be able to use the subtheory for H-D confirmation' (p.454). The trouble Glymour 1980 alleges is that (classically) 'any true evidential consequence of T confirms almost any hypothesis in theory T with respect to some theory'. For (in Glymour's symbolism) if $T \rightarrow e$, $T \rightarrow H$ and not $(H \rightarrow T) \rightarrow e$ then e confirms H with respect to 'subtheory' $H \rightarrow T$. However $H \rightarrow T$ is not an appropriate subtheory, as $(H \rightarrow T) \& H$ is not logically equivalent to, i.e., does not coentail, T. Even modally this is so, as T does not strictly imply $H \rightarrow T$ (with $\rightarrow$ representing strict implication). Consider instead the subtheory $H \supset T$ which T does entail, both modally and relevantly (a case Waters

148 It should not be construed as saying *only when*; we are considering only sufficiency. Otherwise Hempel's 1965 objections to the intimately related 'prediction criterion of confirmation' (p.27 ff) would apply.

149 He remarks, perhaps too quickly, that this 'can only be done case by case' (Glymour 1980, p.323).

should have considered, as Glymour's formulations are classical). Then $(H \supset T) \,\&\, H \nrightarrow T$, by virtue of the relevant failure of disjunctive syllogism. So either way the first problem dissolves.

A second problem, that of trivial confirmation, is similarly resolved. For trivial confirmation T is set as $H \rightarrow E$ or $H \supset E$ where H is any statement not implying $\sim E$ (which is not a consequence of $\sim E$). But where T is $H \rightarrow E$, the argument that T is a true theory depends upon a no longer available paradox of implication, and where T is $H \supset E$, the deduction to E breaks down. The problem is accordingly again defeated by a neat dilemma.[150]

A third problem, that of irrelevant conjunctions, is more vexatious. The problem is that wherever E confirms H it also confirms $H \,\&\, I$ for any irrelevant statement I at all that meets further conditions. Trivialisation of confirmation also ensures, under weak conditions; for if some E confirms any H at all then it confirms $H \,\&\, I$, and so by a consequence condition confirms I, effectively anything. A mere switch to relevant logic cannot evade these problems (as $H \,\&\, I \rightarrow H$), though relevant logical theory provides apparatus—such as that of *use* of premisses in deductions—which can be put to good work in meeting the problem. A version of such a procedure, put into practice through a *partitioning* of information into units, is elaborated to some effect by Waters 1987 (p.457 ff). Partitioning methods are, however, difficult to apply satisfactorily, because of arbitrarinesses in imposing divisions, and difficult to impossible to make work generally (for example Merrill's 1979 attempt to repair H-D, classical attempts to treat reasoning from inconsistency by partitioning each inconsistent corpus). We shall float a different suggestion; thus we dispute Waters' extravagant claim that 'any reconstruction of the H-D method must be carried out in this [his partitioning] type of framework' (p.457). Our suggestion again amounts to building up from weakness, adding such relevant cases of simplication and the like as are genuinely needed and used, rather than cutting down from classical strength, as for instance by partitioning.[151] The suggestion is that relevant *connexive* logic be used in place of relevant logic. As connexive

[150] At the same time, we succeed not only in disposing of Glymour's smart little theorems, which depend essentially on irrelevant implication, but in ridding ourselves of Quine's 1981 trick (pp.26-7) for empirical content, which depends equally on irrelevance (iteration of exportation shuffles, as in the move from $T \wedge O_1 \rightarrow O_2$ to $T \rightarrow (O_1 \rightarrow O_2)$. Both proceed to shunt support from evidence through material conditionals to whole theories, shunting relevance duly curtails. Therewith, with such trickery left behind, we enter new territory in the philosophy of science.

[151] Of course classical strength is just one sort of strength, and a manifestly undesirable form in many respects, such as that of inducing gratuitous puzzles and paradoxes. Strong

logic does not validate $H \,\&\, I \rightarrow H$ for irrelevant conjuncts I, the problem of irrelevant conjunctions is automatically removed. Perhaps the problem is excised too neatly and easily, as some will quickly allege. To try to meet the charge we shall gradually draw together an historical and systematic case for taking the connexive switch seriously in these contexts. For among other things, it amounted (so we claim) to the presupposed logic of science up to the early 20th century.[152]

As a bonus, relevant reformulation also voids other objections to 'the crude hypothetic-deductive account of scientific inference'. According to Salmon 1975,

> One can hardly improve upon Russell's classical example. From the hypothesis, "pigs have wings" in conjunction with the observed initial condition, "pigs are good to eat", we can deduce the consequence, "some winged things are good to eat". Upon observing that such winged creatures as ducks and turkeys are good to eat, we have a hypothetico-deductive confirmation of the hypothesis, "pigs have wings" (pp.34-5).

The classical deduction involved depends on the relevantly invalid principle of Factor,

$$(p \rightarrow q) \rightarrow (p \,\&\, r \rightarrow q \,\&\, r),$$

specifically on the quantification form (in standard relevant symbolism),

$$\forall x(Px \rightarrow Wx) \rightarrow (\exists x(Px \,\&\, Gx) \rightarrow \exists x(Wx \,\&\, Gx)).$$

Thus "some winged things are good to eat" is not a relevant consequence, and the example given is not a relevant counterexample to H-D methods. As it happens, Russell was making a substantially different point (in the passage Salmon refers to) from the work Salmon tries to commandeer the example to do. Russell was making fun of 'the ordinary procedure of induction: "if p, then q; now q is true; therefore p is true". For example, "if pigs have wings, then some winged animals are good to eat, now some winged animals are good to eat, therefore pigs have wings. This form of inference is called "scientific method"' (Russell 1948, p.149). As a counterexample to induction, however, Russell's example is even less promising. Those who contended that induction was some sort of converse

---

relevant and strong connexive logics can provide rival forms of strength through incorporation of classically trivialising principles; see RLR I, section 3.6.

152 *See Chapter Five for further elaboration. [Editors]*

of deduction, always supposed that a *good* deductive relation was required. But Russell's "if ..then" is enthymematic, omitting all information about some pigs being good to eat, and accordingly does not even supply a *strict* implication (implicitly the modally inadmissible principle of exportation has been applied). However there are several other more persuasive examples telling against deduction-conversing induction, to which we shall subsequently turn in considering the intimately related and vexed converse consequence principle of confirmation.

The problems that beset the H-D method, and also converse consequence, bear a striking resemblance to those that beset early attempts to formulate an adequate verification principle. One of Ayer's early attempts at formulating a verification principle (Ayer 1946, pp.11-13) went as follows:

VP. S is meaningful if, for some observation statement O and adequate auxiliary statements A such that $A \not\rightarrow O$, $S \,\&\, A \rightarrow O$.

The resemblance of VP to the H-D schema in the critical case of confirmation is readily apparent.[153] In VP simply replace "meaningful" by "confirmed by E" (ignoring the relativization of confirmation to auxiliary theory A) and replace "observation statement O" by "evidence E". In looking for a verification principle of 'meaning' Ayer was looking for ways in which a statement could be linked to possible reports of experience (or observation). H-D theorists do much the same thing; they too require that the entailed reports of possible experience be true, as observation statements are supposed to be, and in some way confirming of their hypotheses through deductive linkage. Given this close resemblance between the H-D methods and many formulations of the verification principle, problems which come to

153 The pattern of resemblance is richer still, and takes in, as well as VP and H-D, the D-N (deductive-nomological) account of *explanation*. According to the D-N account a law-like sentence H explains a true statement E only if the inference "$H, T \,/\, E$" is deductively sound, with T now a suitably restricted statement of 'antecedent conditions'. Since the inference "$H, T \,/\, E$" is deductively sound iff the entailment $H \,\&\, T \rightarrow E$ obtains, the basic format of D-N explanation is just that of H-D methodology, and open to similar objections and anomalies. Blum's 1989 recent anomaly in the D-N account, for example, amounts to a clever version of the second problem, here that of trivial explanation. It too is automatically resolved relevantly; that is, the anomaly disappears if deduction proper is required and not a classical surrogate. For Blum's anomaly depends squarely upon features of material implication to guarantee his $L' \supset E$ (Blum 1989, p.366) and then on the rule form (rule γ) of disjunctive syllogism to ensure deductive soundness of an inference enlarging enthymematically upon "$L', L' \supset E \,/\, E$", namely "$L, L' \supset E \,/\, E$" with suitable L and E (see Blum 1989, p.366). A very similar, similarly defective, mechanism will shortly recur with Goodman's 'new riddles'.

infect the one are bound to be similar to problems which infect the other (for a survey of several of these defects, see Scheffler 1964, p.150 ff).

Unremarkably then, elementary counterexamples, like those which now plague H-D methods, quickly appeared. For example, if N is any 'nonsense' statement and if for A we take $N \rightarrow O$ (or $N \supset O$), then N is meaningful by VP. Plainly this counterexample is like the (second) problem of trivial confirmation over again, and can be similarly resolved relevantly. Ayer's subsequent revision of VP (Ayer 1946, Introduction) also succumbed to related problems. An argument of Church, applying the statement

C_1. $(\sim O_1 \,\&\, O_2) \vee (O_3 \,\&\, \sim S)$,

has gained much exposure and variation. Church's argument involves repeated application of disjunctive syllogism. For example, the implication

$$O_1 \,\&\, C_1 \rightarrow O_3$$

used in showing that C_1 is indirectly verifiable, applies that relevantly flawed principle, as does the further step, $S \,\&\, C_1 \rightarrow O_2$, showing that an arbitrary statement S is indirectly verifiable. The same flawed principle is used in variations upon Church's argument attacking repairs to repairs to VP, which apply statements such as

C_2. $(S \,\&\, O_1) \vee \sim O_2$, and

C_3. $S \,\&\, (O_1 \vee \sim O_2)$.[154]

The instrumental implication, $O_2 \,\&\, S \,\&\, (O_1 \vee \sim O_2) \rightarrow O_1$, using C_3, just is augumented disjunctive syllogism. However once again the relevant switch does not dislodge problems of irrelevant conjunction. For example, in VP, if S passes the test then so does $(S \,\&\, I)$, for far too many irrelevant and perhaps nonsensical I. Again a connexive switch (or parallel partitioning strategy) averts this problem. It is a serious question whether however to take this further step, this time into the logically unknown. No doubt increasingly many of us have made, or always had, relevant commitments; but connexivism, though ancient, is something else. For it is not as if verificationism is something we are entirely keen to restore or preserve.[155]

---

[154] For these forms see Ullian 1965. Pretty much these forms will reappear with Goodman's 'new riddles' of confirmation and induction.

[155] Interestingly, Wright's recent vigorous defence of on-going attempts to repair a verification principle (a vehicle that is still out on too many philosophical roads despite its unroadworthiness) circuits through—what is *different* from issues of meaning and meaningfulness (despite a certain formal isomorphism)—confirmation and corroboration (see

It is one thing to remove straightforward paradox as undermining instantial confirmation and hypothetico-deductive methodology, more positively determining so much as adequacy conditions for a logic of confirmation is another and more difficult thing.

10.4 Towards basic relevant logic and semantics, geared towards instantial confirmation: conditional confirmation

The cluster of confirmation paradoxes surrounding ravens comprised but the first round of hurdles that Hempel set up in his difficult obstacle course for theories of confirmation. Another connected series of obstacles to be confronted involved, and still involves, the sorting out of adequacy conditions for confirmation logic. It quickly turns out that certain perhaps expected and certainly rather popular conditions lead, in certain often admitted combinations, to local triviality of theory, in the sense that any statement confirms any other. A striking in-house example derives, using transitivity, from the combination of an entailment condition with a converse consequence condition. By the latter A confirms $A \,\&\, B$, given A is a consequence of $A \,\&\, B$; by the former $A \,\&\, B$ confirms B, given $A \,\&\, B$ entails B; whence, by transitivity, A confirms B, for any statements A and B whatsoever. While we have the advantage of operating with relevant technology which can rapidly put a stop to much anything-confirms-anything nonsense, that alone does not give sufficient control. Like Hempel, Goodman and many others before us, we shall have to proceed with due caution. We in fact proceed, as indicated, from comparatively uncontroversial weakness, arriving at at basic system in which all three of transitivity, entailment and converse consequence are optional extras. The problem then is which to add, in what strength.

An initial rock-bottom logic can begin from the first adequacy condition to which Hempel 1965, chapter 1 draws attention, the *equivalence condition*.

Wright 1989, p.621). There is of course a long and entertaining history of unsuccessful attempts to repair the verification principle, perhaps culminating to date in Wright's 1989 revision of an earlier failed proposal. While Wright is still on an old-style cut-down-from-strength trip, he *has* moved inside strict implication with his 'compact entailment'. In general, however, cutting-down is an unlikely approach compared with building up from sure ground (only by rare good fortune, when something is got right, will top-down and bottom-up directions coincide). So far Wright's approach through 'compact entailment' does not guarantee paradox-removal and relevance. For example, where $A \rightarrow B$ is compact, so is $A \rightarrow (B \,\&\, t)$, for any tautology t in new parameters. So therefore is the paradoxical $A \rightarrow t$, by direct verification (not by transitivity, which fails). Thus whatever rescue Wright's sort of approach might hold out for the beleaguered verification principle, it does little for the complex of issues we have investigated, of which the vicissitudes of that principle are but a small part.

That is readily satisfied relevantly where *equivalence* is construed as provable co-implication, i.e., as what systemically represents coentailment or codeducibility. But that is exactly what Hempel intends equivalence to formulate, as his case for the condition (p.13) shows.

'*Equivalence condition*: Whatever confirms (disconfirms) one of two equivalent sentences, also confirms (disconfirms) the other', i.e., in our symbols:

$$A \leftrightarrow B, C \leftrightarrow D \,/\, (A \rightarrowtail C) \leftrightarrow (B \rightarrowtail D).$$

To connect with extant logical theories, of conditionals in particular, this rule is advantageously broken down into two parts:

REA. $A \rightarrow B, B \rightarrow A \,/\, (A \rightarrowtail C) \rightarrow (B \rightarrowtail C)$
(antecedent replacement)

REC. $C \rightarrow D, D \rightarrow C \,/\, (A \rightarrowtail C) \rightarrow (A \rightarrowtail D)$
(consequent replacement).

In making the breakdown certain minimal features of a background entailment logics are assumed; in particular $A \leftrightarrow B =_{df} (A \rightarrow B) \& (B \rightarrow A)$, and simplification and adjunction of connective &. A range or neighbourhood-style semantics is rcadily supplicd for thc logic comprising REA and REC as adjoined to an underlying implicational logic (for the relevant version, see PLI; for an irrelevant version, Chellas 1975, from whom the labels are adapted; Chellas regards the logic as a minimal conditional logic). But the logic and semantics, while acceptable enough, do little to distinguish $\rightarrowtail$ as confirmation; it could be any number of other propositional relations, including even preference or betterness (see PLI). To obtain a more distinctive logic, it is worth looking at strengthening one or other of REA and REC, and further pursuing the analogy with conditionality—already well in the general act—where a similar strategy pays off.

There are two ways the analogy can straightforwardly be taken: first, as a (sort of) *converse* conditionality, the secondly, simply as a sort of *direct* conditionality. The converse approach is suggested by a several considerations; the traditional idea of induction as the converse of deduction, the appeal of converse consequence conditions (discussed, for example, by Hempel), hypothetical-deductive practice, and so on. The assumed correspondence of confirmation with (a sort of) converse conditionality delivers the connection

CC. A confirms B iff B accounts (conditionally) for A, i.e., (for suitable restricted conditionality) iff if B then A; or in symbols, $A \rightarrowtail B$ iff $B > A$.

Then rewriting the schemata for basic conditional logic (of PLI), set within the framework of relevant logic, gives the following basic logic of confirmation. The system $CCL{\rightarrowtail}$ adds to an underlying relevant logic L the two rules:

REC. $A \rightarrow B, B \rightarrow A \,/\, (C \rightarrowtail A) \rightarrow (C \rightarrowtail B)$
(consequent replacement)

RSC. $A \,\&\, B \rightarrow C \,/\, ((A \rightarrowtail D) \,\&\, (B \rightarrowtail D)) \rightarrow (C \rightarrowtail D)$
(complex suffixing).

The complex suffixing rule can be seen as a condensation of two principles, which are easier to assess; namely rule suffixing (got by equating A and B)

RS. $A \rightarrow C \,/\, (A \rightarrowtail D) \rightarrow (C \rightarrowtail D)$,

and an &-distribution principle. We have already travelled far enough to see that this approach is a dead-end. For applied to simplification, $A \,\&\, B \rightarrow A$, RS yields,

DEL. $(A \,\&\, B \rightarrowtail D) \rightarrow (B \rightarrowtail D)$,

which is patently wrong. For it would allow what may be critical confirming information to be simply deleted. Such a suppression-of-confirming-information principle also carries us well along the fatal route to where anything confirms anything, and confirmational triviality. For suppose A really confirms D, but that we can attach a piece of minor further information B which does not detract from A's success. Then by DEL, B confirms D, without A!

Much more promising is a direct connection

DC. $A \rightarrowtail B$ iff $A > B$, for some sort of conditional $>$.

There is, after all, a considerable analogy between confirmation and conditionality—both, for instance, offer grounds or reasons for something with a view to arriving at, establishing, strengthening something; both afford a sort of partial sufficiency. The analogy is sufficiently good to make it worth investigating whether the further conditions for a basic conditional logic also hold for confirmation. The basic logic LC, which has an elegant semantics (as PLI will show), consists of the following two rules, which are added to underlying logic L:

REA. $A \to B, B \to A \,/\, (A \rightarrowtail C) \to (B \rightarrowtail C)$
(antecedent replacement)

RPC. $A \,\&\, B \to C \,/\, ((D \rightarrowtail A) \,\&\, (D \rightarrowtail B)) \to (D \rightarrowtail C)$
(complex prefixing).

Complex prefixing may be analysed into a straight prefixing rule

$$A \to B \,/\, (D \rightarrowtail A) \to (D \rightarrowtail B)$$

(which may in turn be replaced by an axiom scheme coupled with REC) and the scheme,

&-Dist. $((A \rightarrowtail B) \,\&\, (A \rightarrowtail C)) \to (A \rightarrowtail (B \,\&\, C))$

the main connective of which may be strengthened to a coimplication.

Call this basic logic LC, conditional confirmation logic.[156] This basic conditional confirmation logic is conditional in two senses: that it is based on a direct analogy with the relevant conditional, and that it is provisional (in more than one respect). For of course a basic logic is not a final resting point, only a starting point. There are further principles, regularly taken to hold for the statemental logic of confirmation, to add to the logic LC and to model semantically. These include, as well as transitivity, $((A \rightarrowtail B) \,\&\, (B \rightarrowtail C)) \to (A \rightarrowtail C)$, both the following:

$A \to B / A \rightarrowtail B$, i.e., where A provably implies B, then A confirms B, and

$((A \rightarrowtail B) \,\&\, (C \rightarrowtail D)) \to ((A \,\&\, C) \rightarrowtail (B \,\&\, D)$,
a praeclarum principle enlarging &-Dist.

Both of these are appealing; both are easily modelled.

Already with &-Dist, however, we may have encountered trouble. For Carnap 1962 claims to have supplied counterexamples to it: cases where

[156] There is a ready-made worlds semantics for the basic logic. The key semantical rule is essentially that for relevant conditional logic, namely $I(A \text{ confirms } B, a) = 1$ iff $c(a, [A]) \subseteq [B]$, i.e., "A confirms B" holds at situation a iff B holds at every situation selected as confirmationally relevant at a to the antecedent A. Confirmational relevance may in turn be spelt out in various ways, for instance in forms of confirmational similarity, nearness, etc., as for conditional theory. (Technically, $[B]$ is the range of B, i.e., the class of worlds where B holds, $\{d : I(B, d) = 1\}$; and c is a choice function from pairs of worlds and sets of worlds (ranges) to sets of worlds. For more details see PLI).

A confirms B and also confirms C, but where A does not confirm $B \& C$, but even disconfirms it. If these counterexamples stand up, any such direct comparison of confirmation with conditionals is excluded. But Carnap's counterexamples turn technically upon his probabilistic-relevance analysis of confirmation, an analysis itself deservedly controversial, as it both guarantees paradox and breaks down expected adequacy conditions.[157] However, outside that setting the sorts of counterexamples offered are perhaps not so decisive. Consider the situation of his 'example for 3a' (p.394). The setting is a tournament, where 5 of the 10 players are locals, 5 are juniors, and among the 5 male players 3 are locals, 3 are juniors, but there is only 1 local junior. A, B and C are respectively the statements: "a man wins", "a local players wins" and "a junior wins". Carnap contends, on number-crunching grounds, that A confirms B and that A confirms C, but that A does not confirm $B \& C$, i.e., that a local junior wins. Much turns however on how confirmation claims are reached and supported. If, for instance, only probabilistic-relevance counts, Carnap is no doubt correct; but if other considerations enter, such as arguments from other confirmation claims (for example, by composition principles themselves), then such examples appear to fail.

The enlargement of composition (i.e., &-Dist) to praeclarum offers enough leeway, however, to produce counterexamples based directly on Nicod's criticism. Goodman 1965 has such an example (though he sees it not as a counterexample to the principle, but as further indication of the need to take account of total evidence, pp.71-2). The statement that a first thing (C say) is black confirms, by Nicod, that every thing is black; the statements that a second thing, (C say) is non-black confirms that every thing is non-black; but the compatible evidential antecedents do not confirm the inconsistent predication of the statement that everything is black and non-black. More generally, suppose that (as with the examples of section 2) e_1 confirms h, but e_2 disconfirms h, i.e. confirms $\sim h$ on standard accounts (for example Carnap 1962, p.479); then, by praeclarum, we should have evidence $e_1 \& e_2$ which confirms the contradictory hypothesis $h \& \sim h$. Should we want, as we do, to travel without a burdensome total evidence requirement (which would undercut the premisses of such counterexamples), we may have to abandon praeclarum.[158] Fortunately, it is evident that such examples do not extend to

[157] It would take us too far afield to assess probabilistic-relevance and its relation to confirmation. But, as classically-based, it has little to do with relevant logic, and is, as further investigation would reveal, no satisfactory alternative to relevance. Rather it ensnares us both in the paradoxes we are trying here to escape and in other paradoxes.

[158] Once again there are other possible ways out of the seeming impasse. For instance, we could mess with the logical 'identification' of confirmers (see Goddard 1977; also

touch &-Dist; for one thing, independent evidential statements are involved; for another &-Dist does not permit direct derivation of praeclarum because augmentation principles like $(A \rightarrowtail B) \rightarrow (A \,\&\, C \rightarrowtail B)$ fail.[159]

Several options are open, specifically these:

- *dig in*, namely keep some extension of basic conditional logic without praeclarum, and develop a fuller case against Carnap and probabilistic-relevance analyses;
- *retreat* to some extension of minimal conditional logic without &-Dist;
- *proceed* ahead to further logics and further paradoxes not so far examined.

We shall pursue the last option, without however thereby closing off the other options.

Chihara 1981). There are significant logical relations, commonly glossed over (but sometimes presented as evidence conditions) at the *front end* of the Nicod relation. Consider what is used in reaching: a CD confirms "all A's are B's" (as, for example, in a white shoe confirms "all non-blacks are non-ravens") from an AB confirms "all A 's are B's". Namely, implications "C implies A" and "D implies B", functioning rather like meaning postulates, are applied. These implications (which reflect informational decrease) cannot be upgraded to identities, so more logic than mere 'substitution of identicals' is involved. There is also the important question of the logical form of confirmers, and whether standard representation through extensional conjunction is adequate, or, as Goddard 1977 argues, not.

159 There remains room for doubt about &-Dist, doubt which may motivate us to step down further. For example, suppose statement A describes three pieces of data (z_1, z_2, and z_3) which fits two functions f_1 and f_2 which are incompatible, and B and C describe these functions. Given that A confirms B (for example, f_1 is like so) and A confirms C, do we want to say that A confirms $B \,\&\, C$. Much of the inclination to say that it doesn't derives from the drive, in logic especially, to avoid inconsistency, or even the scent of inconsistency. It is primarily on this basis that Hempel imposed his consistency condition of adequacy: 'if A confirms B and A confirms C then B is consistent with C' (Hempel 1965, p.33, p.49).

Against this condition Carnap duly objected, so Hempel reported. 'One and the same observable phenomenon may well be accounted for by each of two incompatible hypotheses, and the observation report describing its occurrence would then normally be regarded as confirmatory for either hypothesis' (cf. Carnap 1962, section 87).

To this Hempel in turn responded: 'though the point does [...] carry considerable weight [...], the consequence condition has to be given up along with the consistency condition. Otherwise a report confirming each of two incompatible hypotheses would count as confirming any consequence of the two, and thus any hypothesis whatsoever' (Hempel 1965, p.49 and footnote). But that is only so if the consequence relation is not suitably paraconsistent. If we use a relevant entailment, we can retain *both* confirmation of inconsistents *and* the consequence condition. As we shall.

10.5 Proceeding within conditionality: containment confirmation

Should Carnap's counterexamples stand up, we are obliged to retreat *inside* conditionality, in one way or another. On the face of it, confirmation appears to involve conditionality and something else, i.e., $A \rightarrowtail B$ takes the form $A > B \,\&\, F(A, B)$, with F some further functor relating A and B. Fortunately we already have a fair idea of how such logics go, and of what F amounts to, from investigation of relevant containment logics.[160] In such containment logics, containment implication is analysed through an implication and a priority ordering, i.e.,

$$A \twoheadrightarrow B \text{ iff } A \rightarrow B \,\&\, A < B,$$

where $A < B$ says that A is prior to (or precedes) B, in the case of a ponible containment implication (i.e., one that satisfies modus ponens: $A \twoheadrightarrow B, A \,/\, B$). Now confirmation is *not ponible,* i.e., $A, A \rightarrowtail B \,/\, B$ fails; so confirmation cannot be carried by a ponible implication such as $\rightarrow$. But there are certainly kinds of conditionals which are not ponible; and in fact nothing in the *basic* logic of $>$ requires or implies ponibility. The way to an analysis of confirmation avoiding (and not contesting) Carnap's counterexamples accordingly seems clear;[161] we exploit the pretty connection:

$$A \twoheadrightarrow\!\!\!\!\!\!\rightarrowtail B \text{ iff } A > B \,\&\, A < B,$$

i.e., (that) A confirms (that) B iff both B on condition that A and A is evidentially prior to B.[162] Carnap's counterexamples are avoided because $A < B \,\&\, A < C$ may not imply $A < B \,\&\, C$. Call the notion symbolised using ↣↠ *containment confirmation.*

An immediate issue is the logical structure of the ordering relation $<$. Relation $<$ is at least an order, i.e., transitive, whence, given that the conditional $>$ is also transitive (not an automatic matter), confirmation is, as expected, transitive, i.e.,

$$A \twoheadrightarrow\!\!\!\!\!\!\rightarrowtail B, B \twoheadrightarrow\!\!\!\!\!\!\rightarrowtail C \,/\, A \twoheadrightarrow\!\!\!\!\!\!\rightarrowtail C$$

in rule form; and perhaps a stronger implication form also holds, viz.

$$((A \twoheadrightarrow\!\!\!\!\!\!\rightarrowtail B) \,\&\, (B \twoheadrightarrow\!\!\!\!\!\!\rightarrowtail C)) \rightarrow (A \twoheadrightarrow\!\!\!\!\!\!\rightarrowtail C).$$

---

160 *See Chapter Seven. [Editors]*

161 Contextually it is *the* way indicated. Naturally it is not the only way, as we have seen and shall see again subsequently.

162 *Evidential priority is symbolised as '—o' in Chapter Six. [Editors]*

Given Carnap's counterexamples $<$ is not a full lattice ordering. An important question is how much (semi-)lattice structure does $<$ retain? But a preliminary question is whether $<$ is even a partial order. Does reflexivity hold, i.e., $A < A$? As it happens, there is again an ambiguity both in priority, and matching it in confirmation, to sort out and resolve. There is that ambiguity already encountered between

1. A helps to make B firm; then $A \rightarrowtail\!\!\!\!\rightarrow A$ may be in.
2. A helps to make B firmer; whereupon $A \rightarrowtail\!\!\!\!\rightarrow A$ is out.

Whereas 1 can satisfy the entailment condition, 2 fails it. Let us assume for the present that 1 obtains.

Then given $A > A, A \rightarrowtail\!\!\!\!\rightarrow A$. So all of $\rightarrowtail\!\!\!\!\rightarrow$, $<$ and $>$ are partial orders (in *a* standard sense). Both $>$ and (hence) $\rightarrowtail\!\!\!\!\rightarrow$ are nonponible, i.e., in the latter case A and $A \rightarrowtail\!\!\!\!\rightarrow B$ does not guarantee B. Which is as it ought to be. What further logical structure $\rightarrowtail\!\!\!\!\rightarrow$ has hinges of course on its components, especially $<$. In line with the containment idea the full lattice condition $A < A \vee B$ fails, whence so also does $A \rightarrowtail\!\!\!\!\rightarrow A \vee B$, i.e., Addition is out. While full lattice conditions fall down for containment, lattice conditions need not in entirety—whence *containment confirmation logic*, that obtained by imposing a certain lattice structure on $<$ and $\&$. One result of this might be $A \& B \rightarrowtail\!\!\!\!\rightarrow A$; another $A \vee B \rightarrowtail\!\!\!\!\rightarrow B \vee A$, $((A \& (B \vee C)) \rightarrowtail\!\!\!\!\rightarrow ((A \& B) \vee (A \& C))$, etc. The main ingredients for a logical and semantical treatment for various containment confirmation theories have already been supplied in previous work.[163]

There are immediate large, if perhaps unexpected, benefits in stepping down to containment confirmation. Containment logics characteristically remove tack-on conditions such as Addition, $A \rightarrow A \vee B$.[164] Given this, and the failure of disjunctive syllogism in relevant settings, we are already in a position to defeat Goodman's arguments concerning *grue*, i.e., to rout entirely another formidable batch of confirmational paradoxes. Some other problems are also conveniently removed along the way. Here are a couple of example problems assembled by Armstrong 1983 in his opportunistic use of the woes of confirmation theory against rival analyses of laws of nature (they are drawn, slightly altered, from pp.43-4):

(A) "There are snakes less than 1m" entails "There are snakes less than 10m".

163 *See especially Chapter Seven. [Editors]*
164 *See Chapter Seven, section 2. [Editors]*

> Now "Here is a 3m snake" confirms the consequent. Therefore, on the converse consequent condition [in the form $A \to B, C \rightarrowtail\!\!\!\!\twoheadrightarrow B \,/\, C \rightarrowtail\!\!\!\!\twoheadrightarrow A$], "Here is 3m snake" confirms "There are snakes less than 1m".

That is no doubt absurd (but is nonetheless no serious problem for orthodoxy which qualifies converse consequence). However the supporting entailment (A) relies on what amounts to $A \to (A \vee B)$ and accordingly fails for relevant containment, and so for containment confirmation. For it depends upon

s is less than 1m $\to$ s is less than 1m or anything longer,

from which it results by legitimate quantificational argument.

A second example from Armstrong is similarly despatched. It relies on the implication

(B) "Unicorns exist" implies "Animals exist".

Now the presence of a zebra confirms that animals exist, so the presence of a zebra, or that zebras exist, confirms, by converse consequence, that unicorns exist. But principle (B) is typically reached by expansion—by the disjunction "unicorns or zebras or ...", for all types of animals, which can then be contracted to "animals". So (B) too fails for relevant containment and so too for containment confirmation.

10.6 Goodman's paradoxes, gruesome problems and logical resolutions thereto

It is an important observation that the new riddle Goodman introduces 'does not really turn on the introduction of bizarre predicates like "grue". ... if the confirmation of the raven hypothesis by white crows or red herrings is genuine, so is the confirmation of the hypothesis that emeralds examined before t are green' (Rosenkrantz 1979, pp.78-9). Essentially Goodman's paradoxes are of a piece with those Hempel assembled. They extend them in two notable respects: in exploiting positional predicates (like "examined before t"), and in using different (relevantly dubious) principles in the derivation. The derivation proceeds as follows: Green emeralds found or examined before t are presumably, by a standard information-adjustment inference, non-blue emeralds examined before t. Then, by the Nicod criterion (with no restriction to exclude positional predicates), instances of the latter type confirm all non-blue emeralds are examined before t.

But that hypothesis is logically equivalent, by antilogism, to

all emeralds not examined before t are blue.[165]

Accordingly, by the equivalence condition, it too is confirmed, by the same instances. In sum, then, green emeralds examined before t confirm "all emeralds not examined before t are blue".

The key logical principle involved, antilogism, a nontrivial extension of contraposition and direct analogue of disjunctive syllogism, is however no friend of relevance; indeed it yields implicational paradoxes like $A \,\&\, {\sim}A \rightarrow {\sim}{\sim}B$ immediately from simplification $A \,\&\, {\sim}B \rightarrow A$, and collapses relevant logic to modal logic. Because antilogism fails, Goodman's extensions of Hempel's paradoxes are automatically relevantly voided. But it is worth mentioning that *these* extension paradoxes (unlike those based definitionally on grue-type predicates) also fail in modal conditional logics, which jettison antilogism along with contraposition.

However Goodman's arguments are normally presented in a different, definitional way. The new predicate "(is) grue", the first of many similar Goodmanesque predicates, serves to legitimize features of the argument otherwise in deeper doubt, such as illicit use of positional predicates and appeal to a conditionally dubious principles. For the definitional strategy enables more to be accomplished simply through substitution of (supposed) equivalents. Nonetheless, the arguments concerning *grue* depend, essentially it appears, on two principles, both rejected in relevant containment logics, and both (long) under suspicion for confirmation logic, namely, Addition and Disjunctive Syllogism. Conveniently Bass 1987 has recently exposed the dependence of Goodman's argument on these principles. Consider then,

> a small but troubled corner of confirmation theory, the obdurate and much analyzed *inductive paradox* first propounded by Nelson Goodman *to dramatize the New Riddle of Induction*—why some

165 This sort of puzzle was substantially anticipated by Russell 1948, p.411, as was Goodman's prime inference therefrom (these historical insights we owe to Nick Griffin). According to Russell 1948, 'this consideration destroys the whole force of the inductive argument for general theories, unless the argument is reinforced by some principle such as the uniformity of nature is supposed to be. Here again, we find that induction needs the support of some extra-logical general principle not based upon experience' (pp.411-2). But in a relevant setting these conclusions appear unwarranted. With relevant confirmation a way past such principles as uniformity of nature, total evidence, and logical omniscience looks viable.

predicates are projected into the future and others are not. Very briefly: Before time T (T_1), Smith observes emerald a to be green, but Jones claims that the emerald is grue. It turns out that an object is grue just in case it is green and observed before T or blue and observed after T (T_2); and it seems that Smith must agree that the emerald in question is indeed grue as well as green. The evidence, as first interpreted by Smith, offers inductive support for the hypothesis "all emeralds are green", and hence, for the prediction that emerald b observed after T will be green. On the other hand, and with seemingly equal justification, the observed evidence inductively supports the hypothesis "all emeralds are grue", and hence, the prediction that emerald b observed after T will be grue; and hence, blue. Why should Smith project "green" rather than "grue"? How choose between the contrary predictions about the colour of emerald b?

Embedded in the above account are two distinct inferences, each essential to the working of the paradox. Smith agrees that emerald a is also grue only if he accepts the validity of

(1) $(Ga \,\&\, T_1 a) \rightarrow ((Ga \,\&\, T_1 a) \vee (Ba \,\&\, T_2 a))$;

and he *predicts blueness* for emerald b observed after T only if he accepts the validity of

(2) $({\sim}(Gb \,\&\, T_1 b) \,\&\, ((Gb \,\&\, T_1 b) \vee (Bb \,\&\, T_2 b))) \rightarrow (Bb \,\&\, T_2 b)$.

(Bass 1987, pp.32-3; italics added, symbolism modified.)[166]

There are the following elements in Goodman's riddles:

- A truth functional reduction of conditionality. Given his attachment to Addition, Smith should concede that

> a is grue, i.e., green and observed before T or blue and observed after T, i.e., adapting Bass's symbolism: $(Ga \,\&\, T_< a) \vee (Ba \,\&\, T_> a)$.

However Smith need not, ought not, concede that

> if a had not been green and observed before T then it would have been blue and observed after T.

'After all, the greenness of emerald a observed before T provided the *only* reason Smith had for agreeing reluctantly that the emerald was also

[166] Nice though the argument is, really the analysis does not reach deep enough. Goodman's argument also turns upon strict substitution principles and truth-functional reduction. And, in any case, it requires more unravelling than Bass offers.

grue' (Bass 1987, p.34). What the point reveals, though Bass does not finish explaining it, is that Goodman's argument here relies on the blatant paradox $A\,/\sim A > B$. Of course this is achieved through the manifestly defective $(A \vee B) \rightarrow (\sim A > B)$.

• The argument uses a consequence (entailment) condition as well as the Nicod criterion: observation of a green emerald before T confirms all emeralds before T are green and therefore its consequence that all emeralds are grue.

Alternatively, the point can be argued as follows: observation before T of a green emerald is confirmation (*not* observation, presumably) of a grue emerald which in turn confirms that all emeralds are grue. Then use transitivity of confirmation.

• Disjunctive syllogism (DSyll) enters after it has been established that observation of a green emerald before T confirms that all emeralds are grue. Without DSyll, the inference to: "an emerald observed after T will be blue" is blocked. Let d be an emerald observed *only* after T, i.e., $T_{>}d \,\&\, {\sim}T_{<}d \,\&\, Ed$. By the result that all emeralds are grue, $\forall x(Ex > (Gx \,\&\, T_{<}x) \vee (Bx \,\&\, T_{>}x)$, whence instantiating and detaching $(Gd \,\&\, T_{<}d) \vee (Bd \,\&\, T_{>}d)$. Since ${\sim}T_{<}d$, ${\sim}(Gd \,\&\, T_{<}d)$ (by an analogue of Addition). Hence by DSyll, $Bd \,\&\, T_{>}d$, so Bd.[167] Without DSyll and its mates, the inductive paradox incorporated in the New Riddle of Induction vanishes.

It is important to observe that, *given* truth functional reduction, the inductive paradox involved does *not* depend essentially upon Addition principles. Thus we can defeat Goodman's arguments without containment logics. It is unnecessary, for the main run of confirmational paradoxes, to go so far afield as containment confirmation.

The paradoxes generated by terms like "grue", "blite" and "emerose" all take a similar form, all depend on equivalences based on principles scrapped in relevant containment logics, and all are defeated by a straightforward relevant switch. Consider emeroses. For any object x, emerose x iff emerald x & examined x or rose x & $\sim$ examined x. The orthodox assertion that examined emeroses equal, or are equivalent to, examined emeralds depends on the equivalence, $p \,\&\, ((q \,\&\, p) \vee (r \,\&\, {\sim}p))$ iff $p \,\&\, q$, where p represents examined x, q emerald x, and r rose x. Evidently this equivalence requires disjunctive syllogism one way (left to right) and addition the other way to succeed, and in particular the critical inference

167 Variant principles are really used here, but relevantly invalid ones. Note that DSyll is required, not γ, because detachment is from a *nonlogical* principle.

from examined emeroses to examined emeralds uses the relevantly invalid disjunctive syllogism. Consider Hempel's ornithological variant of "grue", namely "blite" defined "examined before t and black or not examined before t and white". Look at Hempel's claim that 'any raven observed before t and found to be black affords a formally confirming instance, in the sense of Nicod's criterion, of the hypothesis "All ravens are blite"' (Hempel 1965, p.50). Of course it does not, without significant logical transformations: without those such an observation confirms at most that all ravens observed before t are black, i.e., antilogising and skating, that all ravens are either black or not examined before t. There are various logical routes, all of which require what amounts to disjunctive syllogism. One transformational route, by logical equivalence (coentailment) replacement, uses the equivalence $((T_{<}x \,\&\, Blx) \vee ({\sim}T_{<}x \,\&\, Whx)) \,\&\, T_{<}x$ iff $Blx \,\&\, T_{<}x$, to transform (as with emeroses) the formally confirming instances into those of blite ravens. Another route follows out consequences of what is confirmed, namely that examined ravens are black, applying the entailment conditions and classical transformations to arrive at "all ravens are blite". What Hempel goes on to claim is also relevantly defective, namely that '... no matter how many such instances may have been collected, they lend no support or confirmation to the hypothesis; for the latter implies that all ravens not examined before t ... are white' (Hempel 1965, p.51). For that implication (from "all ravens are blite" to "all ravens not examined before t are white") involves a direct applications of disjunctive syllogism.

10.7 Converse consequence and connexive confirmation

Already, with the progression to relevant containment linkage, the straightforward entailment condition fails, and therewith a direct consequence condition. For while $A \rightarrow .\ A \vee B$, $A \rightarrowtail\!\!\!\rightarrow .\ A \vee B$ does not hold. (But an analogous condition, based on containment implication, may persist.) Now the entailment condition (which here allowed the tacking on of irrelevant disjunctions) has been a major obstacle in the way of the widely appealing converse condition, that confirmation follows the converse of consequence (that, more sweepingly, induction is the converse of deduction). For the two, entailment and converse consequence, in combination had the untoward outcome that any statement confirmed any other (see Hempel 1965, p. 32). But it is not the only obstacle; an important component of the entailment condition, that supplied by simplification, is enough to wreak serious damage.

For suppose we obtain from simplification (for $\rightarrow$ or $>$), in one way or another,

$$A \,\&\, B \rightarrowtail\!\!\!\!\twoheadrightarrow A;\; A \,\&\, B \rightarrowtail\!\!\!\!\twoheadrightarrow B \text{ (Confirmation simplification)}$$

By the converse consequence condition, or the special entailment case of it, namely

CC. $\quad C \rightarrow D, E \rightarrowtail\!\!\!\!\twoheadrightarrow D \,/\, E \rightarrowtail\!\!\!\!\twoheadrightarrow C,$

given $E \rightarrowtail\!\!\!\!\twoheadrightarrow D$ we can obtain $E \rightarrowtail\!\!\!\!\twoheadrightarrow C \,\&\, D$. But by simplifying, $C \,\&\, D \rightarrowtail\!\!\!\!\twoheadrightarrow C$, whence by rule transitivity of confirmation, $E \rightarrowtail\!\!\!\!\twoheadrightarrow C$. What this shows is that evidence, any evidence, will confirm any arbitrary conclusion.

No doubt more is involved in this argument than simplification, namely a one-step application of a transitivity rule. There are however independent reasons for reassessing simplification. Irrelevant conjunctions supplied by simplification play an entirely analogous role to tack-on by additions and they effect similar damage for all of confirmation, verification and hypothetical-deductive methodology.[168] It is not going too far then to insist that containment confirmation is but a half-way house. It disposes of tack-ons by addition, of the introduction of irrelevant content into consequents, but it does not deal symmetrically with antecedents, with irrelevant conjunctions by simplications. To recover traditional elements of induction—such as induction as the converse of deduction, converse consequence—we shall need to go the full distance, to confine irrelevant introductions in antecedents as well as in consequents.

If converse consequence *is* to be retained, the obvious (and inevitable) move is to reduce structure, in particular to remove unqualified confirmation simplification. This leads towards connexive confirmation, so called because the logical strategy closely resembles that of connexive logic, which as one of its distinctive features, qualifies simplification. The other distinctive feature of connexive logics is the presence of specifically connexive principles, beginning with Aristotle's principle, in implicational form $\sim(A \rightarrow \sim A)$. The matching confirmation principle is $\sim(A \rightarrowtail\!\!\!\!\twoheadrightarrow \sim A)$, which could be guaranteed by either $\sim(A > \sim A)$ or $\sim(A < \sim A)$. Both alternatives are semantically feasible.

---

168 For instance tack-on by addition features in Hempel's criticism of verification principles. But entirely analogous criticism applies simplification, which equally admits non-paying passengers.

More problematic to ensure is CC itself, a principle which, like connexive principles, is powerful in its exclusive features, in drawing other principles out. On the working analysis we have been applying, it requires both

a. $C \to D, E > D \,/\, E > C$ and

b. $C \to D, E < D \,/\, E < C$.

While b can be obtained given the plausible $C \to D \,/\, D < C$, simply reading $<$ as containment-in-content, a is more difficult, but not impossible, since if $\to$ were $\leftrightarrow$, a complication, it would hold.

Accordingly we advance the substantive claim that

- converse consequence will work for connexive confirmation.

There are historical reasons for considering the claim promising. Namely, many traditional logicians who supported converse consequence (or, more sweepingly, induction as the converse of deduction) were implicitly operating within a connexive setting.[169] It was within such a setting that converse consequence principles gained their plausibility.

It is also within such a setting that the principles stand up against counterexamples. The substantive claim appears to deliver the right kind of results with regard to examples, as the following case illustrates. Consider the pair of statements: "Alphonse kicks some cat" (A_c) and "There are cats" (C_I). Now it said, A_c entails C_I. But that there is a cat hardly confirms that Alphonse kicks some cat. It is not hard to see that A_c does not connexively imply C_I; for it turns on an application of simplification. Consider the orthodox formal transcriptions of the statements. They take the respective forms: "for some x, x is a cat and Alphonse kicks x", i.e., $\exists x(Cx \,\&\, Ax)$ for short, and "for some x, x is cat", i.e., $\exists x Cx$. The orthodox argument to the entailment involves generalisation and quantifier distribution from the simplificational form: $Cx \,\&\, Ax \to Cx$; specifically, $\forall x(Cx \,\&\, Ax \to Cx)$, whence $\exists x(Cx \,\&\, Ax) \to \exists x Cx$. Better transcriptions no doubt yield a similar outcome: essential dependence on a connexively invalid form.

Connexively-based confirmation theory, *connexive confirmation*, is essentially a traditional approach (differing from later traditional logic only in removing some controversial reduction proposals). Like traditional theory, it avoids damaging weakness by heavy deployment of traditional logic, most notably syllogistic, which is remarkably flexible in its formalisational capacities.

---

169 For some of the substantial evidence for this controversial historical theme, see NC, and Chapter Five.

10.8 Interim conclusion: reinstating theories of confirmation and inductive reasoning

Relevant investigations give good promise of dislodging the very negative conclusions that have recently been reached, prematurely from a relevant viewpoint, concerning the prospects and status of theories of confirmation and inductive reasoning. Seldom have the prospects been more negatively presented than as follows: '... there is very little left of that putative philosophical subject, the theory of confirmation (and even less of its ancestor, the theory of induction)'.[170] But the bulk of the case for this indictment we have already dismissed, after relevant investigation of the arguments. Much the same applies to other indictments like it, which are based on the intractability of confirmational paradoxes and puzzles—from an assumed classical standpoint. From a different relevant standpoint all that changes, many puzzles dissolve, a fresh start can be made upon investigations of logics and theories of confirmation.

We do not pretend to have taken such investigations any real distance; rather we have simply indicated where, more or less, they lie. What we have so far done is to uncover some main forms of relevant confirmation: conditional, containment, and connexive, and also variants thereon. The choice between these—so far as it needs to be made, so far as these are competing forms, which do not simply capture different preanalytic notions (as, for instance lending support, making firm and making firmer variants do)—turns substantially upon how they deal with preanalytic examples and requirements. Particularly critical are converse deducibility requirements which swing the focus away from conditional-style confirmation to little studied connexive types. It should be evident that our presently favoured field of confirmational investigation is elliptical, with two foci: conditional and connexive forms. Beyond confirmation lie many other intimately connected notions we have simply separated out, not addressed: credibility, plausibility, acceptability and the logical makings thereof.

---

[170] Thus van Fraassen 1983, pp.328-9. But rather similar sentiments have been expressed by several others, typically following in the steps of Popper and Goodman. We doubt that van Fraassen's more positive sketch (p.328), of what we should do instead without confirmation or induction or indeed any theory, will withstand much serious investigation.

10.9 Parallel sympathetic enterprise, notably that of Paul Weingartner and collaborators

Despite the premature announcement, from North America, of the death of relevant logical enterprise and relevance investigations, these activities and their applications to a wide variety of puzzles and paradoxes are proceeding apace in many centres. One significant centre, outside American hegemony, is Salzburg, where results similar to those we have assembled have been reached independently by Paul Weingartner and co-workers (notably Gerhard Schurz). While we build on the recent relevant logic tradition, adopting previous work by ourselves, Waters 1987 and Bass 1987, there is an allied (and in other applications, longer-standing) approach which adds relevance criteria to some irrelevant logic, typically nowadays classical logic. Such an approach is taken by apparent rivals to relevant logic, such as relational logics, and in semantics for other rivals, such as Parry's analytic implication and British non-transitive logics (all described in Chapters Four and Five). It is also essentially the approach pursued, in heavily syntactical form, by Weingartner and Schurz (hereafter W&S).

Whereas we operate in the text with different types of relevant logics, three in fact, W&S work with different relevance criteria, three commonly; and one of these criteria, the A-criterion, matches precisely one sub-type of relevant logics, namely Parry logic within the containment class. It is easy to see, in principle, that the correspondence could be extended, with more criteria (for example a B-criterion, for Belnap, concerning sharing of parameters, to match relevant affixing logics) and with more logics (for example R-logics, for Salzburg, to match the R-criterion).

The three criteria W&S develop to 'solve', as they say,[171] a wide range of paradoxes—criteria for *relevant* valid *implication*, relevant consequence, or relevant deduction, be it noted—are the A-, K- and R-criteria. Let us look at the basic forms of the first two. For only the A- and K-criteria (along with a plethora of subvarieties of these) are applied in defusing paradoxes of confirmation. Under the A-criterion, a valid implication $A \rightarrow B$ is relevant only if every propositional variable (or predicate parameter) in the consequent B also occurs in the antecedent A. To illustrate, addition $A \rightarrow (A \vee B)$ is not A-relevant; and puzzles like those of Ross in deontic logic are instantly derailed. Evidently A-relevance functions like relevant containment. According to the tough K-criterion, $A \rightarrow B$ is relevant,

[171] In Weingartner and Schurtz 1986, the most accessible of their many essays on the topics.

K-relevant, if no predicate (or propositional) occurrence, in the antecedent or consequent, can be replaced by any other predicate (of the same arity, including 0) preserving validity (or equivalently, without changing logical content, or again, upon replacing a subformula by its negation). This K-criterion performs analogous work to relevant connexive logic, relevantly invalidating simplification, $H \& I \rightarrow H$, and its mate, augmentation, $A \rightarrow B \,/\, A \& C \rightarrow B$. So it similarly appears to enable restoration of a converse consequence condition and traditional induction.[172] It should already be evident, then, how similar relevant logical and W&S relevance criterial approaches actually are.

Although logical and criterial approaches are often cast as rivals (including by W&S), really they are not in serious competition, but advance a similar overarching strategy: Namely, that a *relevant* implication or deducibility be used in properly evaluating such critical relations as confirmation, verification, verisimilitude and the like. Both approaches are committed to the theme that, whatever the merits of classical and other irrelevant logics, *classical consequence, and irrelevant deducibility relations more generally, are inadequate for confirmation and other connected notions*, but are indeed the source of a myriad of paradoxes that appear to afflict these notions. Both are similarly committed, in resolving or evading paradoxes, to operating with relevant implications in conjunction with familiar connectives such as those of conjunction, disjunction and negation, that is, to operating in a broadly relevant logical setting.

They are thus mutually reinforcing approaches, which could profitably join forces. For instance, the approach through relevant logic could learn from the criterial as to how to get to grips with verisimilitude (where relevant activity halted prematurely), and conversely the criterial from the logical in reaching a superior treatment of confirmation, which dissolves Hempel's paradox.

However, under pressure from a familiar differential imperative, W&S try to distance and differentiate their approach from that of relevant logic (see for example Weingartner and Schurz 1986, p.2). Fortunately critical features of their differentiation effort do not withstand much detailed investigation. They advance three prime points of differentiation 'in order to avoid misunderstanding':

---

[172] There may be difficulties stored up for K-relevance, not touching connexive confirmation. For K-relevance permits irrelevant tack-on. Such forms as $A \rightarrow (A \vee (B \vee \sim B)$ are K-relevant, though not A-relevant and not connexively valid.

(1) Purposes are different: 'the main aim of our relevance criteria is to avoid paradoxes not only in one area [...] typical paradoxes of implication [...] but in different areas' covering a wide sweep of philosophy, comprehending paradoxes and puzzles in epistemology and philosophy of science, throughout ethics and elsewhere. A fine aim. But—where W&S go wrong—that has also been a goal of relevant logical enterprise, virtually from its inception. Relevant logic was seen not merely as a means for resolving many puzzles and paradoxes in philosophical logic, but as 'a universal paradox solvent'.[173]

(2) The role of principles like addition, $A \rightarrow A \vee B$ and its mates, is different.

> [...] it is not sufficient to concentate on implication because there are also paradoxes which are dependent on disjunction, conjunction and negation. For instance, the formula $p \rightarrow (p \vee q)$ which is relevant in all well-known systems of ... relevance logic is irrelevant according to both of our criteria. The reason is that this formula is responsible for a persistent series of paradoxes in different areas. (Weingartner and Schurz 1986, p.4)

Once again we do not want to dispute these claims, so much as to reorient them, to reduce the appearance of separation distance. Systems of analytic implication, which eschew addition, are commonly accounted (broadly) relevant logics, and indeed virtually stole the show at the one and only international relevant logic conference in 1973 (as documented in DRL). Relevantists do not deny that relevant containment logics, including analytic implication systems, have significant application to paradox and puzzle dissolution: quite the contrary (see Chapter Three, section 2 and Chapter Seven). So much so far for diminishing differences. But a few things remain astray. A minor one is that rejection of addition may turn eventually on features of implication, as it does both in semantics for analytic implication (where extensional connectives behave classically and implication is different because of a parameter containment clause) and similarly under the connected A-criterion. More major is divergence on the question of what is wrong in the paradoxes, when something is wrong. (Is 'Ross's paradox' a paradox? Or Hempel's paradox: Weingartner and Schurz 1986, p.33?) Is addition the real culprit in Goodman's paradoxes, as W&S suppose (Weingartner and Schurz

[173] See, for example, JB, p.903 ff, and p.813 where applications to many of the areas W&S list are claimed. See also RLR I and DRL. Early on, Anderson was concerned with applications to deontic logic and motivational and epistemic theories; more recently Dunn has been making significant applications in metaphysics (see Dunn 1990).

1986, p.35)? Or is it, as we say, what W&S give a clean bill of logical health, disjunctive syllogism? To these issues we shall return.

(3) The attitude to classical logic is different. '[...] whereas the usual system of entailment and relevant logics are non-classical our criteria are based on classical logic' (Weingartner and Schurz 1986, p.2). Again the 'difference', and its 'importance' are exaggerated. For so far as much of the present paradox-resolving exercises are concerned, the relevant logics applied could almost as well be extensions of classical logic (in the style of the 'classical axiomatics' of RLR I, chapter 5). While such 'unusual' formulations of the logics would interfere drastically with their paradox-resolving application where paraconsistent logics are required, these applications are not here material (but they are part of the larger case against persisting with a W&S criterial approach). While certainly, such formulations of the logics would upset some of our smooth motivation, they would accord better with the classical logical commitments of one relevant logic camp (on the divisions within relevant logic, see DRL, conclusion). The point to emphasize then is that already within relevant logic there are underlying ideological differences (especially as to the adequacy and merits of classical logic) of the sort that are alleged to separate relevant logical and criterial approaches—differences furthermore that may not effect presently relevant parts of the operational superstructure. Similar points apply, *mutatis mutandis*, to the criterial approach (which can be based readily enough on amenable logics other than classical): use of the approach need apply no substantial commitment to a classical ideology. For instance, that logic is merely used as convenient simplistic basis, and taken as giving a workable approximation for certain contexts (that it cannot mess up too much). No doubt classical logic is good for what it is good for; tautologically, classical logic is classically correct for classical validity.

Having emphasized the parallels in overall approach, we wish now to record some (further) differences in *detailed* treatment of paradoxes of confirmation.

- Disjunctive syllogism and its mates create a significant watershed. While the omission of these is what really distinguishes relevant logics proper, the basic forms, such as disjunctive syllogism itself, are validated under both A- and K-criteria. So W&S are powerless to defuse paradoxes where disjunctive syllogism and its mates play a crucial role, such as inductive paradoxes in the vicinity of Goodman's riddles. Unless, that is, further criteria emerge, which their somewhat opportunistic practice with criteria indicates W&S would not be averse to deploying. However disjunctive syllogism has proved decidedly resiliant against criterial syntactical approaches designed to reveal its deep irrelevance. As W&S have offered nothing so far to show it and its mates up (it may even be that they suppose these principles are relevantly correct), they have yet to dispell variants on Goodman's riddle like those Rozenkrantz, Russell and others have propounded (whether in our logical fashion, or some other). Accordingly their removal of paradoxes of confirmation remains incomplete, as well as unsystematic.

- Over Hempel's paradox and its variations there is divergence too. The criterial approach W&S are pursuing offers no way of resolving the main paradox (published by Hosiasson-Lindenbaum however called after Hempel). It is disappointing to find W&S, when they discover they cannot exclude the paradox, falling back on the classical refrain, which can be sung to most of the paradoxes they are trying to solve, that the paradoxical upshot 'cannot be regarded in general as paradoxical' (Weingartner and Schurz 1986, p.33). But of course it is widely regarded as paradoxical, as a veritable forest of literature confirms. It is not that W&S are blocked off from all escape routes. Instead of simply assuming that a universal material implication adequately formalizes the universal judgement involved, "all ravens are black", they could, for example, follow down the route we felt obliged to take.

- The motivation W&S offer for how they proceed, which criterion they apply, or which new one they invent for the purpose (for example Weingartner and Schurz 1986, p.20, p.27), strikes us as leaving much to be desired. So far accusations of opportunism and *ad-hoc*ery look as if they would stick. While we appreciate our own logical procedure allows scope for improvement, we anticipate, in part because of the pedigree of the main logic upon which we justify variations, that we are substantially less vulnerable. Part of the trouble is that W&S present no argument for which criterion, if any, should apply, or as to how the particular one cited is supposed to resolve the type of paradox for which it is listed. For example, for the 'Paradox of Good Samaritan', one of the first paradoxes treated (p.9), we are quoted three different criteria for three different principles, and nothing more at all. No clues are given as to why to assume these principles here, not there, or as to why these should operate where and how they do. Still less are we informed how to design a worthwhile logical theory, no doubt incorporating all three, in which we are no longer threatened with such paradoxes (contrast the relevant logical resolution of the same paradox, in MD).

Once again we do not claim that such shortcomings are in any way fatal. All can be surmounted. The criterial approach, despite its severe syntactical limitations, has resources which W&S have not yet exploited. We encourage them to push ahead with their significant program.[174]

[174] *As noted at the beginning of the chapter, this essay appeared in a festschrift for Paul Weingartner. It was followed by an essay by G. Schurz, which includes a reply to the claims made here by Sylvan and Nola; see Schurz 1991b. [Editors]*

11 Conditional Probability as the Probability of a Conditional

This essay reproduces a manuscript marked "Working Draft 1982". From marginal notes within the manuscript, it is clear that Sylvan intended to revise it, and the draft table of contents for Relapse show that he planned a chapter on the topic dealt with herein. However, no later versions were found in his papers. We can offer no explanation of why the project was left untouched for so long, but have taken the view that this manuscript, though old, was worthy of reproduction. [Editors]

It is an appealing thesis, which has rightly tempted many authors,[175] that *probabilities of conditionals are conditional probabilities*, in symbols,

(6) $P(A > B) = P(B, A)$ for every wff A and B,

for a (suitable) conditional connective $>$.[176] Call such a conditional a *probability conditional* (for function P). Hardly surprisingly, such a conditional must be different from the material conditional, for $P(A \supset B)$ and $P(B, A)$ coincide generally only for extremal values. Rather more surprisingly, the connective $>$ must differ from most other studied conditionals. However we should expect to 'discover a good deal about' the logic and semantics of a probability conditional 'just by using what we know about conditional probabilities'. Exactly this obvious procedure, which Lewis suggests (p.298), will be applied to locate a probability conditional.

[175] Some of the (former) adherents are listed in Lewis 1976, p.298, footnote 3. Another who remains tempted by these is van Fraassen; see van Fraassen 1976, and the attached correspondence with Stalnaker in Harper and Hooker 1976, p.302 ff. All unprefixed page references in the chapter are to Lewis 1976.

[176] In principle several conditions could be involved: as it turns out one is enough. Subsequently exactitude will be required as to whether P, which is serving in (6) as both a one and two place function from wff to the closed interval $[0, 1]$, is a partial or total function, and if total as to how it is to be defined when denonimators are zero, for (6) when $P(A) = 0$. The labelling of schemata such as (6) corresponds, for easy cross-reference, to that of Lewis 1976.

11.1 Locating a probability conditional

Lewis has claimed that such a search cannot succeed: 'the quest for a probability conditional is futile'. More precisely, *Lewis's thesis* is

> that there is no way to interpret a conditional connective so that, with sufficient generality, the probabilities of conditionals will equal the appropriate conditional probabilities (p.298).

Lewis's thesis, which has been widely, if sometimes reluctantly accepted, has significant corollaries.[177] So it is not a merely idle matter that Lewis's thesis can be rejected, that a probability conditional of 'sufficient generality' can be found.

Lewis's thesis is supported at bottom by triviality results. But these results lack the comprehensiveness they have been assumed to have, and required to sustain the thesis. In particular, as will emerge, Lewis has managed to give the *first* triviality result (pp.299–300) a bogus generality, for example by identifying a probability conditional—already characterised (as above) much more generally—with a so-called 'universal probability conditional', into which however dubious closure conditions are built, and by treating one way of determining the class of probability functions to be considered as it were *the* way. Let us begin then by considering, and sharpening, the *second* triviality result, where the closure assumptions made lie near the surfaces.

The key principle, upon which the triviality results from is not (6)—or what (6) comes to in fully binary form

(6*) $P(A > B, t) = P(B, A)$

with t some (arbitrary) tautology—but

(7) $P(A > C, B) = P(C, A \& B)$.

Lewis assumes (6) only where $P(A) > 0$ and so argues his way to (7) only where $P(A \& B) > 0$. While there are accordingly places where we shall have to take account of these restrictions or ensure that they would make no significant difference, it is also convenient for some of the argument to remove them and use total binary probability functions. This can be achieved simply by adding to some formulation of conditional probability (in the style

[177] For extravagant statements as to what the result achieves, see for example Adams 1975, Ellis 1979 and Stalnaker (cited above). Contrast however the enthusiastic reception of Lewis's result in Adams 1975 with the reluctant acceptance in Ellis 1979 (for example pp.74–5).

of Popper[178] and others), a further two-place connective, $>$, satisfying the formation rule: where A and B are wff so is $A > B$. Since then $P(B,A)$ and $P(A > B,t)$ are always well-defined, there is no need, and (should (6) hold in the qualified form) little reason, to restrict (6) by requiring that $P(A) > 0$, i.e., in effect that $P(A,t) > 0$. Similar points apply as regards (7), (6) being a *special* case of (7), since $A \,\&\, t$ is strictly equivalent to A.[179]

The hard core of Lewis's triviality results is that any probability theory containing a conditional conforming to (7)—whether qualified or not—is simply trivial. Such a theory (or class) is *simply trivial* if all its probability functions are, where a probability function, whether conditional or absolute, is *simply trivial* if it takes only the extremal values 0 and 1 where it is defined. A function is *Lewis trivial* if it is restricted to at most 4 values, i.e., it has only values, $1, r, 1-r, 0$, for some r with $0 > r > 1$. A probability function is *finitary* (or is *finitely trivial*) if restricted to finitely many values. While universal finitariness need not be death to a probability theory in quite the way that simple triviality, or for that matter Lewis triviality, is, it is nonetheless, as with a propositional logic,[180] a serious defect. Unfortunately for classical probability theories with (over-) *strong* conditionals the situation is worse than mere finitariness.

Proposition 1. Any classical probability theory[181] containing also a conditional connective $>$ conforming to (7) is simply trivial. Nor is simple triviality removed by requiring of (7) that $P(A \,\&\, B) > 0$.

Proof From (7), by substitution.

(9) $P(A > C, C) = P(C, A \,\&\, C) = 1$
(from Popper 1972, fn 1, p.353, (28′)) and

(10) $$P(A > C, \overline{C}) = P(C, A \,\&\, \overline{C}) = \begin{cases} 0 & \text{normally} \\ 1 & \text{abnormally} \end{cases}$$

Explaining and establishing (10) involves a detour:-
To begin with,

---

178 See Popper 1972, appendices, and, for more recent formalisation, Morgan 1982.

179 The fact that (6) is a special case of (7), and (7) a generalisation *of a sort* of (6), does nothing to transfer correctness from (6) to (7). Contraposition ($A \rightarrow B \rightarrow . \sim B \rightarrow \sim A$) is a special case, of a sort, of Antilogism ($A \,\&\, C \rightarrow B \rightarrow . \sim B \,\&\, C \rightarrow \sim A$), but while the first is relevantly correct the second is not, but directly induces paradox.

180 For the need for the infinitariness requirement, see RLR I, chapters 1 and 12.

181 Theories are employed because of their generality. For (closure) conditions on theories, see RLR I, chapters 2 and 4.

$$
\begin{aligned}
P(C, A \,\&\, \overline{C}) &= P(C \,\&\, \overline{C} \,\&\, A, A \,\&\, \overline{C}) \\
&\qquad \text{(cf. Popper 1972, fn 1, p.353, (30'))} \\
&= P(f \,\&\, A, A \,\&\, \overline{C}), \quad \text{where } f \text{ is } C \,\&\, \overline{C} \text{ say} \\
&= P(f, A \,\&\, \overline{C}) \times P(A, A \,\&\, \overline{C}) \\
&\qquad \text{(by Popper 1972, p.349, B2)} \\
&= P(f, A \,\&\, \overline{C}) \text{ since } P(A, A \,\&\, \overline{C}) = 1 \text{ as above.}
\end{aligned}
$$

But $P(f, A \,\&\, \overline{C}) = P(\overline{A \,\&\, \overline{C}}, A \,\&\, \overline{C}) = 0$ (normally) or $= 1$ (abnormally). Firstly, $P(f, B) + P(t, B) = 1 + P(\overline{B}, B)$ (Popper 1972, p.352, (64)), since $P(t, B) = 1$ and $(P(t, B) = P(\overline{B}, B)$. Secondly, $P(\overline{B}, B) = 1$ iff $P(\overline{B}, A) = 1$ for every B, and otherwise $= 0$ (see Morgan 1982 and references cited there). That completes the detour.

Now by expansion of cases for any wff D,

(11) $\quad P(D) = P(D, C) \times P(C) + P(D, \overline{C}) \times P(\overline{C})$.[182]

Hence with D taken as $A > C$, using (11), (9) and (10),

$$
\begin{aligned}
P(A > C) &= P(A > C, C) \times P(C) + P(A > C, \overline{C}) \times P(\overline{C}) \\
&= 1 \times P(C) + 0 \times P(\overline{C}), \\
&\qquad \text{in the normal case (where } P(A \,\&\, \overline{C}) > 0) \\
&= P(C).
\end{aligned}
$$

Otherwise in the abnormal case, $P(A > C) = P(C) + P(\overline{C}) = 1$,[183] so simple triviality is immediate.

It remains to show in the normal case, where

(12) $\quad P(C, A) = P(C)$,

that P is simply trivial. Suppose otherwise that, for some real number r with $0 < r < 1$, $P(E) = r$ for some E. Then $P(\overline{E}) = 1 - r$ with $0 < 1 - r < 1$.

---

[182] The argument for this from Popper's theory is rather more complex than using the (limited) connection $P(D, C) = P(D \,\&\, C)/P(C)$. However by (96) $P(D, C) \times P(C) = P(D \,\&\, C)$, and similarly with $\overline{C}$, so the RHS of (11) reduces to $P(D \,\&\, C) + P(D \,\&\, \overline{C})$. But since D is provably equivalent to $D \,\&\, (C \vee \overline{C})$ and so to $D \,\&\, C \vee D \,\&\, \overline{C}$, the LHS of (11) expands to $P(D \,\&\, C \vee . D \,\&\, \overline{C})$. But then by (79) $P(D) = P(D \,\&\, C) + P(D \,\&\, \overline{C}) - P(C \,\&\, \overline{C} \,\&\, D) = P(D \,\&\, C) + P(D \,\&\, \overline{C})$, since $P(C \,\&\, \overline{C} \,\&\, D) = 0$. For $P(C \,\&\, \overline{C} \,\&\, D, t) = P(\overline{t}, t)$, as in the text above. But since absolute probability is not trivial, for some F, $P(F, t) \neq 1$, so by (63), $P(\overline{t}, t) = 0$.

[183] Since $P(C) + P(\overline{C}) = P(C, t) + P(\overline{C}, t) = 1 + P(\overline{t}, t) = 1$, since as in the preceding footnote $P(\overline{t}, t) = 0$. That absolute probability can take values distinct from 1 follows from (26), i.e., for some D and E, $P(D, E) \neq 1$. For $P(D \,\&\, D) = P(D, E) \times P(E)$, by (96), so $P(D \,\&\, E) < 1$.

Thus, for example, $P(E, \overline{E}) = P(E \,\&\, \overline{E})/P(\overline{E}) = 0$ (by (97), Popper 1972, p.355). Hence by (12), $P(E) = 0$, so $r = 0$ contradicting assumptions.

The use of partial conditional probability functions defined in terms of absolute probability functions by the bridge principle

(1) $\quad P(A, B) = P(A \,\&\, B)/P(B)$ with $P(B) > 0$,

as in Lewis's triviality arguments, simplifies the initial argument above. For it follows directly that

(10) $\quad P(A > C, \overline{C}) = 0$ where defined, and

(11) reduces at once to $P(D) = P(D \,\&\, C) + P(D \,\&\, \overline{C})$ which is easily proved. However use of partial functions also excludes the final steps of the argument above. For as non-zero denominators are required, (9) and (10) are only well-defined where $P(A \,\&\, C) > 0$ and $P(A \,\&\, \overline{C}) > 0$ (where also $P(A) > 0$, $P(C) > 0$ and $P(\overline{C}) > 0$. Thus application of (12) to $P(E, \overline{E})$ is ruled out.

Suppose instead then, that for some D and E, $P(D) = P(E) = r$ where further $D \supset E$ demonstrably. So provably $D \,\&\, E \equiv D$, whence $P(D \,\&\, E) = P(D)$. Then, using (12),

$$P(E, D) = P(E) < 1 = P(D)/P(D) = P(D \,\&\, E)/P(D) = P(E, D)$$

which is impossible. Most simply let E be D. The downright implausible equation

(12*) $\quad P(A > C) = P(C)$,

which has the probability of a conditional dependent only on the probability of its consequent, and independent of the probability of its antecedent, none too surprisingly includes triviality.

(12*) is false, but (12*), like its equivalent (12), follows by classical probability theory given only (7). So, arguing classically, *(7) is false*. There are moreover grounds independently of proposition 1 for concluding that (7) is false. For one thing it follows from (7), by (6) (and so by (7) *itself*, given a definition), that

(7*) $\quad P(B >. A > C) = P(A \,\&\, B > C)$.

Since the equality holds for every function P,

(7†) $\quad A \,\&\, B > C \equiv. B >. A > C$,

for some equivalence of at least material (and indeed strict) strength. For if the two sides of (7†) were to differ in the truth-value we could find (using, for example, the semantical methods of Leblanc and Morgan) probability functions reflecting this difference and so countering (7*). Now (7†), which combines principles of exportation and importation, though it holds for material (and intuitionistic) implication, fails for proper conditionals. Counterexamples to exportation abound; nor is importation unproblematic (counterexamples to both are assembled in RLR I, chapters 1 and 3). Moreover recent theories of conditionals offered by Stalnaker, Chellas and others, including Lewis himself, one and all reject (7†). Counterexamples to (7) do not generally extend however to counters to (6) which yields *at most*

(6†) $\quad A > B \equiv . \, t >. \, A > B.$

There are, it should be observed, excessively classical features beyond (7) and (7†) implicated in proposition 1. Indeed the proof of proposition 1 breaks down for probability theories which are not classical, such as relevant and quantum probability theories. For in such theories the principle (11) of expansion by cases fails. (The principle depends for its derivation on the equivalence $D \leftrightarrow . \, D \,\&\, C \vee . \, D \,\&\, \overline{C}$ which is rejected, though for different reasons, in both relevant and quantum logics.) Naturally principles like (9), and so also (7), also fail for relevant probability theories; for no relevant logic has as a thesis what delivers (9) recouched in terms of a relevant implication $\rightarrow$ for all probability functions, namely the paradox $C \rightarrow . \, A \rightarrow C$. Lewis's argument to his conclusions—which is in itself decidedly curious, making no use of properties of implication at all, except that implication admits of (second degree) nesting—depends essentially then on features of classical probability, and breaks down for relevant probability functions.

Thus Lewis's thesis and argument can be blocked without the 'manoeuvres' he subsequently notes (p.304)—as if alternatives were always merely *ad hoc* and never independetly motivated approaches. Lewis's sole response to a differently motivated theory of probability, which defeats his results, is: 'I do not see what is to be gained by insisting on calling them "probabilities"' (pp.304–5). The short answer is, Much: a fuller answer (begun upon in JB, Appendix 1) goes into the curiosities and paradoxes of classical probability theory, the divergence of its claims from ordinary ones (and also quantum claims), and the way it impedes a decent theory of such things as probabilities of conditionals. When pressed, Lewis's point turns into an argument from meaning reminiscent of Quine's specious argument that the meaning of negation is the classical one, and that there cannot be an alternative connective properly called 'negation'.[184] But we shall not need to venture quite so far into non-classical territory as non-classical probability theories to defeat Lewis's argument.

As seen, it simply does *not* follow that because (7) is false, its special case (6) is false—any more than it follows from the fact that the principles of arithmetic are false as applied to all objects that they are false when suitably restricted to medium-sized solid entities. There are ways however of trying to bridge the gap between (6) and (7). The most important of these, exploited by Lewis, is by conditionalisation of probabilities.

Define $P_A(B,C) = P(B,C \,\&\, A)$. Then $P_A(B) = P_A(B,t) = P(B,A)$ as usual, except that the restriction that $P(A)$ be positive is removed. Alternatively, where restrictions to non-zero denominators are imposed, $P_A(B)$ may be defined as $P(B \,\&\, A)/P(B)$ and $P_A(B,C)$ as $P_A(B \,\&\, C)/P_A(C)$, i.e., $P(B \,\&\, C \,\&\, A)/P(C \,\&\, A)$, i.e., $P(B,C \,\&\, A)$. The function P_A is sometimes said to *come from* P *by conditionalising on* A. A class of functions is *closed under such conditionalising*—C *closed*, to be neutral—iff any function that comes from conditionalising on one in the class is also in the class. In brief, it is C closed iff whenever P is in Γ then P_A is in Γ for every A (for which it is defined). It is taken that such a class not only originates from, but is a class of probability functions, the assumption fails for the *focal class* of classical (conditional) probability functions—or FCC probability functions—conforming to (6), that is for the class of what are (from the present perspective) *focal* (probability) *functions*.

Proposition 2. (1) The class of merely classical probability function is C closed.
(2) The class of focal probability functions (i.e., classical functions conforming to (6)) is not C closed.
(3) A class of focal probability functions is C closed iff (7) holds generally for each function of the class.

Proof. (1) is a matter of detailed, but straightforward, verification that members of the class conform to the postulates of classical probability theory.

As to (3), a focal function exceeds classical requirements only in satisfying (6). Accordingly, given (1), a class Γ of focal functions is C closed iff the conditionalisation of (6) with respect to such wff holds for each function P in Γ, i.e., iff, for every B and every P in Γ,

---

[184] Quine's well-known argument against 'deviant logics' which focuses on negation is analysed and taken apart in NC. A similar critique can be made of attempts, like Lewis's to restrict probability to classically-defined probability.

$$P_B(A > C) = P_B(C, A).$$

But (6C), the (B) conditionalisation of (6), is just tantamount to:

$$P(A > C, B) = P(C, A \,\&\, B),$$

upon elimination of definitions. Under the alternative restrictive definitions some computation is required:- Suppose firstly (6C). Then $(A > C, B) = P_B(A > C) = P_B(C, A) = P(C, A \,\&\, B)$, as shown above. Suppose conversely (7). Then $P_B(A > C) = P(A > C, B) = P(C, A \,\&\, B) = P_B(C, A)$. again as shown above. Given (3), (2) results from modellings given below which falsify (7) but not (6).

Proposition 3. (Lewis's second triviality result, as amended.) Where Γ is a C closed class of probability functions with a probability conditional (i.e., for some connective $>$ (6) holds generally for every function P in Γ) Γ is simply trivial (i.e., every probability function in P is simply trivial). That is, where Γ is a focal probability theory which is C closed, Γ is trivial.

Proof. By Proposition 2.3, (7) holds generally in Γ. Hence by proposition 1, is simply trivial.

Proposition 3 seems to point to the conclusion that C closure is neither to be expected nor desirable. Lewis argues otherwise: at least—to impose a heavy restriction—he argues otherwise where [focal] theory Γ_B comprises 'the entire class of all those probability functions that represent possible systems of belief' (p.302), that is, so Lewis thinks, the class of 'all possible subjective probability functions' (see pp. 301-2).[185] The argument that Γ_B is C closed (*thesis* LC), which is really crucial to Lewis's case (since otherwise the corollaries Lewis wants to draw from proposition 3 do not emerge), is as follows:-

L1. 'The [rational?] change of belief that results from coming to know an item of new evidence ... take[s] place by conditionalising on what was learned' (p.302).

[185] So long as 'subjective' probability functions are classically constrained, the assumption that 'probability functions ... represent people's systems of belief' is quite evidently false, and open to evident objections. In particular, beliefs concerning statements and their negations do not vary inversely in the way that their classical probabilities do. A person may believe (also give due credence to) neither A nor $\sim A$, but both B and $\sim B$. The point is argued in Routley and Routley 1975; a corollary to be drawn is that relevant probability functions do a *much* superior job to classical ones in representing systems of belief.

L2. 'Rational change of belief never can take anyone to a subjective probability function outside the class' (p.302).

These premisses are both palpably insufficient for the intended conclusion LC and at best dubiously true. Firstly, the middle terms of the argument are not the same, yet equivocation on the middle term appear to be relied upon in the argument. For change of belief, as distinct from *classically-rational* change of beliefs, can certainly take a person to a function outside Γ_B (even rationally: see previous footnote) so violating L2; yet it is not evident that conditionalising on data always affords a classically-rational change of belief. Since no argument is offered for L2, and it is indefensible without the restrictions (and *then* is defended where Lewis points to for defence of L1), let us similarly restrict L1 (as the bracketed insert suggests). Secondly, what is required for the argument is the *converse* of L1 (obtained by inserting "only"), namely,

L1(C). Every case of conditionalising on probability functions supplied by Γ_B is a case of rational change of belief.

Whatever plausibility L1(C) may enjoy in certain individual cases (not in *any* case) vanishes when we consider every case, and in particular cases required for proposition 1 which conditionalise on both C and $\overline{C}$. Lewis quite illicitly restricts L1 to what is learned (or what one comes to know), so excluding the cases his triviality argument required.

Lewis smuggles the unwarranted C closure condition into his first triviality argument, thereby giving it a bogus universality. Here is how the trick is pulled off. Recall that conditional $>$ is a probability conditional for P iff (6) holds generally for that function P (see p.299). Call $>$ *a sham universal* probability conditional iff $>$ is a probability conditional for every merely classical probability function P. The conditional is a sham universal, and not fully universal in the way Lewis's alternative choice of terminology would suggest, because of the limitation to merely classical probability functions. Varying the class, either broadening it or focusing it, can sacrifice C closure, and so void the following results:

Proposition 4. (Lewis's first triviality result as amended). Any classical probability theory with sham universal probability conditional ($>$ say) is simply trivial.

Proof. By definition, (6) is required for every merely classical function. But by Proposition 2(1) the latter class is C closed. Hence by Proposition 3, the theory is simply trivial. (Note that the focal theory involved is that of merely classical probability functions for which (6) is imposed.)

What Lewis brings out is not that *Lewis-upsetting* probability functions (those for which (6) holds) are all trivial—which is what his triviality theorems have been taken to show—but that Lewis-upsetting functions are not nearly as common (on the ground) as might have been classically expected, and in particular are not to be expected among more mainstream apparatus.

Lewis's results concern C closed probability functions, and requisite use of these functions effectively involves portation (i.e., combined exportation and importation). Now while portation holds for classical and intuitionist logics, it is generally undesirable and fails for the conditionals and implications of both modal-style and relevant-style logics. It looks initially then as if we might expect to find non-trivial Lewis-upsetting functions through modal implications, for example with $A > B = \Box(A \supset B)$ for some modal connective $\Box$. But unfortunately with connectives of this sort with the right-properties (for example, those in systems in the vicinity of $S1$ and beyond), the modal approach breaks down.

Indeed *strict implicational analyses, and modal analyses, more generally fail.* Firstly, there are no strict implicational probability conditionals which are not entirely trivial. The result is formulated more exactly in the following two propositions.

Proposition 5. Where $-\!\!3$ is the strict implication of $S4$ or any extension of $S4$, the probability conditional $-\!\!3$ yielded is entirely trivial.

Proof. It turns on this strict equivalence of $S4$:

$$A \multimap B \multimap . \, t \leftrightarrow . \, A \multimap B, \text{where } t \text{ is some (any) tautology.}$$

Suppose $-\!\!3$ is a probability conditional, so $P(A, B) = P(B \multimap A)$. Then, since in any system of strict implication extending $S1°$, $\Box D \leftrightarrow . \sim D \multimap D$,

$$\begin{aligned} P(\Box D, t) &= P(\sim D \multimap D, t) \\ &= P(t \multimap . \sim D \multimap D) \\ &= P(\sim D \multimap D) \\ &= P(D, \sim D). \end{aligned}$$

But $P(D, \sim D)$ is entirely trivial (namely, 1 iff $P(D, D) = 1$ for every C, and 0 otherwise: see theorem 3 of Morgan 1982). Hence as $P(A \multimap B) = P(\Box(A \supset B), t)$, $-\!\!3$ is entirely trivial.

Proposition 6. Where $\dashv$ is a strict implication (any extension of Feys system $S1°$ say), the probability conditional $\dashv$ is entirely trivial.

Proof. Like the previous result except that a line is skipped. For $P({\sim}D \dashv D, t) = P({\sim}D \dashv D)$ simply in virtue of the connection of conditional and absolute probability.

The point of the longer first proof is that it connects with *a* way one might go about trying to *prove* in some system that an implication is a probability conditional through $S4$ principles. In fact such proofs are difficult to come by (as van Fraassen 1976 helps demonstrate), and it is much easier to look for modellings for probability conditionals. However a proof (of the worthwhile sort) would be rather more satisfying than a mere modelling (of almost whatever sort). One attempt at a direct proof using these two equivalences (which can be modelled in a weird extension of $S4$):

F1. $A \to (A \to B) \leftrightarrow . \, A \to B$ and

F2. $(A \to B) \,\&\, A \leftrightarrow . \, A \,\&\, B$

runs as follows (cf. van Fraassen 76, p.278).

$$\begin{aligned} P(A \to B) &= P(A \to . \, A \to B) && \text{by F1} \\ &= P(A \to B, A) && \text{by (6) for} \to \\ &= P((A \to B) \,\&\, A)/P(A) && (\text{with } P(A) > 0) \\ &= P(A \,\&\, B)/P(A) && \text{by F2} \\ &= P(B, A) && \text{by (1) again.} \end{aligned}$$

The problem with this sort of 'proof' is that it only proves (6) by assuming it in the course of the argument.

11.2 Boolean models for probability conditionals

A classical probability function is essentially a real valued measure on a Boolean algebra. But, by Stone's representation theorem, every Boolean algebra is isomorphic to a field of sets. Accordingly, without serious loss of generality, we can restrict consideration to certain measures on fields of sets.

In the finite situation at least, where the field can be divided into a set of n exclusive elements (atoms), the matter can be represented as diagrammed:

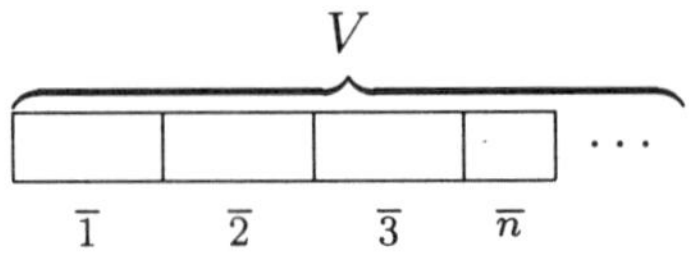

More generally, the universal set V is the union of elements, or cells, $\overline{1} \ldots \overline{n}$: $\overline{n}$ need not be finite. The components of the algebra consists of arbitrary unions of these elements, so the Boolean algebra B will have 2^n components. Represent the components of B as a, b, c,etc.

Now absolute probabilities can be freely assigned to cells, provided their sum equals 1 (or some suitable finite value); i.e., $0 \leq P(\overline{i}) \leq 1$ for each i, and $\sum_{i=1}^{n} P(i) = 1$. Then with any Boolean union is correlated the arithmetic sum of the probabilities of the arguments, with any Boolean complement the arithmetic subtraction from 1, $P(\Lambda) = 0$ and $P(V) = 1$. Conditional probability is defined as follows:-

$$P(a, \Lambda) = 1, \text{ and } P(a, b) = P(a \cap b)/P(b), \text{ provided } P(b) > 0.$$

The latter proviso we shall generally guarantee by assigning non-zero absolute probabilities to cells. But in case $P(b) = 0$ for all b, set $P(a, b) = 0$ when $a \cap b = \Lambda$, and $P(a, b) = 1$ otherwise. These Boolean models provide models of classical conditional probability, i.e., all the postulates of that theory, as given for instance by Popper (in 1972, Appendix iv), are satisfied.

To model also an FCC probability function it suffices to find an operator $>$ with assignments in B such that $P(b > a) = P(b > a, V) = P(a, b)$. This can be achieved if all probability values are among the values of $P(a, V)$, i.e., $\{r : \text{for some } a, b,\ P(a, b) = r\} \subseteq \{r : \text{for some } a,\ P(a) = r\}$. It sometimes can be achieved, as will now be shown to make it clearer what is going on and what is at issue. First the two element four-valued Boolean model is presented, and used to show the following

Proposition 7. There are FCC functions which are not simply trivial.

Proof. Apply the two-element Boolean model, with elements in $\{\overline{1}, \overline{2}\}$ and components or values $1 = \{\overline{1}, \overline{2}\} = V$, $2 = \{\overline{1}\}$, $3 = \{\overline{2}\}$, $4 = \Lambda$. The matrix for conjunction, tantamount to intersection can be read off the lattice diagram:

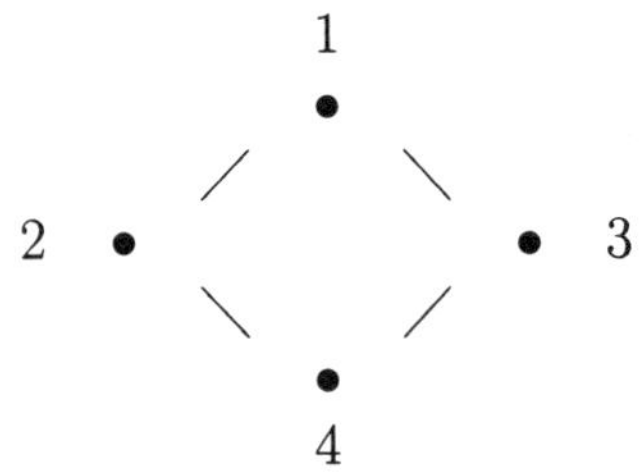

Matrices resulting are then:

$\cap$	1	2	3	4	$-$
1	1	2	3	4	4
2	2	2	4	4	3
3	3	4	3	4	2
4	4	4	4	4	4

	$P(a,b)$	1	2	3	4 [b]
	1	1	1	1	1
[a]	2	r	1	0	1
	3	$1-r$	0	1	1
	4	0	0	0	1

Absolute probability is determined by the assignment $P(\overline{1}) = r$ with $0 < r < 1$. Then $P(\overline{2}) = 1 - r$ with $0 < 1 - r < 1$ also. Since then for each cell b, $P(b) > 0$, conditional probability is entirely determined by the equations $P(a, 4) = 1$ and $P(a, b) = P(a \cap b)/P(b)$. Simple computation then delivers the matrix for $P(a, b)$, and hence for $P(b > a)$. Since $P(a > b) = P(a > b, 1)$ the matrix for $a > b$ is uniquely determined by the first column of the $P(a, b)$ matrix, and can be read off using the correspondence $\left\{\begin{matrix} 1 & r & 1-r & 0 \\ 1 & 2 & 3 & 4 \end{matrix}\right.$. So result these matrices:

	$P(a>b)$	1	2	3	4 [b]
	1	1	r	$1-r$	0
[a]	2	1	1	0	0
	3	1	0	1	0
	4	1	1	1	1

$>$	1	2	3	4
1	1	2	3	4
2	1	1	4	4
3	1	4	1	4
4	1	1	1	1

Then P is an FCC probability function, since it is classical and has been modelled to satisfy $P(a > b) = P(b, a)$. But P is not simply trivial because it takes four values.

Corollary P is not C closed.

For if it were P would be simply tivial, by the second Lewis result as amended. In fact the leading equations in the Lewis argument are straightforwardly countermodelled. To refute:

$$P(A > C, B) = P(C, A \,\&\, B), \text{ where } P(A \,\&\, B) > 0 \quad \text{and}$$
$$P(A > C, C) = P(C, A \,\&\, C), \text{ where } P(A \,\&\, C) > 0$$

assign B and C value 2 and A value 2. Then $P(A \,\&\, C) = P(2) = r > 0$, but $P(C, A \,\&\, C) = 1$ and $P(A > C, C) = P(1, 2) = r \neq 1$.

Although the connective $>$ is not a modal connective,[186] its logical behaviour is known from the investigation of relevant logic, where it has been studied as a, somewhat degenerate, conditional. Degenerate it may be—though little more so than the connectives of Stalnaker, Lewis and others that pass for conditionals—conditional it also certainly is while degenerates get admitted. In fact the connective $>$ is closely connected with that of first degree entailment, and coincides with it for an important class of cases.

Although there is then a pretty four-valued logic for a probability conditional, this is not going to satisfy many probability buffs. Probability is more than four-valued, even with roving third and fourth (coupled) values. The probability function involved is still Lewis trivial, and accordingly inadequate for a range of probabilistic purposes. What one would expect is that by taking larger Boolean models one could find FCC (functions admittedly not C closed) which are not Lewis trivial. This expectation proves wrong: *there are no larger finite Boolean models for probability conditionals.*

What goes wrong can be seen by looking at the next example with $n = 3$, there are three cells, $1, 2, 3$. Define values $1, ..., 8$ as follows (the lattice diagram can then be read off):

$1 = V = \{\bar{1}, \bar{2}, \bar{3}\}$	$8 = \Lambda$
$2 = \{\bar{1}, \bar{2}\}$	$7 = \{\bar{3}\}$
$3 = \{\bar{1}, \bar{3}\}$	$6 = \{\bar{2}\}$
$2 = \{\bar{2}, \bar{3}\}$	$5 = \{\bar{1}\}$

The standard eight-valued Boolean matrices then follow:

186 The matrices for set $\{\&, \sim, >\}$ invite comparison with those of C.I. Lewis for $\{\&, \sim, \dashv\}$. Only the implication matrix differs crucially, and then not in designation pattern.

$\cap$	1	2	3	4	5	6	7	8	$-$
1	1	2	3	4	6	6	7	8	8
2	2	2	5	6	5	6	8	8	7
3	3	5	3	7	5	8	7	8	6
4	4	6	7	4	8	6	7	8	5
5	5	5	5	8	5	8	8	8	4
6	6	6	8	6	8	6	8	8	3
7	7	8	7	7	8	8	7	8	2
8	8	8	8	8	8	8	8	8	1

For the most general probability assignments on $B_{(3)}$ assign absolute probabilities as follows: $P(\{\overline{1}\}) = n_1$, $P(\{\overline{2}\}) = n_2$, $P(\{\overline{3}\}) = n_3$ where $0 < n < 1, 0 < n_2 < 1, 0 < n_1 + n_2 < 1$ and $n = 1 - (n_1 + n_2)$. Then the matrix for conditional probability is as follows:

$P(a, b)$	1	2	3	4	5	6	7	8
1	1	1	1	1	1	1	1	1
2	n1+n2	1	n1/n1+n3	n2/n2+n3	1	1	0	1
3	n1+n3	n1/n1+n2	1	n3/n2+n3	1	0	1	1
4	n2+n3	n2/n2+n1	n3/n3+n1	1	0	1	1	1
5	n1	n1/n1+n+2	n1/n1+n3	0	1	0	0	1
6	n	n2/n1+n2	0	n2/n2+n3	0	1	0	1
7	n3	0	n3/n1+n3	n3/n2+n3	0	0	1	1
8	0	0	0	0	0	0	0	1

Now as $P(b > a) = P(b > a, 1)$ all values of $P(b > a)$ are represented in the first column of the conditional probability matrix. But there is *no way in general* in which the values in the first column can comprehend all values in the conditinal probability matrix, as would be required to model a FCC function. Only for quite special restrictions on n_1 and n_2 will this be possible. The base problem is that the initially given arithmetic class of values is not closed under division. (This statement renders more exact the intitial problem encountered in finitary situations as here, where there are eight values (or propositions) at most, but more than eight different probabilities get assigned to pairs of these eight values.) Indeed except for the case where the values collapse to four, because $n_1 + n_2 = 1$, so $n_3 = 0$, there is no assignment of rational numbers to n_1 and n_3 such that all values in the conditional probability matrix occur in the first column. It is not difficult to see why (at least in outline).

Proposition 8. There is no three element Boolean model for FCC functions which is not Lewis trivial.

Let $n_1 = k_1/m_1$ and $n_2 = k_2/m_2$ in lowest common form in the general case. Then $n_2 \neq 1 - n_1$, $0 < k_1 < m_1$, $0 < k_2 < m_2$, all elements being integral, and

$$n_3 = 1 - \left(\frac{k_1}{m_1} + \frac{k_2}{m_2}\right),\ n_1 + n_2 = \frac{k_1}{m_1} + \frac{k_2}{m_2},\ n_1 + n_3 = 1 - \frac{k_2}{m_2},$$
and $n_2 + n_3 = 1 - \frac{k_1}{m_1}$.

These provide all the values in the first column. The denominator class comprises $\{m_1, m_2, m_1m_2\}$. Now consider what happens upon division. Take $n_1/n_1 + n_2 = k_1m_2/k_1m_2 + k_2m_1$ after reductions. Then $k_1m_2 + k_2m_1$ exceeds m_1 and exceeds m_2 , in virtue of arithmetic requirements, so $k_1m_2 + k_2m_1 = m_1m_2$, or else the values of the first column are exceeded (given the assumed reduction of n_1 and n_2 to lowest common form, i.e., no common division except 1). But now $1 = k_1m_2/m_1m_2 + k_2m_1/m_1m_2 = k_1/m_1 + k_2/m_2 = n_1 + n_2$, contradicting assumptions. The argument from the three-element case is now generalised to any finite n-element case.

Proposition 9. There is no finite Boolean model for FCC functions which is not Lewis trivial.

Proof. Let $r_1 = k_1/m_1, ..., r_{n-1} = k_{n-1}/m_{n-1}$ be the probabilities assigned to elements, with integer k_i and m_i having no common divisor except 1. The probability of the remaining element r_n is determined by the relation $r_1 + ... + r_n = 1$. For each i, $1 \leq i \leq n$, $0 < k_i < m_i$. All remaining probability values in the first column are sums, without repetition, of the r_i's. The denominator class of all values in the first column is accordingly $\{m_1, ..., m_{n-1}, m_1m_2, ..., m_{n-2}m_{n-1}, m_1m_2m_3, ..., m_1m_2...m_{n-1}\}$; that is, it is all combinations combined by multiplication of $\{m_1, ..., m_{n-1}\}$ taken one at a time, two at a time, ..., all together, but, since there are no repetitions, powers of m_i do not occur (except insofar as $m_k = m_j$ for some k and j).

Although finite models for FCC probability functions which are not Lewis trivial are excluded—essentially because probability values outrun propositions—*infinite* models are not thereby ruled out.

One known infinite model—the interval model—for classical probability theory which will serve also to model FCC functions goes as follows:
The values are the intervals I (possibly discontinuous) within the interval $V = [0, 1]$. A valuation function $v : wff \rightarrow I$ mapping wff to intervals is defined as follows:

$v(P) \in I$, i.e., sentential parameters are assigned intervals (say, continuous intervals).
$v(A \& B) = v(A) \cap (B)$; $v(A \vee B) = v(A) \cup V(B)$; $v({\sim}A) = V - v(A)$, i.e., the interval comprising the set-theoretical complement of that interval assigned to A.

Also $v(t) = V$ and $v(f) = \Lambda$, where t and f are (arbitrary) tautologically true and logically false propositions respectively. Given this Boolean structure, define absolute probability thus:

$P(A) = \ell(v(A))$, where $\ell(a)$ for $a \in I$ is the length of a, with $\ell(\Lambda) = 0$ and $\ell(V) = 1$.

Finally, define conditional probability thus:

$P(A, B) = P(A \& B)/P(B)$ where $P(B) > 0$, and when $P(B) = 0$, set $P(A, B) = 1$.

Given that the interval model satisfies the postulates of classical probability logic, it remains to assign values to $A > B$, or $B > A$ for convenience, so that $P(B > A) = P(A, B)$. This can be achieved by assigning to $B > A$, for given assignments $v(B)$ to B and $v(A)$ to A, an arbitrary x in I for which the length of x just is the conditional probability assigned for the pair $v(A)$ and $v(B)$, in symbols $v(B > A) = \xi x \in I(\ell(x) = P(A, B))$.[187] For the latter, $\ell(x) = P(A, B)$, to be non-null it suffices that every probability is represented in the values of $P(C, V)$ (i.e., in the first column of the matrix to be indicated next): but this is guaranteed by the modelling.

Now $P(B > A)$, i.e., $P(B > A, V)$, is the length of the element x assigned to $v(B > A)$, i.e., $\ell(\xi x(\ell(x) = P(A, B)))$. (For given $v(A) = x$, $P(A, V) = \ell(x)$.) But this length just is $P(A, B)$. Hence $P(B > A) = P(A, B)$.

---

187 '$\xi x A$' *represents some particular* x *such that* A. *[Editors]*

12 Cause as an Implication

Co-authored with Newton da Costa, this essay originally appeared as Sylvan and da Costa 1988. [Editors]

The notion of cause, having fallen from favour in the heydays of logical positivism, has enjoyed a contemporary resurgence. But despite its fashionability now, especially as a major foundational element in epistemology, the logical and structural properties of causation remain quite insufficiently examined. In this situation, who knows whether the foundations will carry the philosophical castles being built (they are never complete, and invariably ramshackle)? Our preliminary investigation of causal implication suggests they will not; like structurally and materially short-supplied high-rise buildings, they will come tumbling down.

In treating *cause* as like a conditional (or implicational type) there is a familiar problem of getting in a satisfactory way from natural language locutions, for example of the form "α causes β", to conditional forms which typically couple sentences. A variety of expressions plug into the causal form, not just event subjects but gerundives such as "smoking"—but not significantly sentences. One difficulty in working with "α causes β" as primitive is that not all the usual sentential connectives are particularly well-defined on the relevant substituents; for example, negation becomes problematic with event clauses, though not unintelligible as with proper names.

Still, it is not so difficult to make out what negation is doing applied to event clauses, to construe for instance $\sim$(mowing the lawn), as not mowing the lawn. Let us allow in *whatever* can be made to fill out variable places, and let variables correspondingly range over such makable-out values. The constraints will then be *imposed* by the formation rules, not fixed in advance. Connectives which combine schematic variables A, B, C, ... delimit what can be made out. For comparison with familiar logics, let these connectives be, to start with, just &, $\vee$, $\sim$, giving the zero-degree wff. Then we add the causal connective $\ni$, read in context "(the) obtaining of ... causes (the) obtaining of ...". It can be left open for the present to what extent $\ni$ combines with extensional connectives &, $\vee$, $\sim$; for instance, whether an entire higher degree results or only a first degree part. This is a matter of what can be made out. Making out itself can eventually be taken up through significance.

In this way we arrive, hopefully then, at a sufficiently well-behaved form $A \ni B$, to begin logical investigations, a form where &, $\vee$, $\sim$ are well-defined on combinations of A and B, and also on uniterated forms $A \ni B$. For example, $A \ni B$ may read "A causes B", with A and B not sentences, or "should A happen it would cause B to happen", or it may sometimes be read "A's happening causes B's happening", or it may even be read "the truth of A *causes* the truth of B" (whatever that really means). The essential point is that we don't shift out of the causal idiom at the outset (as for example Burks 1951, Zinov'ev 1967, and others do), and so prejudge or prejudice several issues, such as whether causal idiom is modal, whether we are working with something like partial sufficiency or not, and so on.

In this way also we can wend through a minefield of research-paralysing objections (such as those laid by Davidson 1980, for example p.161) as to what the grammatical and logical forms of causal statements really are. In particular, we don't say, or require, that the relation of causation is represented through a sentential connective. It is enough that $\ni$ forms sentences on certain terms, which are sometimes propositional or fact-like or rendered such by happening or occurrence functions (so delivering a partial higher degree). Most conveniently it is much the same with *implies* as it is with *cause*, and as it is with bridge verbs such as *means* and *intends*. All these two-place connectors are sentence-forming on many linguistic items other than propositional or fact-like expressions. For instance, "that flood caused a famine"; is parallelled by "that flood meant a famine"; "the failure of the sprinkling system caused that fire" is matched by similar sentences built around "meant" or "implied". What is this implication-like causal connection like?

The method of the reflective equilibrium approach we then adopt is this: we try to reach, by working through a standard list of key logical principles for implication, a core logic and some important rejections for the notion of cause. We certainly leave open the option that various types of cause can be characterised through satisfaction of extra principles, for example reverse causation, proximate causation, logically extended causation, etc. In reaching a core we are much helped if we separate out various notions that have been confused with cause, such as conditionals. Having begun to move to a core causal logic, and associated typology, we begin to cast about for some logical explanation of the core, preferably of a semantical cast or semantically adaptable. In the presentation we borrow symbolism and terminology from RLR I.

12.1 Apparent principles, and rejections, of causal implication

Clearer principles of a logic of cause, formulated in $\ni$, include these:

Transitivity: $$\frac{A \ni B \quad B \ni C}{A \ni C}$$ (at least in rule form)

(In fact Conjunctive Syllogism, (CS) $A \ni B \,\&\, B \ni C \rightarrow .\ A \ni C$, looks alright.) It is a little theorem, given appropriate epilogical apparatus, that:

$$\frac{A_1 \ni A_2 \quad A_2 \ni A_3 \quad \ldots \quad A_{n-1} \ni A_n}{A_1 \ni A_n}$$

Irreflexivity: $\sim(A \ni A)$.

Here the logic of cause diverges, sharply and distinctly it seems, from the logic of implication (under usual construals and interpretations). It also departs from the logic of explanation. For one thing, explanation does not require existence. The scientific creation, the initial Big Bang, may lack a cause, because nothing existed before it, but it does not lack an explanation. For another, self-explanation is not ruled out logically. Explicit irreflexivity separates causal implication from what is now dubbed non-circular reasoning; for that, while rejecting Identity leaves its negation open.[188]

Modus Ponens (MP): $$\frac{A \quad A \ni B}{B}$$

(Like substitutivity, for example where Γ is null, this exposes A; for in the first premiss A isn't embedded in a context of form $\ni$). MP appears to settle against *mere* partiality of cause: the connection really is invariant. The principle also helps distinguish *cause* from *reason*.[189] Similarly, where it makes sense (with A and B so exposed),

Modus Tollens: $$\frac{\sim B \quad A \ni B}{\sim A}$$

The principles arrived at so far are enough to justify the honorific title 'implication' (though some would want to insist that it is an exceedingly weak, or decidedly rum, implication). The linguistic connections seem to be more or less in order. We can move from $A \ni B$, through "If A *then* B, as a matter of causation", through "A causally implies B", and back again,

---

188 Marketed non-circular reasoning, as in Martin and Meyer 1983, diverges from causal implication, and perhaps from genuine non-circular reasoning, in its heavy endorsement of *Exported* Syllogistic principles (for example, symbolised in $\rightarrow$ form, $A \rightarrow B \rightarrow .\ B \rightarrow C \rightarrow .\ A \rightarrow C$ and its permuted mate).

189 *See further Chapter Six. [Editors]*

naturally enough, for many substitutions upon A and B. The logical issues as to whether some sorts of containment connections (those of *im-plicatio*) can be induced we shall come to: but to anticipate, the answer is that they can.

From here on, however, principles get less and less clear. For consider

Contraposition: $\dfrac{A \ni B}{\sim B \ni \sim A}$ (which yields Modus Tollens from MP)

This seems wrong; for example, no lung cancer doesn't *cause* no heavy smoking, though it *implies* in some sense that there's been none. Cause induces an implication connection that *extends* it, $\mapsto\!\!\ni$ say. Then in place of Contraposition the following appears to hold:

$$A \ni B \,/\, A \mapsto\!\!\ni B \qquad \text{and} \qquad A \mapsto\!\!\ni B \,/ \sim B \mapsto\!\!\ni \sim A$$

The *direction* of cause also appears wrong for Contraposition. The matter of directionality of causation is of course an important, and not uncontroversial, one.

Substitutivity: $A \ni B, B \ni A \,/\, \Gamma(A) \ni \Gamma(B)$ (in familar form)

where $\Gamma(A)$ is a compounding (at least truth-functional) of A, fails. Normally its success would be tied to that of Contraposition and analogous compounding principles. But in causal cases, joint satisfaction of the premisses is excluded. If both premisses were to obtain then by transitivity $A \ni A$, violating irreflexivity. So substitutivity fails by default.

Connexivity: $\dfrac{A_1 \ni A_2 \quad A_2 \ni A_3 \ldots A_{n-1} \ni A_n}{\sim(A_1 \ni \sim A_n)}$

with special cases

Aristotle: $\sim(A \ni \sim A)$

Rule Strawson: $A \ni B \,/ \sim(A \ni \sim B)$

Causation is, so to say, strongly consistent; for with causal consistency defined, $A \diamond B = \sim(A \ni \sim B)$, then $A \diamond A$, etc. Such connexive principles tend to make the theory strongly resistant to known logical technology.

Composition: (&-form) $A \ni B, A \ni C \,/\, A \ni B \,\&\, C$
(∨-form) $A \ni C, B \ni C \,/\, A \vee B \ni C$

But that is around about where familiar &-∨ behaviour terminates. By contrast, other lattice principles almost all fail. Thus, with $\dashv$ representing in the usual way rejection, $\dashv A \,\&\, B \ni A$, $\dashv A \ni A \,\&\, A$, $\dashv A \vee A \ni A$,

$\dashv A \ni A \vee B$; etc. Distribution, however, may hold where the same variables figure appropriately. Consider next

Augmentation: $\dfrac{A \ni B}{A \,\&\, C \ni B}$ and $\dfrac{A \ni B}{A \ni B \vee C}$

Such principles are decidedly doubtful. They are in doubt for the same sorts of reasons that Augmentation is problematic, at least, for what is closely related to causation, counterfactual conditionals. For the tacked-on additions are doing no *casual* work, and more important may *interfere* with causal connections.

Maybe instead what hold are the implicational extensions:

$$\frac{A \vdash\!\ni B}{A \,\&\, C \vdash\!\ni B} \quad \text{and} \quad \frac{A \vdash\!\ni B}{A \vdash\!\ni B \vee C}$$

Factor: for instance in the &-form, $A \ni B \,/\, A \,\&\, C \ni B \,\&\, C$,

is similarly dubious, because the tacked-on components typically do no causal work, and may indeed interfere.

The *higher degree*, insofar as it can be made out (or makes sense), is still more problematic. One principle that looked good to one of us at first is:

DC: $\dfrac{A \ni B \qquad A \ni (B \ni C)}{A \ni C}$

But another principle that went along with DC is *much* less evident, namely

Contraction: $\dfrac{A \ni (A \ni B)}{A \ni B}$, whence

n-place Contraction: $\dfrac{A \ni (A \ni \ldots (\ldots (A \ni B) \ldots)}{A \ni B}$

Perhaps Contraction is a *defective* case of the DC principle; it ordinarily would be derived using $A \ni A$, no principle of causation.

As is by now evident, there are two directions of primarily first degree development feasible:

(1) embedding the logic in an entailment logic, as with principle (CS) of Conjunctive Syllogism above.

Then principles like $\dfrac{A}{\sim\sim A}$ and $\dfrac{\sim\sim A}{A}$ are automatically given in.

(2) elaborating the logic of its own. Then the pleasant weakness of inferential principles can become a serious initial difficulty.

12.2 Beginning on a typology

Various types or subvarieties of cause can be distinguished. As indicated, *augmentable* causation can be distinguished, either as that subclass also satisfying augmentation principles or, more satisfactorily, as the closure of the notion under augmentation rules; *a reversible* causation as the closure under contraposition; and so on. Different is actual (or realised) causation. The emphasis on what is actual is especially strong with notions like cause, but so far we have been operating, in a way resembling conditionals, with hypothetical or putative causes. *Actual* causes, ontologists want to insist, must happen. Let's distinguish actual causes then, for instance by the principle:

$$A \ni B \rightarrow . A \& B$$

Then actual cause fails various principles that hold for (putative) cause, such as Modus Tollens; and would accordingly make a difficult starting point. In general, actuality requirements damage the smoothness and generality of a logic, here as elsewhere (for example, with existential requirements and loading in quantification logic). However given (hypothetical) causation, actual causation can be defined as holding *where* the antecedent is realised, i.e., in extensional approximation, $A \ni^E B =_{df} (A \ni B) \& A$. Then the distinguishing principle follows, where a suitable implication is available.

Quite different 'types' again are those of *partial* cause—not really a cause—and *local* and *proximate* causes. A partial cause is a causal factor in Mill's sense, one significant component in a cause. As a first shot it can be characterised enthymematically: A is partial cause of B iff A is part of some cause C which causes B. Local or immediate cause is a cause, delimited by a neighbourhood requirement (or even more narrowly, in the seventeenth century, by a juxtaposition requirement). Local cause contrasts with remote cause or action at a distance. Provided "neighbourhood" is not too narrowly construed, cause is a transitive closure of local cause. And local cause is of course cause restricted to a neighbourhood, for example $A \ni_L B$ iff $A \ni B \& (\text{for some } n)(A \subseteq n \& B \subseteq n)$, or some such. Some topology can be pulled in to clarify neighbourhood behaviour. In physical theory action at a distance was made local by way of fields. Is there any holistic analogue (some sort of causal flux or 'field') that can be profitably introduced into causal logic? These are all questions we can postpone until the more pressing logical work has advanced, until we have gained some semantical grip upon the core implicational notion.

12.3 An organising process account

A process analysis both has much appeal and promises a neat semantical theory for causal implication—at least at the first degree, and presumably it admits of *some* higher degree extension. The basic connection, given in translation form,[190] rather than in semantical form, is as follows:

P1.	$A \ni B$	iff	$A \rightarrow B$	&.	$A < B$
	process:	=	*relation:*	+	*direction:*
	causal		sufficiency		initiation, priority,
	conditional		conditional		e.g. time order

While there is nothing inevitably first-degree about this connection (unless priority is narrowly construed), suppose for the present that causal imlication is first-degree, i.e., $\ni$ is not nested; this is after all the usual intuitive setting. Now the first-degree semantical analysis of implication or conditional, $\rightarrow$, is essentially a *constant conjunction* one; $A \rightarrow B$ is true iff for every situation where A holds so does B. The situations involved are in no way restricted to those that actually occur, so there is no serious problem about accidential 'conjunction'. But the association may still not be causal, for instance B could be a logical consequence of A or A an effect on B. The problem is resolved, it is supposed, by a priority ordering, which gives a direction. Commonly it is assumed that the priority ordering is a temporal one, but it strikes us as illegitimate to rule out time-reversing causation, such as is alleged to occur in some psychic phenomena, in this crude way. So we generalise time ordering to 'action priority'; the time of A may in principle be later than B, yet A has initiation action or priority to B; causally A is the *prior moment*. Those who find this idea paradoxical (for example, because initiations or actions fix times) can however construe $c(A, a)$, the causal moment of initiation of A at situation a, just as the time of A at a.

The priority ordering $A < B$ will be interpreted then, at a, through the precedence ordering $c(A, a) < c(B, a)$. Writing $c(A)$ for $c(A, T)$, i.e., the action of A at actual (truth determining) situation T, then $A < B$ will be true when $c(A) < c(B)$, i.e., roughly when the time of A precedes that of B. The conditional, $A \ni B$, is interpreted in the standard relevant and strict way in terms of formal (truth-)value preservation, i.e., as $I(A, a) = 1 \supset_a I(B, a) =$

[190] Connections of this sort appear in the literature; for example P1 (below) is suggested in essentials by von Wright (who also introduces the notion of action, but unfortunately conflates it with manipulation). Indeed, any combination of a conditional with a tense-logical ordering will yield something like P1.

1, where $I(A,c) = 1$ reads: A holds at (situation, world) c. Thus, at the first-degree, we can arrive at:

P2. $A \ni B$ is true iff $(a \in K)(I(A,a) = 1 \supset I(B,a) = 1) \,\&\, c(A) < c(B)$,

i.e., iff, whatever apposite situation a, B holds at a when A does and A is initiated before B, i.e., iff *B is constantly conjoined with A which is the prior.*[191] Delineation of the class K of *apposite situations*—perhaps certain situations neighbouring or similar to that where evaluation is taking place—is of course critical in a full theory and in many applications. In the further clean-up of P2, we may want to restrict situations also to those that conform to various constraints, for instance (at some risk of circularity) to natural laws. But those are important details that we can here comfortably leave in the background.

With but few assumptions, we can now substantially confirm initial intuitive principles. It will be assumed—what can be cogently argued—firstly that implication ($\rightarrow$) conforms at least to the principles of first degree relevant implication (thus it could be a strict implication, etc.), and secondly that the order $<$ is transitive and irreflexive (and so, antisymmetric).[192]

Proposition 1. Causal conditional $\ni$ is an implication satisfying Rule Transitivity, Irreflexivity, Modus Ponens, Modus Tollens but failing Identity.

Verification uses P1 and properties given.

To proceed the further intuitive distance, it will be assumed, thirdly, that at least for extensional connectives in the same components priority is determined componentwise, that is $c(\sim A) = c(A)$, $c(A \,\&\, A) = c(A)$, $c(A \vee A) = c(A)$. This leaves open, so far, what is said as to $c(A \,\&\, B)$, and so on, for different components.

---

191 As we found, on subsequently browsing relevant literature, we have been thoroughly anticipated—this is no deficiency here, but grist to our mill—in a semantical account of general form P2, in particular by Mackie 1980, who proposed a conjunction of necessity-in-the-circumstances with causal priority (see also the discussion of Mackie in Leman 1984, pp.255-6). We agree with Mackie and Leman that a conditional or qualified-necessity analysis of causation needs to be supplemented by a further element, affording a foundation for causal asymmetry. But we disagree with Leman that this foundation, provided by priority (or fixity), has to be epistemological (for many reasons, some of which Leman himself advances, p.256).

192 *The following claims concerning the causal conditional mirror claims made ealier concerning "reason for". See Chapter Six, section 3. [Editors]*

Proposition 2. Causal conditional $\ni$ is a connexive implication satisfying Aristotle but failing Contraposition, Simplification, Addition (and elementary lattice conditions generally), and very important, failing (SI) Substitution.

Verification is again elementary, but some examples are instructive. Consider Aristotle, which expands to $\sim(A \rightarrow \sim A) \vee \sim(c(A) < c(A))$, and is carried by the latter disjunct. If Simplification held, then $A \,\&\, A \ni A$ would hold, implying $c(A) < c(A)$. Rule Contraposition fails because $<$ is antisymmetric. Etc. As the example of Aristotle reveals, some very curious negative statements are also verified. Let $\triangle_1(A)$, $\triangle_2(A)$ be any extensional functions just of A. Then $\sim(\triangle_1(A) \ni \triangle_2(A))$.

To fill out the first degree picture, it remains to impose appropriate connections between $c(A \,\&\, B)$ and the pair $c(A)$, $c(B)$. A natural approach to the issue of connections is to let actions (or times) pluralise, so the actions of $A \,\&\, B$ are those of A and of B, i.e., $c(A \,\&\, B) = c(A) + c(B)$, where $+$ is some sort of union operation (a semi-lattice operation suffices). Then, since any truth-functional combination $\delta(A_1, ..., A_n)$ can be defined in terms of just connectives & and $\sim$, $c(\delta(A_1, A_2, ..., A_n)) = c(A_1) + c(A_2) + ... + c(A_n)$. For instance, $c(A \vee B) = c(\sim(\sim A \,\&\, \sim B)) = c(\sim A \,\&\, \sim B) = c(\sim A) + c(\sim B) = c(A) + c(B)$ (given semi-lattice conditions and either suitable definitions of truth-functions or the principle $A \leftrightarrow B \,/\, c(A) = c(B)$ for A and B truth-functional). It follows, what may be a little surprising, that action reduces to that of sentential variables, at least in truth-functional cases. (Clearly, such distribution principles fail, or *should* fail, for such functors as the tense-logical F, it will be the case that: $c(A) \neq c(FA)$ for many A.)

Proposition 3. Causal implication satisfies various Composition principles, conforms to Augmentation only in restricted form, and fails Factorisation principles unless also restricted.

Verification and elaboration. Consider for instance &-Composition. Assuming $A \ni B, A \ni C$ it follows that $A \rightarrow B$, $A \rightarrow C$, $c(A) < c(B)$, and $c(A) < c(C)$. Then, by the first pair $A \rightarrow (B \,\&\, C)$, while by the second $c(A) < c(B) + c(C) = c(B \,\&\, C)$, whence, assembling, $A \ni (B \,\&\, C)$. Factorization by contrast fails, for there is no guarantee that when $c(A) < c(B)$ that $c(A \,\&\, C) < c(B \,\&\, C)$, i.e., $c(A) + c(C) < c(B) + c(C)$. What happens with Augmentation, which similarly fails in general, is interesting. Consider the form: $A \ni B \,/\, A \,\&\, C \ni B$. The critical issue, given implication sustains augmentation, is when $c(A) < c(B)$ guarantees $c(A) + c(C) < c(B)$. Presumably generally only if $c(C) < c(B)$; roughly C is causally relevant to B. In short, causally relevant additions can be tacked on, but not arbitrary ones.

12.4 Running out initial semantics

The account given lends itself to ready semanticization. The semantics extends, or can be pushed, to the higher degree, i.e., for full sentential systems—given in our first attempt the implausible assumption that $c(A, a)$ is always determined through initial components (i.e., in pure calculi, variables). We shall offer semantics for full systems, though the significance of nested occurrences of "$\ni$" is often in doubt, to say the least, beyond the second degree. As we shall see, relevance itself provides some natural restrictions, and enables reduction of implausibility.

We take over semantics for relevant logic formulated in terms of connectives $\{\rightarrow, \&, \vee, \sim\}$. We view the semantics as fashioned for some deeper relevant system, for instance for a relevant theory of conditionals such as that of Hunter 1983 (not really for the excessively strong relevance systems of Anderson and Belnap or the analogous strict systems of Lewis, though they can be used for all of them). A relevant model structure m.s. $\langle T, K, O, R, *\rangle$ comprises a set K of situations (or worlds), a subset $O \subseteq K$ of regular situations (where theorems hold), a base (actual) situation T in K), a three-place relation R on K (modelling implication) which collapses to a three-place relation in irrelevant logics, and a reversal operation $*$ (which takes care semantically of negation). As it happens, the regular situations will come to play a significant role in relevant causal implications, as in deeper relevant logics. We add to such relevant m.s. (further explained in RLR I), a set Ca of *moments* for each situation a in K, and an operation $+$, on each Ca, which is commutative, associative, idempotent (i.e., $\langle Ca, +\rangle$ is a semi-lattice). The moments or actions of set Ca (which will be construed canonically just in terms of classes of sentential variables, whatever these objects do) are open to a variety of interpretations, here as initiation or priority or fixing stages, but elsewhere—for different readings of "$\ni$" as grounds, information bits, contents. Order relations are defined for α, β in Ca as follows:

$$\alpha \leqslant \beta \text{ iff } \alpha + \beta = \beta; \ \alpha < \beta \text{ iff } \alpha \leqslant \beta \,\&\, \alpha \neq \beta.$$

The modelling itself takes weaker or stronger forms, depending upon whether causal implication is to be fully relevant or not. We present the stronger, more simplistic form first, and then give the variations demanded by relevance.

A CI m.s. $\langle T, K, O, R, *, C, +\rangle$ adds to a relevant m.s. just a function C from situations to sets of moments and a semilattice operation $+$. A CI (causal implication) *model* adds to a CI m.s. two valuation functions, I and c. Valuation I, from sentential parameters and situations to holding values 1 (on) and 0 (off), subject to a hereditariness requirement, is as in relevant semantics. Valuation c is also from sentential parameters and situations but maps to situationally-associated moments, i.e., $c(p, a)$, with c representing the causal moment or initiation or time (or differently, information or content) of p at a, belongs to Ca.

Both I and c are extended from initial sentential wff to all wff inductively. The evaluation clauses for I, as applied to connectives &, $\vee$, $\sim$, $\rightarrow$, are just those of relevant semantics. The further clauses are as follows:

$c(A, a) = \sum_{j=1}^{n} c(p_j, a)$, i.e., $c(p_1, a) + ... + c(p_n, a)$,
for $p_1, ..., p_n$ exactly the sentential parameters of A.
$I(A < B, a) = 1$ iff $c(A, a) < c(B, a)$
$I(A \ni B, a) = 1$ iff $I(A \rightarrow B, a) = 1$ & $I(A < B, a) = 1$,
i.e., iff for all b, d, $(Rabd$ & $I(A, b) = 1 \supset. I(B, d) = 1)$
& $c(A, a) < c(B, a)$.

To make axiomatisation and demonstration of its adequacy run smoothly further clauses for $\leqslant$ and $\approx$ are added:

$$\begin{aligned} I(A \leqslant B, a) &= 1 \text{ iff } c(A, a) \leqslant c(B, a); \\ I(A \approx B, a) &= 1 \text{ iff } c(A, a) = c(B, a). \text{ Then,} \\ I(A < B, a) &= 1 \text{ iff } c(A, a) \leqslant c(B, a) \,\&\, c(A, a) \neq c(B, a), \\ \text{i.e., iff } I(A \leqslant B, a) &= 1 \neq I(A \approx B, a). \end{aligned}$$[193]

Validity and other semantical notions for CI are defined as usual for relevant logics.

For a matching stronger axiomatisation of causal implication, CI, we add to any relevant-based affixing logic in connectives ($\rightarrow$, &, $\vee$, $\sim$)—the *carrier* logic—the following postulates for further connectives $\ni$ and $\leqslant$ of CI, where $A \approx B =_{df} A \leqslant B \,\&\, B \leqslant A$ and $A < B =_{df} A \leqslant B \,\&\, \sim(A \approx B)$:

1. $A \ni B \leftrightarrow. A \rightarrow B \,\&\, A < B$ (i.e., P1)
2. $A \leqslant B \,\&\, B \leqslant C \rightarrow A \leqslant C$ (i.e., $\leqslant$ is transitive)
3. $A \leqslant C \,\&\, B \leqslant C \rightarrow. A \,\&\, B \leqslant C$ (i.e., $\leqslant$ composes)

193 Part of the apparent circuity in argument here involved can be traced back to the relevant behaviour of negation (see the final stages of the completeness proof, section 4). In irrelevant settings such as strict implication such circuity is easily avoided.

4(I). $D \rightarrow. A \leqslant B$, where the sentential parameters of A are a subset of those of B, i.e., in symbols $v(A) \subseteq v(B)$.

The irrelevant form 4(I) immediately yields

4. $A \leqslant B$, where $v(A) \subseteq v(B)$,

as well as much irrelevant junk, such as $q \rightarrow. p \leqslant p$.

In irrelevant logics which offer some control (over classical licence), D of 4(I) can be restricted in form, for instance to $\Box D$ in modal logics and $D \ni D$ in intuitionism (and in I-systems); and those logics of course enable the derivation of 4(I) thus restricted from 4. It is the quest for relevance, and thus for replacement of 4(I) and its variations by 4, that leads to weaker and deeper axiomatisations of causal implication.

It is a simple matter to reaxiomatise in terms of $<$ (and $\approx$) as primitive (or in terms of a single union operation). Replace 2, 3 and 4 by

2′. $A < B \,\&\, B < C \rightarrow. A < C$

3′. $A < B \,\&\, B < C \rightarrow. A \,\&\, B < C \vee. A \,\&\, B \approx C.$

4′. $A < B$, where $v(A) \subset v(B)$.

A main reason for some indirection is that 4′ complicates semantical validation, excluding (correctly) some finite representations of $c(A, T)$. To work directly with $<$ it is enough to require that for p_i distinct from p_j, $c(p_i, a) \neq c(p_j, a)$. Introduction of $\leqslant$ is however required for the type of completeness argument adopted.

Although the carrier logic is relevant-*based* in the sense of admitting a relevant logic, it may be decidedly irrelevant, for example in classical or intuitionistic directions. Thus any of a huge range of carrier logics may be chosen (including indeed, with minor variations in the argument, many not based on relevant systems).

Some theorems used in showing completeness are recorded.

T1. $A \leqslant A$; by 4.

T2. $A \leqslant A \,\&\, B$; $B \leqslant A \,\&\, B$, by 4.

T3. Where $A_1 \leqslant A_2$ then $A_1 \,\&\, B \leqslant A_2 \,\&\, B$
For, as $A_1 \leqslant A_2$ and $A_2 \leqslant A_2 \,\&\, B$, by T2, $A_1 \leqslant A_2 \,\&\, B$. But, by T2, $B \leqslant A_2 \,\&\, B$ also; whence by 3, $A_1 \,\&\, B \leqslant A_2 \,\&\, B$.

T4. Where $B_1 \leqslant B_2$ then $A \,\&\, B_1 \leqslant A \,\&\, B_2$. Similar to T3.

T5. Where $A_1 \leqslant A_2$ and $B_1 \leqslant B_2$ then $A_1 \,\&\, B_1 \leqslant A_1 \,\&\, B_2$. By T3, T4.

T6. $D \rightarrow. B \leqslant B$, $D \rightarrow. B \leqslant A \,\&\, B$, $D \rightarrow. A \leqslant A \,\&\, B$; applying 4(I).

Stronger adequacy result. A wff A is a theorem of CI iff A is CI-valid. (In fact the argument will yield strong completeness).

(I) Soundness. The following properties of relevant semantics much simplify verification of postulates:- $I(C \rightarrow D, T) = 1$ iff for every situation a, where $I(C, a) = 1$ then $I(D, a) = 1$. Similarly $I(C \leftrightarrow D, T) = 1$ iff for every situation a, $I(C, a) = I(D, a)$. Then new postulate 1 is immediate from the evaluation clause for $\rightarrow$.
ad 2. $\leqslant$ is transitive on Ca, for each a.
ad 3. $c(A \& B, a) = c(A, a) + c(B, a)$; but where $\alpha \leqslant \gamma$ and $\beta \leqslant \gamma$ then $\alpha + \beta \leqslant \gamma$, by semi-lattice properties.
ad 4(I). Suppose $v(A) \subseteq v(B)$. Then $c(A, a) = \sum c(p_A, a) \subseteq \sum c(p_B, a) = c(B, a)$, where p_A ranges over variables of A and p_B over those of B. Hence $I(A \leqslant B, a) = 1$, and thus where $I(D, a) = 1$ for arbitrary wff, so classically does $I(A \leqslant B, a) = 1$.

(II) Completeness. To the canonical m.s. for the carrier logic, defined as usual but on *non-null* situations (see RLR I), are added further details for C and $+$. The requisite canonical semi-lattice is arrived at by standard methods for Lindenbaum algebras. An equivalence on wff, $\tilde{a}$ is defined for each situation a thus: $A \ \tilde{a} \ B$ iff $A \leqslant B \in a \& B \leqslant A \in a$. Then $\tilde{a}$ is reflexive and transitive by postulates 4 and 2 and symmetric by the definition. Let $|A|_a$ be the equivalence class of A under $\tilde{a}$, i.e., $\{B : B \ \tilde{a} \ A\}$. Then Ca is the class of all these classes, i.e., $Ca = \{|A|_a : A \text{ is wff}\}$; and correspondingly $|A|_a + |B|_a = |A \& B|_a$. The latter definition is admissible, because, by virtue of T5, $|A \& B|_a$ is suitably independent of the choice of A and B.

Canonical valuations are defined as usual: $I(p, a) = 1$ iff $p \in a$ and $c(p, a) = |p|_a$, for every sentential parameter p and every situation a. These interconnections are extended inductively to every non-definitionally supplied wff A. Connectives $\approx$, $<$ and $\ni$ can be construed as definitionally supplied. The details beyond those for the carrier logic are these:-

- $c(A, a) = |A|_a$, for every wff A and situation a.

The generalising step is as follows, where $p_1, ..., p_n$ are all the sentential parameters of A:

$$\begin{aligned} c(A, a) &= \sum c(p_j, a), \text{ by its evaluation rule} \\ &= \sum |p_j|_a, \text{ by given basis} \\ &= |p_1 \& ... \& p_n|_a, \text{ by iteration of } |p \& q|_a = |p|_a + |q|_a \\ &= |A|_a. \end{aligned}$$

It is in this final step that use of non-null situations and of 4(I) (or a modal-style variant upon it) is crucial. By 4(I), $D \rightarrow .\ p_1 \& \ldots \& p_n \leqslant A$ and $D \rightarrow .\ A \leqslant p_1 \& \ldots \& p_n$. As a is non-null choose some wff D in a. Then by $\rightarrow$–closure of situations, $p_1 \& \ldots \& p_n \leqslant A \in a$ and $A \leqslant p_1 \& \ldots \& p_n \in a$, whence $p_1 \& \ldots \& p_n \ \tilde{a}\ A$, vindicating the final step.

- $I(A \leqslant B, a) = 1 \quad \text{iff} \quad A \leqslant B \in a\,.$

In view of the evaluation rule for $\leqslant$ it suffices to show $c(A,a) \leqslant c(B,a)$ iff $A \leqslant B \in a$. Now

$$\begin{aligned} c(A,a) \leqslant c(B,a) \quad & \text{iff} \quad |A|_a \leqslant |B|_a, \text{ as above,} \\ & \text{iff} \quad |B|_a + |A|_a = |B|_a, \\ & \qquad \text{by semi-lattice definition of } \leqslant \\ & \text{iff} \quad |B|_a = |A \& B|_a, \text{ by definition of } + \\ & \text{iff} \quad B\ \tilde{a}\ A \& B, \text{ by canonical definitions} \\ & \text{iff} \quad B \leqslant A \& B \in a \;\&\; A \& B \leqslant B \in a, \\ & \qquad \text{by canonical definitions} \\ & \text{iff} \quad A \leqslant B \in a, \text{ as is now shown.} \end{aligned}$$

Suppose $A \leqslant B \in a$. Since $D \rightarrow .\ B \leqslant B$ by T6 and a is non-null, $B \leqslant B \in a$. Hence, by 3, $A \& B \leqslant B \in a$. Also since $D \rightarrow .\ B \leqslant A \& B$ by T6, $B \leqslant A \& B \in a$. Suppose the converse; the operative conjunct is $A \& B \leqslant B \in a$. By T6, $D \rightarrow .\ A \leqslant A \& B$, so $A \leqslant A \& B \in a$, whence $A \leqslant B \in a$ by 2.

Further steps for positively-defined connectives are straightforward. For example,

- $I(A \approx B, a) = 1 \quad \text{iff} \quad A \approx B \in a\,.$

It suffices to show $c(A,a) \approx c(B,a)$ iff $A \approx B \in a$, i.e., $c(A,a) \leqslant c(B,a) \,\&\, c(B,a) \leqslant c(A,a)$ iff $A \leqslant B \in a \,\&\, B \leqslant A \in a$. But this follows from the previous case. Similarly a derived step for $A \ni B$ is straightforward, given that for $A < B$. But the latter, negatively-defined wff provides problems in relevant settings. For while $I(A < B, a) = 1$ iff $A \leqslant B \in a \,\&\, A \approx B \notin a$ and $A < B \in a$ iff $A \leqslant B \in a \,\&\, {\sim}(A \approx B) \in a$, nothing guarantees $A \approx B \notin a$, i.e., ${\sim}(A \approx B) \in a^*$, is equivalent to ${\sim}(A \approx B) \in a$ unless modal collapse ensures $a = a^*$. So we avoid $<$ in the canonicalness argument, as (given its definitional eliminability) we are entitled to do. The remainder of the completeness argument is standard (see for example RLR I).

12.5 More fully relevant causal logics

While stronger causal implication systems do not enable the derivation of any irrelevant causal implication claims—because the implication conjunct serves to filter out any irrelevance priority may introduce—still the logics are irrelevant in containing theses of the form $C \to D$ where C and D share no variables. Amusingly we could have avoided this problem, and other hassles by basing the theory of causal implication on an irrelevant strict implication or associated irrelevant conditional.[194]

To obtain a more fully relevant logic including both $\ni$ and $\to$, and so on the present approach $\leqslant$, involves reducing postulate 4(I) to 4. But this reduction, like felling a rainforest tree, takes some other things with it. It seems to require a certain withdrawal to O, and restructuring as follows:- Model structures are unchanged, but the interpretation functions concerning c are varied. In particular, for $a \in O$, $I(A \leqslant B, a)$ is as before; for $a \notin O$ however $I(A \leqslant B, a)$ can be determined otherwise, or arbitrarily, by the modelling. So we take this opportunity to remove an obvious deficiency, and also qualify the assessment of $c(A, a)$ in terms of $c(p_j, a)$ to $a \in O$. Otherwise, for $a \in O$, $c(A, a) \in C_a$, as for initial assessments. As a result, intensional wff are no longer assessed, priority-wise, through their sentential parameters. This is a major improvement. With $c(A, a)$ so decoupled, $I(A \leqslant B, a)$, etc., can most conveniently be left intact, as $c(A, a)$ gives requisite arbitrariness.

The effect of the decoupling of $c(A, a)$ gives the right results; validation of 4(I) breaks down. Call the relevant logics with 4(I) weakened to 4, but postulates 1-3 intact and semi-lattice principles of implication strength,[195] DCI systems, deeper causal implication. The modified semantics with $c(A, a)$ decoupled supplies DCI-validity.

Deeper adequacy result. A is a theorem of DCI iff A is DCI-valid.

Argument varies that for CI. In soundness, only the details for 4 are different. What did hold generally in 4(I) now holds only for $a \in O$, which serves to validate 4 but not 4(I). Completeness calls for more changes. For $a \in O$, $c(A, a)$ is coupled to its components; for $a \notin O$ define $c(A, a) = |A|_a$. Then $c(A, a) \in C_a$ as required. The proof given for CI that $c(A, a) = |A|_a$ works only for $a \in O$ in the absence of 4(I): no matter, since the rest is obtained

---

194 Such an irrelevant approach to deeper relevant logics themselves is not entirely excluded.

195 This introduces residual irrelevance, which it would be pleasing to eliminate, in the shape of the principle: 5. $B \leq A \leftrightarrow A \leq A \,\&\, B \,\&\, A \,\&\, B \leq A$.

(not by hard work but) by stipulation. Given $c(A,a) = |A|_a$ generally, proof of $I(A \leqslant B, a) = 1$ iff $A \leqslant B \in a$ is as before.

There remain sharp limits to the extent to which this style of approach, however relevant, can be improved. The reason is, of course, that a semantical elucidation of causal priority through atomic variables is ultimately unsatisfactory.

12.6 Corollaries, problems, further ado

An apparent corollary is decidability in those cases where the carrier logic succumbs to filtration methods. Another apparent corollary is that several philosophical castles will tumble down because their foundations are not adequate. A main reason in that they do not admit a sufficiently broad class of situations to support proper analyses of causal implication and duly distinguish it from other notions, such as mere time-ordered association (see SM).

Among the problems we have left open are some we started with, the extent and shape of the higher degree of causal implication, and the purer theory. The semantical theory we have presented is thoroughly entwined with that of some (sufficiency) implication. But it would be interesting also to know what the pure theory of $\ni$ looks like, and what the $(\ni, \&, \vee, \sim)$ part comprises. We can present a small amount of information,[196] some suggestions, and some conjectures. First, where the carrier relevant logic is near to or at basic affixing relevant logic, the pure $\ni$ logic contains no theorems, but only a rule structure (with rules such as transitivity). For the carrier logic contains only implicational theorems of the form $A \rightarrow A$ (see RLR I) and these causal implication strips off. The $(\ni, \&, \vee, \sim)$ logic is a richer connexive logic, as we know. Determining its extent is a more difficult enterprise, one to which present semantical technology (which works better applied to stronger systems) is still not well adapted. (One reason is that Affixing rules certainly appear to fail for causal implication.) We conjecture

[196] One important thing we *do* know is that there are several notions of logical significance of a similar doppelganger, double-banger, type to causal implication, permitting analysis in terms of a pair of more tractable notions. Containment logics, which have a rather similar semantical theory, are one sort, relational logic another, connexive logics another, logics for reason itself yet another. The connections are made in Chapter Six, section 5 where some of the history of these connected developments is duly acknowledged.

We also now know there are other interesting—but less satisfactory—options for analysing causation, through a relevant conditional (rather than relevant implication), either directly as a sort of conditional or by combining a conditional with a priority ordering.

that algebraic methods however will yield both an axiomatisation, and a conservative extension result. The status of many higher degree principles, which could be added to CI and DCI, remains open and, so far as we know, largely univestigated. Such business is part of the more immediate further ado.

There is always further ado. We have not looked at the quantificational theory; we have not considered the [false] principle of universal causation which obtains a nice nontrivial formulation within the theory, simply as $(UB)(PA)(A \ni B)$;[197] we have not investigated the foundations we suspect the theory affords for 'the logic of conditions'; nor have we glanced at the whole great ocean beyond this pleasant pebbly shore.

---

[197] *As mentioned in Chapter One, the existentially-neutral quantifiers U and P replace the more usual (and, in Sylvan's view, existentially-loaded) $\forall$ and $\exists$ respectively. [Editors]*

13 Part III Fragments

In his proposed monograph, Sylvan intended to include material on non-deductive reasoning. Extant fragments of draft material on this topic are reproduced below as individual chapter sections. They all appear to have been drafted between June and October 1995. [Editors]

13.1 Induction as the converse of deduction: induction and connexive confirmation

The theme that induction is the converse relation of deduction, once widely espoused, has fallen into disrepute. Indeed more, it has been abandoned, though closely related successors have sprung up in its place: notably the hypothetico-deductive method, and latterly that sceptical usurper that would replace it in turn, with some such variant as corroboration, namely deductivism of Popperian and other forms.

Both the demise and criticism of the *converse* theme, historically extremely successful criticism, were based on a fallacy, a sort of equivocation. For the criticism proceeded to replace the traditional notion of deduction—*traditional* deduction for short—by a new fangled, contemporary, and frankly entirely hopeless notion of deduction or consequence—widely known though through a stunning twist of irony, as *classical* consequence.

Traditional deduction was not classical. There is a twofold case. First, a detailed appeal to history. Secondly, the presumption that deduction was classical would render bizarre much history. Consider, for instance, Descartes' sustained and detailed effort to secure mathematics against scepticism, by a deduction from entirely certain basics. He need not have bothered; a simpler semantical argument to their necessary truth cuts through his whole elaborate enterprise. Somewhat similarly classically undermined is Kant's related transcendental deduction.

What was traditional deduction like? The traditional mainstream that is, as there were undoubtedly sidestreams. As the mainstream was never formalized, it remains less than entirely clear. Elsewhere the plausible hypothesis has been advanced that traditional deduction was—except in high medieval times—connexive.[198] One striking piece of indirect evidence for the connexive hypothesis is that the full Aristotelian logical theory, including

---

[198] *See Chapter Five. [Editors]*

the syllogism, emerges intact under a connexive hypothesis. Recall that, by contrast, direct classical translation destroys syllogistic theory. It also destroys induction theory.

Under connexive construal the theme of induction as the converse of deduction makes sense, and even enjoys some plausibility, whereas the theme looks rather ridiculous under 'classical' interpretation—which is why it was abandoned. For example, classically, A would induce $A \& B$, for arbitrary B, whence, given an entailment condition, A would induce B for arbitrary B, thereby trivializing the connection. Connexively *both* steps in this simple trivialization argument fail. Connexive theory can stand up, whereas such a classical theory falls down, absolutely flat.

The present exercise is aimed at standing up the converse theme by taking deduction as nonclassical, specifically as traditional, as connexive. The resulting relation, of connexive induction, is called *induccion*. As regards standing connexive induction theory up again, a few matters should be clarified, above all what is being claimed and what is not. It is claimed that induccion is a viable relation, and that it does bear a striking resemblance to what is accounted induction, enough to have been considered historically a promising candidate for explicating induction. But although connexive theory is seen as far from unreasonable, it is not claimed that connexive theory is correct—or that induccion is induction, or even an adequate part of induction, induction being a many-spendoured, not a monistic, thing. Nonetheless a connexive theory enjoys considerable plausibility; a historical position on induction, construed connexively, deserves more credence than it has normally received recently. There is another thing as well. The theory affords a wedge or, swapping images, a springboard—a stage in developing a logical theory of induction, against considerable recent opposition to any such theory.

The converse theme was widely espoused until the classical deductive paradigm displaced traditional logic. Indeed it was perhaps the mainstream position in the nineteenth century. Typical was Jevons: 'Jevons took induction simply to be the inverse process of deduction'.[199] Jevons is quite clear as to how *induction* is characterised: it *is really the inverse process of deduction*. (Plus, it is extended similarly to probable reasoning by a principle of Laplace.[200]) The basic *technique* consists in inventing hypotheses then checking deductions.

---

199 'As division is the inverse process of multiplication', Reese 1980, p.252.

200 On both see Jevons 1874.

With a suitably tight deductive connection—no irrelevance, no dangling components, roughly traditional connexivism—there does not seem to be any insuperable reason why something like this should not work, more or less.

Explicitly the definition proposed runs: A inducces B iff A is connexively derivable from B, whence by a deduction theorem, iff B connexively implies A.[201]

While there are several nonclassical connections worth investigating in developing improved theories of inductive relations,[202] connexively based formulations merit early investigation for two reasons: Firstly there are the historical linkage already indicated. Secondly, connexive logic includes some principles, critical in traditional reductionistic proofs, not available (because not valid) in relevant nonclassical logics, principles such as Antilogism ($A \,\&\, B \rightarrow C \rightarrow .\ A \,\&\, {\sim}C \rightarrow {\sim}B$) and Disjunctive Syllogism ($A \,\&\, ({\sim}A \vee B) \rightarrow B$).

Let us symbolize the basic induccion relation between A and B, $A \text{—}s\ B$, read also *A connexively supports B*. Many support implications are immediate from connexive logic. For example: $A \text{—}s\ B, B \text{—}s\ C \,/\, A \text{—}s\ C$ i.e., support is (rule) transitive. More generally support is an order relation. Additionally, $A \text{—}s\ B \,/\, {\sim}B \text{—}s\ {\sim}A$, i.e., support is (rule) contraposing. $Aa \text{—}s\ \forall x Ax$, i.e., support is instantial.

It may be imagined that even this much basic theory reveals that induccion is an immediate target for Hempel paradox and like trouble. It is interesting to show why this is not really so. Hempel's paradox concerning any irrelevant objects which are not black and nonravens confirming "all ravens are black" turns upon, among other things, the Nicod principle of instantial confirmation formulated rather classically. Specifically it is assumed for Hempel's paradox that:

NIC. $fa \,\&\, ga$ confirms "all f's are g's",

---

[201] We could *widen* this by using plication or probabilification in place of implication. Then we can handle 'J robbed the safe' *probs* 'his fingerprints are upon it' but inconclusively, because he may have worn gloves. In the first place, however, we adopt only one-way linkage with implication; then the converse gives partial evidence.

[202] Intensional conjunction, $\circ$, of relevant logics also displays many of the advantageous features of connexive conjunction. For example we can represent *much* of the syllogism in relevant logic by using relevant *and,* $\circ$. What we shall not be able to derive is $A \rightarrow I$ *(i.e., A-propositions implying I-propositions [Editors])*, unless we add connexive principles, viz. $(B \rightarrow C) \rightarrow .\ B \circ C$, i.e., in R, $B \rightarrow C \rightarrow .\ {\sim}(B \rightarrow {\sim}C)$.

a generalization Hempel proceeds to formalize as $\forall x(fx \supset gx)$. But it is enough that such universal generalization is formalized in terms of some contraposible implication for the paradox argument to go through as follows:

$$\sim fa \,\&\, \sim ga \text{ confirms } \forall x(\sim gx \rightarrow \sim fx)$$

and therefore confirms its equivalent $\forall x(fx \rightarrow gx)$.

Of course the equivalence principle involved holds for connexive confirmation, i.e.:

$$A \mathrel{—s} B, B \leftrightarrow C \,/\, A \mathrel{—s} C.$$

Nonetheless, Hempel's paradox breaks down for connexive confirmation because principle NIC fails. Very simply, $\forall x(Ax \rightarrow Bx)$ does not imply $Aa \,\&\, Ba$, either connexively or for any decent implication. So it is *not* the case that $fa \,\&\, ga \mathrel{—s}. \forall x(fx \rightarrow gx)$. Moreover it is easy to discover what supplants NIC in connexive theory, namely:

IC. $fa \rightarrow ga \mathrel{—s}. \forall x(fx \rightarrow gx)$.

Nor is it difficult to argue that $fa \rightarrow ga$ (also read "a which is f is g" or "a such that f is g") is a proper connexive formulation of such an instantiation. For recall that the proper formulation of the particular form "some A's are B's", the form required for the complete connexive theory of the Syllogism, is $\exists x(Ax \rightarrow Bx)$.

The worst paradoxes of confirmation (for example, that $\sim fb$ confirms "all f's are g's") are removed under any theory that replaces a material implicational formulation of the universal conditional by one in terms of a more discerning implication. So too they are eliminated by connexive theory. But does not an element of paradox remain? Certainly $\sim fa \,\&\, \sim ga$ does not connexively support $\forall x(fx \rightarrow gx)$, so too a white shoe does not straightforwardly support "all ravens are black". But is it not awkward nonetheless that $\sim ga \rightarrow \sim fa$ connexively supports $\forall x(fx \rightarrow gx)$?[203]

Not only does a rather promising basic theory begin to emerge. Furthermore there are favourable or harmonious concomitant outcomes. One unremarkable feature is the fit with traditional induction theory.

A catalogue of standard inductive principles is to be found assembled in encyclopaedic works such as Carnap 1962 (an excellent, if biassed data base), and has been summarised in many derivative sources. Now what can

[203] Thus connexive confirmation theory does not fare quite as well as the relevant conditional theory discussed in Chapter Ten, where conditional contraposition breaks down.

be shown is that most of the standard principles can be brought out on a connexive theory. Moreover apposite qualifications are suggested.

Consider first some examples.

- Simple enumerative induction:

$$Ax_1 \rightarrow Bx_1, Ax_2 \rightarrow Bx_2 \text{ ... support } \forall x(Ax \rightarrow Bx).$$

Proof is immediate from instantiation. As of course that tiny proof shows, one instant is enough for support. What we are looking at amounts only to very low level partial support, a *basic* support connection. The connexive theory would have to be elaborated—as it could be—to supply information as to *quality* of support, strength or reliability, and so on.

(Notes indicate that more examples were to be inserted here. [Editors])

There is reason to feel vaguely annoyed, and rather scandalized, that, given such apparent clarity last century, the whole inductive enterprise should appear destroyed, or at best in ruins. I felt particularly annoyed when I had a batch of arguments (like those of limits to growth, for example) which I wanted to assess for adequacy, but lacked a background theory.

How did the wrecking occur? There are several components (but I don't yet see the linkages etc. clearly):

1. Displacement of extended traditional logic by classical, the inverse relation of which was hopeless.
2. Discrediting of traditional processes for induction (and deduction), for example as one from less general to more general (and inversely, more general to less). Literally, this doesn't hold of course (though something can be done with *content*).

As it happens, Jevons was quite careful about this, putting deduction more generally in terms of 'developing the consequences'.

1. Attempts to design a new connection, cut loose from a classical inverse, foundered—in paradox, etc. The most famous attempt is through confirmation, but there are others, I suppose: corroboration, reliability out of probability theory, etc.
2. Rubbishing the whole idea, as per Popper. All the same the result looks remarkably like Jevons. Instead of 'inventing hypotheses' for deductive testing, we conjecture theories to which we apply deductive methodology. What has been dropped, viewed this way, is the 'definition' of induction as *an* inverse process.

What has been sketched out—in very crude form—is rather a large project. It has at least two main initial components: analytico-logical and historical. If successful it could be called something like: the wrecking and restoration of induction. Or: induction, its fall and rise. But maybe something else fatally flaws induction? (For example, the whole idea is much too simple-minded, one of the things often said about one of its substitutes H-D methodology, perhaps rightly.)

Critique of connexivism. In some respects the theory is too strong, in others too weak. No doubt in some cases where it is too weak it can be strengthened, but in other cases this could cause trouble. Strength is sometimes more of a problem. For example, transitivity of connexive support looks too strong. Support is not unrestrictively transitive, giving out over chaining (consider too iteration of probabilities).

13.2 Nondemonstrative reasoning and nonponible logics: neglected groups of unlinked logics

There are various readings of connective "if", or associated with conditionality, for which Modus Ponens fails. That is to say, where connective $>$ symbolizes such a 'conditional', the ponibility rule:

MP. $A, A > B \,/\, A$,

does not hold—does not warrant detachment, or appropriately preserve validity—for all permitted assignments of wff A and B. Such logics are called *nonponible*.

It is sometimes claimed that MP breaks down for conditionality itself. Recently such counterexamples to MP as the following have been proposed (McGee 1985, p.463):

> Having learned that gold and silver were both once mined in his region, Uncle Otto has dug a mine in his backyard. Unfortunately, it is virtually certain that he will find neither gold nor silver, and it is entirely certain that he will find nothing else of value. There is ample reason to suppose that:
>
> If Uncle Otto doesn't find gold, then if he strikes it rich, it will be by finding silver.
> Uncle Otto won't find gold.

Since, however, his chances of finding gold, though slim, are no slimmer than his chances of finding silver, there is no reason to suppose that:

If Uncle Otto strikes it rich, it will be by finding silver.

Of course such examples, even if sustainable, only tell against rules like MP *as extendible*, not as devices for generating theorems from theorems. But prevailing logical theory has been fairly slack on distinguishing kinds of rules. While a theory of conditionals which allows for lapses of MP, extendible or not, is readily devised (such as relevant theory as presented in RC), and is not without philosophical consequence, a much bigger syndicate of logical interests than just conditionality is involved here. The aim is to catch a very important class of nondeductive logics, those associated especially with nondemonstrative reasoning.

In these nonponible logics, the *plication* connective $>$ assumes different readings, for instance of a case-*making* or evidential kind, short of conclusiveness. There are important substitutions to be made for parameter p in the reading $A > B$ as: A makes B p-able. For example, *plausible*, *probable*, *possible*, and also *evident*, *relevant*, and so on. Plainly too there are associated readings such as: "A is evidence for / a reason for / relevant to ... B". In none of these cases does MP hold generally. Only when the linkage is made conclusive on the intended antecedent ground (after perhaps enthymematic gaps have been closed) is MP ensured. For this making p-able has to guaranatee making *true*, as making *necessary* does; then the truth-preservation that MP intimately connects with *is* forthcoming.

13.2.1 A more general setting: unlinked logics

Nonponible logics form a subclass of a more general class of nondeductive logics. It is illuminating to bring nonponible logics together with the kinds of other logics, nontransitive and cut-resisting logics, of this general class. (Note that these unlinked logics comprise only one major class of nondeductive logics; other classes are nonmonotonic and nonidentity logics.)

It is a feature of deductive logics that accumulated linkages can be reduced to and components can be cut (down). Thus the chain of steps, inferences:

Ch. $A_1 > A_2, A_2 > A_3, ..., A_{n-1} > A_n$

yields the (big) step $A_1 > A_n$ in which all the intermediate links are cut out, middle terms stepped over. Alternatively, moving from implicational to inferential form, from the chain Ch, the premiss A_1 yields the conclusion A_n.

Syllogistic reasoning, where however the components are confined to very special logical forms, is the traditional exemplar, par excellence, of the first, cutting, procedure. But it is also characteristic of the deductive sciences; theorems, for instance, are typically the result of assembling many small obvious steps into one big usually less obvious step.

A significant, diverse, but little investigated, class of nondeductive logics calls this cumulative procedure into question. Some of the notions involved, which we have already glimpsed call linkage of chaining into question regularly, others upset it occasionally. Among the connectives which cast such linkages into doubt are once again, firstly, conditionals; secondly, evidential and case-*making* connectives such as "A is evidence for B", "A makes plausible B", "A probabilifies B", and thirdly (and overlapping the second) relevantly controlled connectives, such as "A is relevant (relevantly related) to B". A fourth increasingly important class includes default connectives, where an implication or the like which is normally admissible, breakes down because of trouble emerging, trouble such as inconsistency identified as occuring back earlier in the chain.

For want of a better term, these logics, which break the standard removal of linkages, at least sometimes, are called *unlinked* logics. In these the basic procedure:

$$\frac{A > B \quad B > C}{A > C} \qquad \text{or} \qquad \frac{A \Vdash B \quad B \Vdash C}{A \Vdash C},$$

or the like (for example, if the rule is extendible, $A > B \,\&\, B > C \rightarrow. A > C$), is not always admissible. The reason for this is that the data base from which initial inferences are drawn turns out to be—as long chains reveal—*unsatisfactory* in some way. A most conspicuous case of unsatisfactoriness, so it is commonly thought, is inconsistency.

Unlinked logics which block more classical moves from inconsistent data provide a noteworthy group of paraconsistent logics—a group unduly neglected in previous standard texts, notably PL. Negative paradox arguments are rather easily stopped by nontransitivity or nonponability, to consider basic cases. For example, the medieval argument, in conditional form:

$$\text{Ch1.} \quad \frac{A \,\&\, {\sim}A > (A \vee B) \,\&\, {\sim}A, \; (A \vee B) \,\&\, {\sim}A > B}{A \,\&\, {\sim}A > B},$$

depends on transitivity; and the corresponding inference:

$$\frac{\text{Ch1}}{A \,\&\, {\sim}A \therefore B}$$

turns on ponability. Both are stopped without them. Such logics thus offer a direction in which Wittgenstein's *Stop*! rule, and other default rules, can be made good.

To swing back to the initial point, there are *two* closely related groups of unlinked logics: modus ponens violating logics and transitivity or cut violating logics. The forms which are invalidated are, respectively,

$$\frac{\text{Ch}}{A_1 \therefore A_n} \qquad \frac{\text{Ch}}{A_1 > A_n} \quad \text{(or with} \Vdash \text{ for } > \text{).}$$

These forms are equivalent given Modus Ponens itself and the (weak) Deduction Theorem. The forms also reduce to one another in more special settings, A less familiar example is that of the first degree theory (of RLR I) where a zero degree wff A is always tantamount to a conditional $t > A$. Then Modus Ponens

$$\frac{A \quad A > B}{B} \quad \text{just amounts to} \quad \frac{t > A \quad A > B}{t > B},$$

that is, to a version of transitivity. Thus Modus Ponens can be, so to say, eliminated in this special setting. It can similarly be eliminated in favour of transitivity wherever a constant t is available, with characteristic two-way rule property $A \,\|\, t > A$. A more familiar example is that of Gentzenisation, where Modus Ponens is absorbed by Cut, which is but a general form of transitivity. Modus Ponens becomes that special case where sequents are null on one side

$$\frac{\Lambda \Vdash A \quad A \Vdash B}{\Lambda \Vdash B}, \quad \text{i.e.} \quad \frac{\Vdash A \quad A \Vdash B}{\Vdash B} \quad \text{where } \Lambda \text{ is really nullity}$$

But though they can be brought together, as above and especially in Gentzen-style settings and the like, in general the two types of unlinked logics are distinct. In order to clarify the systems already extant in the logical literature, and to bring some order to the scene, it is certainly worth separating these two major types. The second more amorphous type quickly bifurcates into two subtypes: nontransitive logics and cut-resistant logics. For there are cut-resistant logics which include forms, sometimes all forms, of transitivity; and there are cut-permitting logics which reject all but minimal forms of transitivity. The details will not be pursued, as the emphasis in what follows is very much upon nonponible logicistic systems in 'Hilbert' (pre-Jaśkowski-Gentzen) formulation.

There are evidently problems in formulating nonponible logics in straightforward Hilbert-style axiomatisations, since these are typically based on a Modus Ponens rule, commonly as sole rule. Unfortunately for logical variety, such an approach has become highly favoured. However many other rules are available: Prefixing, Suffixing, Transitivity, Replacement, etc., etc. So much Arruda and da Costa have made plain in their pioneering work on J systems. Not so differently from J methods, Gentzen-style methods will yield a rich variety of unlinked logics. Of course these systems have not usually been viewed in quite this way; mostly, failure in obtaining a cut-elimination theorem has been regarded as failure. According to Scott 1973, however, 'in general cut cannot be eliminated. Gentzen's Elimination Theorem holds only for very special relations'. There are untapped riches to be won, from an abundance of cut-resistant systems.

There is fortunately a convenient way around the paucity or awkwardness of the rules (or principles) of some nonponible notions (i.e., notions represented through a nonponible $>$). That is through the device of a carrier logic, an underlying logic, where implication connective ($\rightarrow$ say), supplies the requisite inferential linkages. Mostly it is not pure nonponible notions, but nonponible connectives combined with other connectives such as the standard set $\{\rightarrow, \&, \vee, \sim\}$ at the sentential level, that are of main interests. Moreover, from the larger setting the pure systems can then be disentangled and isolated, where required. In subsequent systematic development of nonponible (relevant) notions, a relevant carrier logic will be presupposed. In fact, formulations like those in terms of Gentzen systems and J systems involve an implicit carrier logic; the sequent $A_1, ..., A_n \Vdash B$, for example, does the same work as $A_1 \& ... \& A_n \rightarrow B$.

Given a carrier logic, it is not difficult to design and model a range of relevant, and irrelevant nontransitive systems. Consider to illustrate a döppelganger approach (the method deployed in Chapter Six). Set $A > B = A \rightarrow B \& A\, r\, B$, where the carrier implication, $\rightarrow$, is some relevant or irrelevant implication of known behaviour. Then MP holds but not Rule Transitivity. (It is *Rule* Transitivity that is the basic form of transitivity that is of importance here. It is easy to design intelligible systems without other forms of transitivity; relevant logics in the vicinity of basic system B are of this sort.) The equation delivers a semantical rule, relativizing to world in the set of worlds K:

$$I(A > B, a) = 1 \quad \text{iff} \quad I(A \rightarrow B, a) = 1 = I(A\, r\, B, a).$$

Here it seems likely that the r rule can be replaced by something neater, for example $c(A, a) \subseteq c(B, a)$ for $I(A\ \ B, a)$, where $c(A, a)$ is (say) the information or content of A at situation a. It is an interesting question to what extent extant nontransitive systems can be analysed in this way; relational logics certainly can be.

13.2.2 A brief survey of extant nonponible logics and their intended roles

Among the early unlinked systems that we so far know of, apart from Gentzen-style systems, is the extendible calculus of Hiż 1959. This is an interesting if bizarre system, which is paraconsistent while containing all theorems of classical sentential logic. The feat is accomplished by failing Material Detachment, and therewith Ex Falso Quodlibet, and supplying instead an alternative less extendible rule structure. The basic system $\overset{\circ}{H}$, in schematic formulation with just connectives $\supset$ and $\sim$, is as follows:

Axiom Schemes	Rules
$\sim(A \supset B) \supset A$	$A \supset B, B \supset C \,/\, A \supset C$
$\sim(A \supset B) \supset \sim B$	$A \supset (B \supset C), A \supset B \,/\, A \supset C$
	$\sim A \supset B, \sim A \supset \sim B \,/\, A$

Plainly it admits of much variations, for example in intuitionistic and other directions. (For example, since all but the first axiom and the last rule are intuitionistically correct, intuitionistic $\overset{\circ}{H}$ comes through varying these.) Perhaps surprisingly, there is no larger purpose, paraconsistent or other, in Hiż's limited enterprise; it is just to show that the exercise is 'not completely trivial', and can be done. It has now been done in many less bizarre ways than with the implicationally-paradoxical axiom-schemes of $\overset{\circ}{H}$. For example, it has been done within relevant logics, such as the system E of 'entailment', which contains all classical tautologies, but not the rule MD (of Material Detachment) i.e., $A, A \supset B \,/\, B$, as an extendible rule. Unlike $\overset{\circ}{H}$ these systems have the advantage of a connective other than $\supset$, with a (perhaps genuine) implication $\rightarrow$ for which Modus Ponens does hold. $\overset{\circ}{H}$ retains however a certain purity and compact Hilbert-style formulation which it would be pleasing to emulate with other systems, such as requisite parts of systems with more generous classes of connectives (for example, a standard set $\{\rightarrow, \&, \vee, \sim\}$ in place of $\{\supset, \sim\}$).

By contrast with Hiż, both Fitch and Arruda-da Costa work with different formulations, so complicating comparisons—Fitch with a natural deduction formulation, and Arruda-da Costa with modified Gentzen systematization. By important contrast too with Hiż, both have much larger objectives in view. A main objective is to design a set theory, or equivalent grand system, which includes (substantially) unqualified comprehension principles. Logical paradoxes are avoided by cutting linkages, by imposing restrictions on Modus Ponens. Arruda and da Costa fail in their purpose, while producing a bundle of interesting systems $J_1 - J_5$, with $J_2 - J_5$ neighbours of $\overset{\circ}{H}$. For as Bunder 1983 shows, these systems do admit a restricted form of Modus Ponens (in $\supset$), which, with unrestricted Comprehension, would allow every theorem of implicational form (i.e., of form $A \supset B$) to be provable. In fact, the J systems, which are based on positive logic, are easily modified to avoid some of these problems. However, as Urbas 1988a shows, a very wide variety of nonponible systems are vulnerable to Curry-style paradoxes. While such Curry objections can no doubt be avoided in relevant variations on the J systems, such systems will typically also admit full Modus Ponens without trouble. Nonponibility is not a promising route to logico-semantical paradox-freedom.

Fitch's systems amount, like the J systems, to variations upon intuitionistic and positive logic. Fitch, who was perhaps the first to deliberately formulate nonponible systems, gives *reasons* for the breakdown of Modus Ponens, in important contrast with other system-building researchers. Though these reasons bear close examination, the breakdown occurs far beyond the propositional stage. Indeed they occur in the regions of self-reference where Curry paradoxes or the like—which undoubtedly involve MP—may be generated. Accordingly Fitch's series of systems are nonmonotonic, in the computer science sense.

The logical systems of Fitch, Hiż, and Arruda-da Costa are nonponible by intention, even if as in Hiż the intended way discloses little deeper purpose. Also abroad are various logics which are nonponible, not by deliberate design, but by accident or adventitiously. Among more recent examples are weak conditional logics (both modal, as in Chellas 1975, and relevant, as in DRL). What happened with these logics was that expected basic modellings did not on their own validate Modus Ponens, which had to be ensured by further conditions. In the orthodox modal case, Modus Ponens is mixed up with other desired principles; but in relevant conditional theory such modal confusion is removed, and Modus Ponens can be considered on its own merits—a desirable situation now that conditional counterexamples to the inference are beginning to appear.

Both in the evidential nexus and in the related probability nexus, protologics are to be found which drop or void Modus Ponens. Examples include Hempel's logic of confirmation, Popper's theory of corroboration, Rescher's logic of plausible reasoning, and various nonmetrical (sub-)theories of probability. For the most part these embryonic logic theories are not presented in suitable plicational (conditional) form, for instance for ready comparison with other sytems. At best such a plication connective is defined, or is definable. It is past time that a direct approach was made, that systems for these notions, considered in their own merits, get discerned.

All the nonponible systems so far explicitly considered (not excluding conditional logics) validate rule forms of transitivity, such as:

RTrans. $A > B, B > C \,/\, A > C.$

Indeed systems like $\overset{\circ}{H}$ depend heavily on RTrans to supply the inferential machinery. By contrast, extant nontransitive systems typically rely upon MP for main inferential purposes. However there are systems, for notions of philosophical interest, which eschew both MP and RTrans. One notion of that sort is that of 'irrational' preference, for example "what A states is preferable to what B does". Another more important notion is that of plausibility, with $>$ construed in terms of making plausible. While plausibility has gained some due logical investigation, notably by Rescher, the work thus far has two severe, and connected shortcomings:- Firstly, the attempted analyses all involve reduction back to the classical logical paradigm, paradoxes, MP, and all. That is no way to go about obtaining satisfactory accounts of distinctly nonclassical notions. Secondly, there is a heavy, distracting emphasis on the problem of inconsistent antecedents, a distorting diversion that would have been avoided with a paraconsistent start to investigations.

13.2.3 Other irrelevant and relevant variations on historic nonponible systems

Much as there are intuitionistic variations upon nonponible classical systems, so there are strict and relevant and other variations. Formulations of pure implicational, such as pure strict, systems, lend themselves especially well to such elaboration, for they commonly come formulated without Modus Ponens, or are easily restructured to omit it (all wff of theoretic interest being implication). The idea is then that we take some suitable nonponible pure implicational formulation and add axioms and rules to yield a full $\{\rightarrow, \&, \vee, \sim\}$ system keeping Modus Ponens out. But so long as such a system remains within the orbit of modal logic, it will violate relevance requirements—though not necessarily offending paraconsistency conditions. Consider the following minimal strict system, MS (part of Boolean lattice logic):

Axiom Schemes		Rules
$A \,\&\, B \rightarrow A, A \,\&\, B \rightarrow B$		$A \rightarrow B, B \rightarrow\ / A \rightarrow C$
$A \rightarrow A \vee B, B \rightarrow A \vee B$		$A \rightarrow B, A \rightarrow C \,/\, A \rightarrow B \,\&\, C$
$A \,\&\, (\sim A \vee B) \rightarrow B$	(DS)	$A \rightarrow C, B \rightarrow C \,/\, A \vee B \rightarrow C$

Though minimal it contains enough apparatus to reproduce the negative Lewis paradox, by the usual medieval argument.

Theorem $\quad A \,\&\, {\sim}A \rightarrow B$

Proof:

$$A \,\&\, {\sim}A \rightarrow A, A \,\&\, {\sim}A \rightarrow {\sim}A$$
$${\sim}A \rightarrow {\sim}A \vee B$$
$$A \,\&\, {\sim}A \rightarrow .\ {\sim}A \vee B$$
$$A \,\&\, {\sim}A \rightarrow .\ A \,\&\, ({\sim}A \vee B)$$
$$A \,\&\, ({\sim}A \vee B) \rightarrow B$$
$$A \,\&\, {\sim}A \rightarrow B.$$

Nonetheless, despite such irrelevant implicational theorems, MS is paraconsistent. To show this, apply the following M_3 matrices (drawn from Hiż: where in effect $\vee$ and & are defined: $A \vee B =_{df} {\sim}A \rightarrow B$; $A \,\&\, B =_{df} {\sim}({\sim}A \vee {\sim}B)$).

$\rightarrow$	1	2	3	$\sim$
*1	1	3	3	2
*2	1	1	1	1
3	1	1	1	1

$\vee$	1	2	3
1	1	1	1
2	1	3	3
3	1	3	3

&	1	2	3
1	1	2	2
2	2	2	2
3	2	2	2

These satisfy all of MS but reject Spread: $A, {\sim}A \,/\, B$.

There are various ways back to relevance, more than one of which is of interest in the present context. The usual relevance approach drops DS. Then in MS–DS we have a beginning on a nonaffixing relevant logics *without* MP. Expected additions would include:

Positive side: $\quad A \rightarrow A; \quad A \rightarrow C, B \rightarrow C \,/\, A \vee B \rightarrow C;$
$\qquad A \,\&\, (B \vee C) \rightarrow .\ (A \,\&\, B) \vee C;$

For negation: $\quad {\sim}{\sim}A \rightarrow A; \quad A \rightarrow {\sim}{\sim}A; \quad A \rightarrow B \,/\, {\sim}B \rightarrow {\sim}A.$

In fact these are still P-style systems, Nonreplacement systems, without MP. There are two important types of rule upgrading, giving various types of Substitutivity:

RS. $\quad A \leftrightarrow B, C \leftrightarrow D \,/\, B \rightarrow C \rightarrow .\ A \rightarrow D$

RA. $A \to B, C \to D / B \to C \to. A \to D$, i.e., (full) Affixing.

In ponible settings these rules enable the development of distinctive types of semantics for each class of systems: nonreplacement, Replacement and Affixing.

It is an open question whether something like these semantics can be replicated for corresponding nonponible systems. Semantical modellings are decidedly desirable, not least because they supply guidance as to how syntactical systems should be elaborated. But such semantical investigation is not easy. In the best known types of systems, Affixing systems, the modelling condition of Affixing itself is rather inextricably bound up with that for MP. In amputated logical settings, algebraic methods often look a better bet than purely semantical ones. But even there technical innovation is called for, partly because standard methods break down interestingly. Standard methods introduce a top element t, subject to the two-way t rule: $A \parallel t \to A$. Since this rule gives back MP in transitive nonponible systems, it is not admissible. (Thus too a different way of saying A is valid, than $1 \leq I(A)$, is required.)

A way forward, circuiting these difficulties, again exploits the valuable device of a carrier logic. Operating with pure nonponible logics is in some respects like trying to communicate by radio means without a carrier wave. A carrier logic is an underlying logic whose inferential signal intended nonponible logics use and modulate. Later, when main technical work is done, requisite features of pure nonponible logics may be unscrambled. A relevant affixing logic is adopted as carrier logic in the detailed investigations that follow (so the theory of RLR I can be taken over). There is no virtue in dragging along the distorting falsehoods of classically-based logics, for example, when they can be avoided. But by straightforward variation the carrier logic can be made irrelevant, and pretty much as modal or classical as desired. The theory developed will not be short on generality.

Several different sorts of carried nonponible logics and corresponding semantical modellings will be offered. One reason for the variety is that there are many logically different notions of interest in the area; another reason is that a critical issue is not yet sufficiently clear: namely, to what semantical type do various philosophically important notions belong?

13.2.4 Replacement and semi-systematic systems

In these systems, nonponible connective $>$ is added to a carrier logic with connective set $\{\to, \&, \vee, \sim\}$. Replacement systems conform to the replacement rule:

R. $A \leftrightarrow B, C \leftrightarrow D \,/\, B > C \rightarrow . \, A > D,$

a rule that may be broken into antecedent and consequent parts:

RA. $A \leftrightarrow B \,/\, B > C \rightarrow . \, A > C$

RC. $C \leftrightarrow D \,/\, A > C \rightarrow . \, A > D.$

Because the carrier logic satisfies replacement conditions, full substitutivity can be proved, namely: $A \leftrightarrow B \,/\, \delta(A) \rightarrow \delta(B)$ and $A \leftrightarrow B, \delta(A) \,/\, \delta(B)$, where usual conditions apply. In semi-systematic systems, one or other of the antecedent and consequent rules is replaced by a corresponding affixing rule. Most important semi-systemic systems are prefixing systems which combine RA with the prefixing rule:

RP. $C \rightarrow D \,/\, A > C \rightarrow . \, A > D.$

For such systems provide the acclaimed rule structure of conditional logic and the apparent rule structure of several significant evidential and probabilistic notions.

Semantical theories for all these sorts of systems are available, in the form of neighbourhood semantics and upgradings thereof. There are no doubt superior semantics for many of these sorts of systems, in the shape of smart relational semantics, but we shall work here with neighbourhood forms. For the prefixing semi-systematic systems on which we shall concentrate, the upgraded forms are pleasant enough and agreeably well-known.

Modus Ponens for $>$ is not a (derived) principle of basic semi-systemic logics. The easily faulted rule: $A, A > B \,/\, B$, is modelled as follows: For every $a \in D$ there is some $x \in O$ such that: $a \in \alpha \supset . \, a \in \underset{\sim}{s}(x, \alpha)$.

For *soundness*, suppose B fails somewhere. Then for some model with some $a \in O$, $I(B, a) \neq 1$. By the modelling condition there is some $x \in O$ such that $a \in \alpha \supset$, $a \in \underset{\sim}{s}(x, [A]) \subseteq [B]$, whence $I(B, c) = 1$, which is impossible.

For *completeness*, suppose for $a \in O$, $A \in a$. It suffices, in the circumstances, to show $a \in \underset{\sim}{s}(x, [A])$ for some $x \in O$, i.e., for all$B, (A > B \in a \supset B \in x)$. Suppose then $A > B \in a$, and show $B \in x$. If both A and $A > B$ hold for every $a \in O$, then both are theorems. Hence by MP, so also is B. So $B \in x$.

13.2.5 Systematic systems

The semantic rule for a connective $>$ is:

$$I(A > B, a) = 1 \quad \text{iff} \quad Tabc \,\&\, I(A, b) = 1 \supset_{b,c} I(B, c),$$

where T is a three-place relation on worlds, i.e., on K^3, satisfying, for the hereditariness lemma, the condition:

when $a \leq b$ and $Tbcd$ then $Tacd$.

The hereditariness step, when $a \leq b$ and $I(A > B, a) = 1$ then $I(A > B, b) = 1$, then follows, as required.

The postulates for $>$ are, as deployed in the completeness argument these:

$>$-Prefixing. $A \rightarrow B \,/\, C > A \rightarrow . \, C > B$

$>$-Praeclarum. $(A > B) \,\&\, (C > D) \rightarrow . \, A \,\&\, C > . \, B \,\&\, D$

$>$-Suffixing. $A \rightarrow B \,/\, B > C \rightarrow . \, A > C$

$>(\vee)$-Praeclarum. $(A > B) \,\&\, (C > D) \rightarrow . \, A \vee C > . \, B \vee D.$

These postulates admit of interesting combination, in various ways, in usually *expected* settings:

a. Prefixing and Praeclarum may be combined in the single rule:

$A \,\&\, B \rightarrow C \,/\, (E > A) \,\&\, (F > B) \rightarrow . \, E \,\&\, F > C$ (Complex Prefixing)

Proofs: Given the complex rule, Praeclarum is immediate from $A \,\&\, B \rightarrow A \,\&\, B$. Then, identifying A with B and E with F, by properties of the carrier logic, $A \rightarrow C \,/\, E > A \rightarrow . \, E \,\&\, E > C$.

Proof of soundness requires extension of results for the carrier logic to the further postulates for $>$. The cases for the Praeclarum schemes are similar. It suffices to suppose $I(A > B, a) = 1 = I(C > D, a)$, and to show $I(A \,\&\, C > B \,\&\, D, a) = 1$ (or $I(A \vee C > B \vee D, a) = 1$). But this is a matter of quantification logic given the rules for $>$ and $\&$ (or $>$). No modelling conditions are required.

ad $>$-Prefixing. Suppose in some model based on $y \in O$, $C > A \rightarrow . \, C > B$ fails to hold. Then for some a, $I(C > A, a) = 1 \neq I(C > B, a)$. Hence for some b, c, $Tabc$ and $I(C, b) = 1 \neq I(Bc)$. As $I(C > A, a) = 1$ and $I(C, b) = 1$, $I(A, c) = 1$. That is $I(A, c) = 1 \neq I(B, c)$. But this counters $A \rightarrow B$ (using the condition that for some $x \in O$, $Rxcc$ and forming a countermodel on base x; see Lemma 4.2, RLR I, p.302).

ad $>$-Suffixing. Similar to previous case. Again no special modelling conditions for T are required; only a condition already given for R.

For completeness, by a canonical modelling argument, the key step consists in showing:

$$I(A > B, a) = 1 \quad \text{iff} \quad A > B \in a.$$

One half of this result tumbles out of an appropriate canonical definition of relation T, as $Tabc$ iff $(A,\ B).\ A > B \in a \,\&\, A \in b \supset .\ B \in c$, for a, b, c suitable theories. Suppose then $A > B \in a$. In order to show $I(A > B, c) = 1$, suppose further, using an induction hypothesis, $Tabc$ and $A \in b$; and show $B \in c$. But this now follows by definition of T. For the converse, suppose $A > B \notin a$. What have to be supplied are suitable theories b and c such that $Tabc$, $A \in b$ and $B \notin c$. The argument is in two stages. First (unprimed) theories b' and c' are defined, with appropriate properties. Second, maximization is applied to reach appropriate b and c. Define $b' = \{D : \vdash_L A \to D\}$. Note that the implication of the carrier logic is used here; use of $>$ would require significantly different (and stronger) conditions upon $>$. Then, in the usual way, $b \in \overline{K}_L$. Plainly $A \in b'$. Define $c' = \{E :$ for some $D,\ D > E \in a \,\&\, D \in b'\}$. Then $\overline{T}_L ab'c'$, and also $B \notin c'$. By $>$-Suffixing, for some $D_1, \vdash_L D_1 > B \to .\ A > B$, whence $A > B \in a$, contradicting the initial supposition. In showing that $c' \in \overline{K}_L$ there are two parts. First for provable implication closure, suppose $E \in c'$ and $\vdash_L E \to F$, and show $F \in c'$, i.e., for some D, $D > F \in a$ and $\vdash_L A \to D$. By assumption, for a D that will do, $D > E \in a$ and $\vdash_L A \to D$. Since $\vdash_L E \to F$, by $>$-Prefixing, $\vdash_L D > E \to .\ D > F$, whence by closure, $D > F \in a$ as required. Second, for adjunction closure, suppose $E \in c'$ and $F \in c'$, and show $E \,\&\, F \in c'$. Then by hypothesis, for some C and D, $C \in b'$, and $D \in b'$, whence $C \,\&\, D \in b'$, and $C > E \in a$ and $D > F \in a$

Thus by $>$-Praeclarum, $C \,\&\, D > E \,\&\, F \in a$, whence $E \,\&\, F \in c'$. To complete the argument, b' and c' are inflated to prime theories, using an analogue of a corollary to the priming lemma (RLR I, p.309, corollary 6); namely: where $a \in K_L$, b', $c' \in \overline{K}_L$, $\overline{T}_L ab'c'$ and $C \notin c$, then there are b, $c \in K_L$ such that $b' \subseteq b$, $T_L abc$ and $C \notin c$. While this yields the desired result, proof of the lemma is not without axiomatic constraints. In particular, the argument uses $>(\vee)$-Praeclarum, $A > B \,\&.\, C > D \to .\ A \vee C > B \vee D$. Finally, the hereditariness modelling condition follows immediately from the inclusion, $a \subseteq b$.

The condition required for Modus Ponens is modestely demanding, and so easy to avoid or defeat:

MP. $\quad A, A > B \,/\, B \qquad\qquad$ for all $x \in O$ there are $y, z \in O$ s.t. $Tyzx$.

Proof of soundness is unproblematic. Suppose in some model, at some $x \in O$, $I(B,x) \neq 1$. Then for some y and z in O, $Tyzx$. Since A and $A > B$ are valid $I(A > B, y) = 1 = I(A, z)$, whence $I(B,x) = 1$, contrary to assumption. For completeness, define y_1 as the class of theorems of the logic and set $z_1 = \{A : (\text{for all } B)(A > B \in y \supset B \in x)\}$. Then $\overline{T}y_1 z_1 x$. To show $z_1 \in O$, suppose on the contrary $\vdash E$ but $E \notin z_1$. Then for some B_1, $E > B_1 \in y_1$ and $B_1 \notin x$. But as $\vdash E > B_1$ by MP, $\vdash B_1$, whence as $x \in O$, $B_1 \in x$. Now, maximize on z and y in turn. Inflating z proceeds as in the completeness argument, and uses preaclarum principles. Inflation of y is more hassle. But Zorning will succeed.[204]

Another condition of implicational logic which fails for evidential notions and those of noncircular reasoning and support is $A > A$. It too admits of interesting modelling:

$$A > A \quad \text{for some } x \in O, Txab \supset .\ a \leqslant b.$$

(Note that the condition $A > A$ is interderivable with $A \rightarrow B \,/\, A > B$ given $>$-Prefixing.)

13.3 Abduction, Explanation and Discovery

'In Aristotle abduction refers to those types of syllogistic inference which fail to bring any certainty with them , ... due to a weak connection between ... terms'.[205] Or at least the Latin rendition, through *abducere* ("to lead away"), does. These ancient non-American origins have been lost sight of in ultra-modern America: 'Peirce (1931-1958) coined the term "abduction" to refer to the formation of explanatory hyptheses'.[206]

Abduction as reintroduced by Peirce, through a single rather special form of inference, hovers between explanation and discovery, between being a method for one or the other.

Peirce's abduction—is it just an example or is it his characterization?—is said to be encapsulated in the following scheme:

The surprising fact C is observed.
If A were true, C would be commonplace [not be surprising].

$\therefore$ A has some probability of being true.

---

[204] Priming lemma will go through when have $A \rightarrow B \,/\, A > B$ i.e., $A > A$. For $(A > B) \vee (C > D) \rightarrow .\ A \,\&\, C > B \vee D$, since both antecedents then yield $A \,\&\, C > B \vee D$.

[205] Reese 1980, p.1.

[206] Thagard 1992, p.53. More accurate dates for C.S. Peirce are 1839-1914.

First observe that the scheme, although suggested as an example of nondeductive reasoning (and regularly presented in that way in subsequent literature), is near to deductive. Its deductivoid character can be exposed by expliciting the conclusion as, for instance: A has an initial positive probability. (Or more quantitatively, initial probability$(A) > 0$.) Now, by modal logic, A is either necessary or contingent or impossible. If it is impossible then, by orthodox logic, it has no prospect of being true. So that case is excluded. If A is necessary, serving for instance to explain some surprising mathematical observation, then A has a high initial probability, namely standardly 1. In the remaining case A is contingent, indeed the context of the argument (with C observed) suggests that this is the intended situation. Now it is conventional wisdom, which can be incorporated into probability logic, that all contingent statements have initial positive probability. Q.E.D. Only this classical argument has the minor drawback that it leaves out A's explanatory role.

Secondly observe that the scheme could be rendered a bit more syllogistic, but at first blush no less plausible, by replacing "be commonplace" by the square bracketed alternative offered. But this replacement (by no means an equivalence) turns problematic should the conditional be contraposible. For then we obtain as premisses:

> C and C is surprising. (The sense of the tautologous first premiss?)
> If C should be surprising, A would not be true.

Which invite the deductively drawn conclusion: A is not true. And that (despite previous conventional wisdom) rather tells against Peirce's conclusion. It rather looks both as if the second premiss does not so contrapose, and as if the *explanatory* role of A has become submerged.

There is a third reason for reflecting that Peirce should be put back at the logical drawing-board. Namely, the inference does not appear to generalize, but to hang upon special features of *surprise,* which need to be brought out. For one thing generality does not obtain. The first premiss, that f-type proposition is verified, presumably paraphrases to:

> C and $f(C)$.

Now take f as some such propositional feature as being always true, or even quite commonplace. Then (conventional wisdom apart again) the inference appears to fail. It hangs on special features of the 'modality' S, "it is surprising that". Functor S is not systematic, and indeed satisfies very few of the usual modal principles. It is hard to get much of a fix upon logically, and indeed is up there with those functors such as "It is trivial that" and "It is informative that" that participate in the 'paradox of analysis'.

As an inferential procedure for generating explanatory hypotheses, what abduction is also explained as, abduction can take other forms, without such operaors as S. Here is a different example (as contrasted with matching terms of induction and deductions):[207]

All the beans in this bag are white
These beans are white

$\therefore$ These beans are from this bag.

Here abduction takes the form of a deductive fallacy—that of conversion—and indeed can be seen as a form of induction! This is not so surprising, as induction too can serve as a means of reaching explanatory hypotheses (which may also be confirming).

To schemes like those offered Thagard 1992 objects that they do 'not look like a possible source of discovery since' the hypotheses to be generated already occurs in effect in the premisses. Two observations are reached. Having explained *abduction* in terms of 'formation of explanatory hypotheses', Thagard switches from explanation to discovery without notice. But discovery only overlaps explanation. Then only by a subterfuge does Thagard himself escape his own objection. Namely he pretends the following premiss is a rule:

$$\text{Rule } \forall x(Fx \rightarrow Gx), \text{ i.e., all } F \text{ are } G, \text{ would explain } G(a),$$

so that F does not itself occur. But "all F are G" is not a rule.

An interim conclusion is tempting: Peircean abduction is neither well thought through nor that promising. Nor are some of its descendents. It is not altogether surprising then that abduction is coming to mean something very different in computing science.

13.4 Outside Standard Deductive Theory

Beyond the confines of standard deductive reasoning it is important to distinguish two distinct types of reasoning:

I. Deductive reasoning,

---

[207] Drawn from Reese 1980, p.420.

involving nonstandard functors or the like, in particular functors concerning probabilistic or statistical notions. For example, many cases of what are presented as probabilistic reasoning can be seen to be, or transformed to be, cases of deductive reasoning with functors that fall outside standard confines. Some examples, with interesting special features, will illustrate the matter.

- *Most* with *probably*; and alleged *inductive inconsistencies*

Most A's are B's
Most B's are C's

$\therefore$ Probably this A is a C

Most A's are D's
Most D's are $\sim C$'s

$\therefore$ Probably this A is $\sim C$

From a mainstream perspective such reasoning is of course highly problematic, since when the premisses are true both Probably P and Probably $\sim P$.[208]

- *Support* or *confirmation*, and alleged *non-deductive reasoning*

One 'well known objection to the idea of non-deductive inferences ... completely ignored by Stove' runs as follows:

> Suppose an inductivist says that a conclusion h may be validly inferred from evidence e provided that the degree of support for h does not fall below some critical level, say 0.9: and suppose that the same evidence e supports a hypothesis h_1 to the degree 0.99, and likewise h_2, ... and likewise h_n, where h_1, h_2, ... and h_n are mutually consistent. Then a rational man who accepts e should proceed to accept h_1 and also h_2, ... and also h_n. But it may very well be that the support provided by e for the *conjunction* of h_1 and h_2 ... and h_n is well below 0.9 and perhaps even zero. Then what is the poor fellow to do? He is supposed to accept each of them considered separately but reject them collectively.[209]

Such arguments *against* 'nondeductive' reasoning do not undermine it, but rather *reveal* its character. Consider the above example used by Watkin against Stove. What it shows, if anything, is that Stove is not entitled to:

[208] Other examples like this feature in work on 'inductive' inconsistencies.

[209] J. Watkins, "Animadversion on David Stove's *Popper and After: Four modern irrationalists by a fifth 'irrationalist'*" *[full reference unknown—Editors]*. The objection lodged is said to have been highlighted many years ago by Kyburg.

$$\frac{e \text{ supports } h_1 \text{ ... } e \text{ supports } h_n}{e \text{ supports } h_1 \,\&\, \text{...} \,\&\, h_n}$$

Or in reduced form: "A supports B", "A supports C" does not yield "A supports B & C". Similarly, presented in non-implicational form, A supports B & A supports $C \nrightarrow A$ supports $(B \,\&\, C)$. Or a bit differently, we can fill it out with *support* in terms of critical levels for example at 0.9. In brief, then: *Support* (or confirmation) is nonadjunctive, and presumably noncompositive. And that is all.

II. Genuinely nondeductive reasoning.

This divides in turn into importantly different classes namely:

II.A. Conclusive reasoning,

that does not conform to deductive requirements, and

II.B. Nonconclusive reasoning.

Part IV

Metaphysics and Epistemology

14 Freedom Without Determinism: decent logics of relevance and necessity applied to problems of free-will

This essay originally appeared as Sylvan 1995. [Editors]

Determinism, although widely espoused especially in the guise of a compatibilism, hardly sports a healthy logical shape. Its more robust forms depend upon vulnerable, speculative reductionisms, based on classical physics, together with a heavy psycho-physicalism. All of these, where they have not already foundered, are really in deep trouble. Most obviously, classical physics has been superseded at the micro-level by a quantum physics which flaunts a sort of indeterminism.[210] It reflects something about the respect for rationality maintained within the philosophy profession that determinism continues to be the inertial position, with free-will seen as struggling hopelessly for its life. (But it had better stay alive, not just for matters of proper responsibility and just punishment, but, just, for all those interminable introductory philosophy courses.)

14.1 A main argument flawed

In fact, however, *arguments* to determinism have been in bad shape logically all along, before newer physics made their debuts and damaging impacts. Until recently the arguments involved were never exposed in sufficient detail. Rather, like religious dogma, sketchy handwaving expositions were simply handed down from above from generation to generation as philosophical

[210] As a result, some hard-line physicalists declare (to the consternation of the philosophical masses) that they are not determinists. Of course the sort of indeterminism that quantum theory can offer affords, like randomness, little comfort to animal free choice (see also below).

gospel. Such shoddy presentation persists, both in popular philosophy and professionally. In both it is imagined that determinism somehow fans out, through laws of nature, from a past, assumed given and fixed, to the present, which is thereby determined.

The basic, naive, argument to determinism assumes the following more detailed sort of form: Consider some proposition B specifying an act, apparently of free choice, such as Eve Modern's picking an apple from a tree. 'But there is a true historical proposition H about the intrinsic state of the world long ago, and there is a true proposition L specifying the laws of nature that govern our world, such that H and L jointly determine [that B]. They jointly imply the proposition that' B thereby determining B (thus Lewis 1981, p.113, for one example among many). Filled out, with A abbreviating the conjunction $H \,\&\, L$, the argument appears to take the logical form:

$$A, N(A \supset B) \therefore NB$$

with the horseshoe symbolising some kind of implication (material implication on usual professional conceptions) and the functor N representing determination (or necessitation). But it is evident, once it is set down decently, that that sort of argument is *invalid*, even classically. For it has not been duly established that H and L are themselves determined. Supplementary arguments, not supplied (whether hard to come by or not), are required to show that NA. For example, Honderich's arguments for determinism, which take the traditional form, are fallacious in precisely this way; they fail to secure requisite determination of initial premisses (see for example his 1979, p.240).

It is evident, moreover, that much that is being assumed in naive arguments to determinism is highly contestable. For instance, how can we be so sure that all the requisite laws of nature are conveniently deterministic? So far in fact we know virtually no psycho-physical laws. To surmount this problem, Honderich 1979, for one, invests much effort to trying to stand up a psycho-physical correlation (which appears however to guarantee little more than a *material* linkage). But whether this succeeds or not, the further argument that Honderich invokes does no more than propose 'a physical sequence going back in time to before the birth of' each relevant agent (p.241). Once again premisses for a valid argument to determinism are not duly addressed.[211]

The old but failed argument, for a necessitation of happenings which excludes free choice, turned on the assumption of universal deterministic

[211] Much else is seriously astray in Honderich's 1979 textbook exposition, some of which we shall come to below.

causation which, given past fact, rendered present happenings necessary. It made 'the mistake of assuming that, since all actions in a deterministic universe are *necessitated by* past events taken together with (necessary) laws of nature, all our actions are themselves necessary ...', *without* however ensuring that 'those past events and laws' are in the requisite sense necessary (Slote 1982, p.5). However a 'new [and obvious] form of argument' patched this 'glaring deficiency' in the old argument 'precisely by insisting that the past events and laws ... are themselves necessary' (ibid. pp.5-6).

What Slote 1982 distills as the 'common argument' for the *new riddle of free-will* is a straightforward variation on the old deterministic argument. Like the old, it

> focuses crucially on the past. Given determinism, there will always be some much earlier set of conditions s that is connected by laws of nature to any given human action a that takes place. But nothing can be done to alter, nothing can be *done* about those laws; and neither, it may be added, can anything be done about s at any time when the doing of a is immediately in question. Since s is thus necessary (for example, unalterable) at the time when a is in question and since a law leading from s to a is similarly necessary (unalterable) at that time, it would seen to follow that a itself is necessary (unalterable, unavoidable) at that time—however ignorant of that fact the agent of a may be. Presumably, then, the agent in question does not act freely in performing a, and, since the argument has been entirely general in its assumptions, one may conclude that no human being every acts freely ... (p.6: 'in a deterministic universe', Slote laboriously adds. It is an important addition for his compatibilist purpose, as will emerge, not for our different libertarian objectives).

Briefly, that past is unavoidable. Laws of a nature link present actions by unavoidable connections to the past. So, present actions too are unavoidable. But the reasoning makes crucial hidden assumptions (in particular, the distribution of the unavoidability functor over laws of nature, however formalised, i.e., essentially $N(A > B) / NA > NB$). Fortunately more exact philosophers have *at last* begun to expose the logical principles assumed in long-standing deterministic arguments. Our commandeered guide among them goes on to contrast the old and new forms 'of deterministic argument against freedom of will' along the following approximate lines:

OLD FORM		NEW FORM
A	with A a statement of earlier events, circumstances or conditions;	NA
$N(A \supset B)$	with $A \supset B$ (an instance of) a law of nature;	$N(A \supset B)$
$\therefore NB$	with B a statement concerning (human) action;	$\therefore NB$

Whereas the old form 'seems clearly fallacious', the new form, according to newer orthodoxy, 'in no way seems fallacious and in fact corresponds to an inference that is valid in the most well known form of alethic modal logic ...' (Slote 1982, p.8).[212]

But that is not at all how it strikes anyone inculcated into relevant thinking. Nor is it altogether how it strikes Slote, whose main thesis turns out to be 'that the modal principle "$(Np.\ N(p \supset q)) \supset Nq$" (is) simply invalid for the sort of necessity involved', whence the new riddles are 'thoroughly undermined' (p.11). But a relevantist is spared the trouble of jumping through all the subtle epistemic hoops that Slote tries to leap through (pp.8-23), and also spared determinism. For the form in fact used, $NA\ \&\ N(A \supset B) \supset NB$, is *already* relevantly invalid!

To sum up thus far. The old riddle of free-will has been seen[213] to rest on an apparently erroneous inferential principle of the form:

SND. $\quad A, N(A \supset B)\ /\ NB$

in which the modal connective N represents some sort of deterministic necessity, *unavoidability* or inevitability. However the old dissolved puzzle has been superseded by a new riddle of free-will which relies on an improved principle of the form:

212 The fallacy of misplaced definiteness is at work. While there are many forms of alethic modal logic where the new form is correct, there are as well *many* lesser known weak forms—more appropriate for a functor like unavoidability—where it is not valid (cf. Appendix 2).

Perhaps a truncated history is also assumed. The 'old' and 'new' forms both appear to be contemporary forms. According to Hacking 1991, '[t]he old problem of free will [pre-1859] had chiefly been the question of whether a person with given motives and states of mind can then choose freely, or whether choice is predetermined by mental state' (p.190). *Then* the new question, asked by Du Bois-Reymond in 1872, was how to find a place for either consciousness or free will given both that mind must be understood solely in physical terms and the Laplacian picture of complete physical determinism (cf. p.190). Quel developpement!

213 For several examples, see Slote 1982, p.5, who conveniently surveys a certain cross-section of the literature. It has been so seen both in these examples and in a succession of recent publications.

ND. $NA, N(A \supset B) \,/\, NB$

(The inferential stylus, "/", is often read "therefore" and taken to correspond to an implicational connection; its exact status is a significant issue.)
The improved principle, ND of N distribution over material implication, $\supset$, at least has the advantage, over SND, of classical correctness. But, *it does not enjoy relevant validity*. For it is none other than an N-covered version of the rule γ (a rule form of the now deposed Disjunctive Syllogism), which (in logics of relevance and necessity) is at best admissible in special closed circumstances.[214] Is another welcome by-product of relevant logic, then, easy dissolution of the riddles, old and new, of free-will?

The elementary idea is not so far-fetched as may at first blush appear. It is part of a systemic method. Application of relevant logic is a powerful paradox and riddle solvent, dissolving puzzles in a wide range of areas, from its implicational and conditional homeland to deontic logic and set theory, and also in regions moving beyond purer logic, most recently in confirmation theory and quantum theory (see Chapter Ten and Chapter Three, section 2). Maybe the elementary idea is worth pursuing?

The elementary idea does further work in dissolving arguments in the freedom-determinism arena. For, a trifle curiously, ND and its classical Formalisation:

CND. $NA \,\&\, N(A \supset B) \supset NB$

---

[214] Forms of Disjunctive Syllogism figure critically elsewhere in the contemporary determinism debate. Consider the 'almost uncontroversial' principle ($A >.\ B \vee C \vee D, A > {\sim}B, A > {\sim}C \,/\, A > D$) that Lewis 1981 applies to reach the apparent (quite absurd) result that if one had done what one was able to do then some law of nature would have been broken (p.114), and indeed uses to structure his whole article. It is nothing but such a form. Naturally it is *invalid* in relevant conditional logic.

There may be a worry that 'mere' absence of Disjunctive Syllogism is insufficient to stop deterministic arguments (but really it is absence of N distributed forms and more). For will not the rule γ hold in the consistent circumstances supposed, and the argument be carried by it? There is no reason to suppose that γ, which holds only in rather restricted logical circumstances can be established, and reason to suppose that it cannot in general. One reason is that consistency, like primeness, cannot always be presumed. Nothing prevents development of paradoxical situations, such as those falling out of the policeman-prisoner exchange. Another reason is just conversion of the deterministic argument itself. The past hereabouts may be duly fixed, and linked law-wise under a good implication to my present effort, but my present scribbling against determinism is not therefore determined (see further Appendix 3). For a similar example, nicely developed, compare Slote 1990, pp.292-3, on the 'natural assignment of truth values'. Of course relevant semantics deliver formal counter-models to modalized Disjunctive Syllogism.

have figured large not (only) in the argument *for* determinism, so much as in a recent libertarian argument *against* compatibilism. For ND is taken to logically formalise the 'consequence argument' (hereafter *C-argument*), which is taken in turn to show that agency and free choice are incompatible with determinism. The core of the C-argument itself is informally expressed as follows:

> If determinism is true, then our acts are the consequences of the laws of nature and events in the remote past. But it is not up to us what went on before we were born, and neither is it up to us what the laws of nature are. Therefore the consequences of these things (including our present acts) are not up to us (van Inwagen 1983, p.56).

For formalisation the crucial condition, *determinism is true*, is often (astonishingly) detached.[215] This accounts, in the main, for the curious double roles of the argument principles, ND and CND. If determinism is true then the very argument demonstrating it is also (apparently) demonstrates incompatibilism, that determinism is incompatible with freedom of choice. After detachment, formalisation can proceed much as for the naive argument to determinism already formalised (following Lewis). Let H represent the true historical proposition, concerning the state of the world before we were born, L the governing laws of nature, A the conjunction of H and L, B a description of some present action, and N the functor "It is not up to us that" or the like (for example "It is not under the given agent's control that").

---

[215] It is not simply deleted by van Inwagen, who explicitly uses the condition. He advances a version of determinism of strict implicational strength $\Box(H \supset . L \supset B)$, inferring therefrom $N(H \supset . L \supset B)$, whence distributing N over $\supset$ and detaching twice, NB. Note however that $\Box(H \& L \supset B)$ certainly does not guarantee that $H \& L$ entails B or that B is deductible from $H \& L$. Should anyone declare that deducibility does obtain, ask *how* proposed deductions run, which laws are used or suppressed? It will be found that in every case where arguments can be produced (not too many!), proposed deductions rely upon laws assuming (or like equations, presupposing) a hypothetical form of less than logical strength, and normally these days of merely material form, whereupon without defective Disjunctive Syllogism there is no deduction.

It might be thought that such an argument admits relevantly correct variation. *Were* we to concede that B is deducible, by applied mathematical calculation, just from H and L (as it may be in certain very favourable cases, with L now *expanded* to include certain mathematical postulates), then $H \& L \to B$. Now apply Ass, $A \& (A \to B) \to B$, with $A \leftrightarrow (H \& L)$ again, and N principles to reach $NA \& N(A \to B) \to NB$, whence detachment willing, NB. For those already locked into depth relevant logic, there is no new problem: Ass is not universally valid. (*See, for example, Chapter Three, section 2. [Editors]*) For those not so committed, while universal determinism hardly threatens (as the concession only holds in certain favourable cases), the problems that Slote confronted do reemerge in a relevant setting.

Assume that consequence is here satisfactorily represented through material implication (a large assumption) and that N duly distributes over all $H \& L$ (a modest assumption, which can be traded in for exportation). Then, gaps plugged in obvious ways, ND formalises the C-argument. So the C-argument of course fails relevantly, just as that it replicates, the main argument of determinism, does. If the case for determinism collapses, there is little point in pursuing compatibilism, which becomes a dead issue.[216] But there is much more critical work to be attempted, such as: Does the elementary idea work satisfactorily? How much determinism still operates in unshackled lives?

Making the elementary observation *work* is a somewhat larger undertaking. What needs to be made out, in particular, to some reasonable level of plausibility, is that enduring deterministic arguments against free choice—those that generate the free-will problem—*do* track through forms like SND and ND, and not in some substantially independent way, and further that these arguments do not permit reinstatement in terms of relevantly valid principles.

In this enterprise a relevantist is soon motivated to investigate issues previously quite insufficiently considered. Of particular importance is the question of the adequacy of formalisation of natural conditionals and laws of nature through the material conditional, $\supset$.[217]

Those formalisations prove inadequate. Laws involve some smarter connective, strengthening a good conditional to natural necessity status. Suppose a basic connective of this sort (for there may be more than one) is symbolised $\Rightarrow$. Simpler law-like conditions will take the form $A \Rightarrow B$, or a restricted universal quantification thereof. Then the challenge to determinism

[216] A serious problem remains for compatibilists, since their precarious position depends both upon fending off CND, as directed to incompatibilism, and retaining CND or similar, as establishing their presupposed determinism. Put otherwise, van Inwagen's strategy, formally refurbishing an old argument against compatibilism, is endorsed:- Suppose determinism is true; then the very arguments establishing this, such as CND, serve also, contrary to compatibilism, to exclude free-will. The strategy also yields a *reductio* of determinism, *given* compatibilism (which permits its assumption). For, continuing the argument, free-will is not excluded but exhibited (according to compatibilism). So free-will is both excluded and also not excluded, which is contradictory. Therefore, by reductio, determinism is false. Naturally, however, there is more ado, which resolute compatibilists used to resist, in demonstrating, given CND, the exclusion of free-will.

There is also an important way out, already glimpsed, and regularly taken by compatibilists: namely distinguish types of necessity or unavoidability. In particular, a distinction is forged between the sort of (physical or even logical) necessity operating in determinism, and the kind of 'selective necessity', such as "It is not up to us that", invoked in parallel arguments against free choice.

[217] *On this issue see already Chapter Ten. [Editors]*

gets represented, in simple form, thus: whether principles guaranteeing an inference of one of the forms,

$$NA, N(A \Rightarrow B) \,/\, NB \quad \text{or} \quad NA, A \Rightarrow B \,/\, NB,$$

can be justified? What would justify them would be schemes of the form $A \Rightarrow B \,/\, NA \Rightarrow NB$ (assuming $\Rightarrow$ ponible), which would render N systemic with respect to $\Rightarrow$ (on the relevant theory of systemic functors, see PLI). We shall argue that there is no such pair of connectives, $\Rightarrow$ and N, for a good law-like connection $\Rightarrow$ and damaging unavoidability functor N, sufficiently generally in view, and that exponents of determinism have done little or nothing to demonstrate that there is.

The arguments do not pretend to be decisive; they do not need to be, to turn the onus of proof tables. They begin with dilemmas for determinism, which are backed up by commonsense counterexamples. A main argument to determinism is broken by the following dilemma: either the argumentative linkage from a rigidly fixed past to present choice holds and would (perhaps) transmit unavoidability, but it is of material order only and is not ponible, or else the linkage sometimes breaks down (as when entailment is tried) or fails to transmit unavoidability (as apparently with natural implication). Another typical dilemma runs as follows: if the past-present linkage *can* be made out, then the connective involved is of insufficient strength to ensure distribution of necessity N, while if a connective is introduced which does warrant such distribution, then there is no assurance that it is not of excessive strength for duly representing past-present linkage. A determinist cannot simply cut through these dilemmas by *claiming* that there is such a connective. Free-will, not determinism, is the commonsense inertial position, and is supported by examples.

Onus to repair the main argument to determinism, for proof thereof, thus lies squarely with determinists and their compatibilist fellow-travellers. For normally we take it, as compatibilists concede, that we sometimes act freely; paradigm examples of such choice and action are readily supplied. Against that normal setting determinism comes very often as a shock. The shock is administered through certain persuasive, but extremely sketchy and slack, arguments. The trouble is, as discussion is beginning to reveal, that those arguments are quite insufficiently detailed. The gaps are becoming apparent, and the arguments are ceasing to be convincing, at least to those who properly demand for validity, tight relevant arguments. It seems fair enough then to issue a challenge to afficiandos of the main argument: Put up a sufficiently detailed sound argument, or back down!

But exponents of freedom can do significantly better than this sort of bravado. For one thing, they can legitimately appeal to many world modellings, in support of a *key contention* already adumbrated: there is no (implicational) connection, suitably stronger than a material conditional, which links past fixed points—an unavoidable past—utterly *ineluctably* with present choice doings and happenings. The leading idea is simple enough.[218] As for relevant implication, semantics for an appropriate connection will be genuinely *many*-world in kind. But in different worlds different outcomes will *evolve* from the same starting points, universes will branch. So a uniform unavoidability functor will not apply to all present happenings in our actual world, because in other worlds some will turn out differently. In short, presumed deterministic *linkages* become unstuck. A now familiar diagram helps display these features.

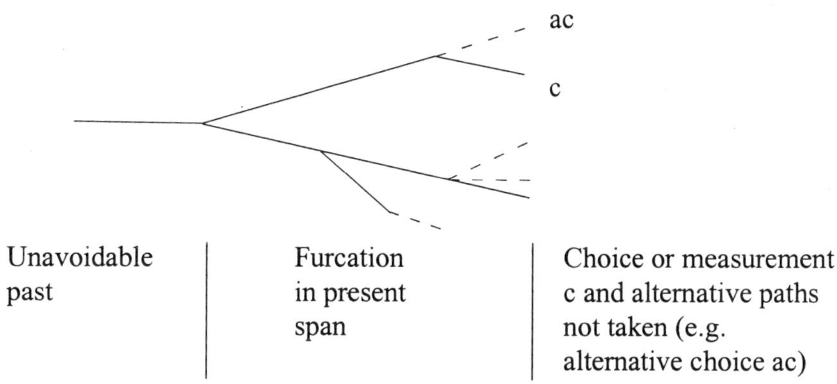

Figure 14.1 Branching worlds or forking times

Choice *c* is not unavoidable because there are other courses that could have been taken, roughly, that are taken in alternative accessible worlds. Similarly for measurement *c*: it is not determined because there are other results that might have emerged (with estimable probabilities in simple frameworks). It is worth emphasizing the rich analogies between agent choice, as set in a many world framework, and quantum measurement, as set within a many-world elaboration of quantum theory (for the common relevant modellings,

[218] It is already nicely exploited in cruder modal environments, for instance by McCall 1994 and also van Inwagen 1983. A major disadvantage with such modal approaches is that Disjunctive Syllogism lives on. A furcation picture is already to be encountered in James's defence, almost a century ago, of free-will.

see CS). As measurement collapses the many worlds (or the wave packet), so univocal choice contracts the alternative worlds to one. Naturally the analogy is imperfect. There is, for instance, no evidently constraining Schrödinger equation or obvious analogue of Heisenberg uncertainty in the choice situation (though a quantum theory of agent's brains is not excluded). Choice in turn has certain intentional preliminaries, which diverge from those of preparation for measurement (both can however be, so to say, accidental). An agent making a choice will possess both a range of information and aspirations or values, and if rational will aim to satisfy aspirations in the light of available information (details of such a satisizing modelling are in MR). With both choice and measurement (which may involve choices of what parameters to try to measure), the future will be opening further, through furcation, *as* alternative-contracting procedures are applied. Such expanding-future (and mind-expanding) pictures are extremely helpful in escaping established fixed-future and future-fixing images, in breaking the hold of old and false problem-generating images (as Wittgenstein stressed).

As initial conditions are assumed fixed, the newer picture focusses attention mightily attention on the character of the controlling laws. If these are quantum laws, then they will not fix a unique future, reducing courses to a single continuation of the past. Indeed, but rarely will systems of laws achieve such a contraction of paths to one. Newtonian particle dynamics, upon which so much philosophical attention has been lavished, is quite exceptional in this respect. So too therefore are Laplacean worlds which much past philosophical discussion has simply assumed. Striking examples are afforded by 'prediction devices' which, helping themselves to a Laplacean setting, supposedly can chart an agent's future behaviour and predict courses of action taken (but part of the argumentative patter, for example 'if the agent will do ... then the machine will predict ...', slides back into logical determinism otherwise). Such devices will fail even for deterministic worlds like chaotic ones, where predictability lapses. Moreover, for worlds governed by less tractable equations, and for indeterministic worlds, things are even worse for cosy Laplaceanism and significantly better nondeterministic modellings. But, remarkably, widespread Laplacean assumptions have tended to make non-Laplacean structures, including those which allow duly for free agency and choice, look anomalous, when it is Laplacean order and control that is exceptional.

Now suppose we try to introduce some connective, $\sqsupset$ say, that duly tracks the law-governed course of actual events through time, through the opening (by branching) and closing (by selection, measurement, etc.) future. Then (taking up the happening of events appropriately), "$A \sqsupset B$" reads something like: "after A then (ineluctably, in the course of events) B". Suppose further that A is placed in the unavoidable past, but B is situated after furcation. Then certainly NA obtains. But NB may well not, because there are accessible alternatives.

14.2 Other determinisms and arguments

It is not enough to flaw just a main argument to determinism. One of the reasons why determinism looms as such a difficult philosophical problem is because it amounts to a tangle of problems. Like most persistent though apparently outrageous philosophical propositions, determinism is a hydra-headed monster. It enjoys furthermore, a kind of mytho-octopusian tenacity: cut off one problem and sure enought another rapidly grows to replace it. There are rather a lot of these argumentative heads to chop off, as quickly appears. But, although there are many argumentative heads, a unifying theme will emerge: *determinism is a logical con-job.* Determinism is in all cases 'established', if not by sketchy classical reasoning (as above), by logical trickery. Where supported by more than 'plausible' speculation (a favourite epistemological method), it is floated on a deviously interwined series of sophistical arguments.

We can *begin* to get to grips with the (mythical) determinism monster by a divide-and-rule technique, of separation through classifications.[219] In this regard there is a need for a *dual* (but interconnected) classification, of:

- *types* of determinism, and of
- kinds of *arguments* for determinism.

Traditional limited classifications tend to conflate these two. *Logical* determinism, for example, is both a strong type of determinism, and a determinism supported by (purely) logical considerations. By contrast, *causal* determinism is both a weaker type of 'determinism', and a form supported by inductive and (quasi-) empirical arguments.

---

[219] Naturally we need not pretend to any sort of complete attack on the monster, as outside help is available. For highly compatible treatments of further arguments (many of them not enduring) in the snarl, see Attfield 1979, chapter 3, and, much earlier, T.H. Reid (whose analysis Attfield develops). A larger exercise would attempt to synthesize all these and other pieces.

14.2.1 On strengthened logical determinisms

These include detachable issues which philosophers sometimes want to set aside as 'views improperly classified as determinist', though they certainly involve determinism; for instance, logical determinism and predeterminism (or predestinarianism). Predeterminism and fatalism have already been sunk (for details see JB, chapter 2, pp.405-409). Logical determinism is said to be 'the doctrine that the future is as fixed and unchangeable as the past: that just as what has been, has been and cannot be altered; so what will be will be, despite anything anybody may do' (Donagan 1973, p.19). But, firstly, this *is* a strengthened determinism, which implies (present) determinism, in the shape of unchangeability of the present train of events and actions. Secondly, there is but little that is logical in the presentation, except for a modal fallacy. For there is a well-known equivocation in the argument from the apparently tautological: what will (inevitably) be will (inevitably) be. Observe that the intended use of *will* must be one that conveys something like inevitability or unavoidability, else a deterministic outcome will not ensue. Symbolise "it will (inevitably) be" suggestively by "$\Box$". Then what we appear to be offered, by some slide of meaning, is transformation of the platitudinous $\Box A > \Box A$ to the damaging $\Box(A > \Box A)$, effectively a form of the fallacious $A \rightarrow \Box A$. Similarly the platitudinous $WBp > WBp$, one tautological construal of the ambiguous "what will be will be", gets transformed to $WB(p > WBp)$, Hollywood determinism. An analogous scope confusion is exploited by Russell to establish determinism. Russell proceeds effectively from $\sim\Box(A \,\&\, {\sim}A)$ to $\sim(A \,\&\, \Diamond{\sim}A)$, i.e., $A \supset \Box A$.[220] As it happens, we can rest confident that such determinism cannot be decisively established logically, for the good reason that alternatives to it, including Aristotle's future indeterminism, can be adequately modelled (see TT).

Strengthened determinisms are not alone in resting upon logical howlers. So, but less transparently at present (because of the prevalence of classical logic), do still fashionable determinisms. Amusingly the very problem of determinism is sometimes so stated that it looks like a modal fallacy: 'By determinist I shall understand ideas according to which that which *is* also (somehow) *had to be*' (von Wright 1983, p.35).

220 See Russell 1954, p.147: Nick Griffin kindly supplied this reference.

14.2.2 On still fashionable determinisms: universal causation, rational decision, and theoretical contraction

Fashionable contemporary determinisms are nonlogical. Typically they depend upon empirical themes (or perhaps synthetic a priori themes), such as universal causation or psycho-physical reduction (including functionalism). No doubt some of the assumptions of logical empiricism may have disappeared, or like those of verificationism have become hidden from general view. But some, important for determinism, remain conventional wisdom: the adequacy of classical logic, one-worldism (there is no world other than the actual (physical) one, so that necessity does reduce to fact), even that the physical is entirely deterministic territory, the only problematic regions being extraphysical (mental, supernatural, and like soft ground). Determined advocates of freedom should concede *none* of these assumptions. Determinism fails all the way down from mental and free spirit reaches, to base metal and bare matter sloughs.[221] Of course failure of determinism, as with quantum indeterminism and other randomism, does not establish any satisfactory sort of freedom. It is usually supposed that what is not decided, or determined, by agents making (potentially rational) choices and is not caused must be random processes (or some such), which do not and cannot qualify as free. For freedom requires, as well as somehow 'realising' one of a sheaf of open alternatives, having the power to select nonrandomly which one of these alternatives is realised (McCall 1994, chapter 9). Outside the process tradition, it is boringly standard to look for, or repudiate, freedom in and only in higher rational reaches, where deliberative choice appears to occur. Let us investigate this higher end.

An immediate challenge to higher freedom derives from universal causation. The drift of the argument is this: if supposedly rational choosers are caused to do what they do, then they are not free to choose otherwise, and their choices are determined after all. There is a semantical analogue. Causation tends to accompany, and enforce, one-worldism. Causal jointing is presumed to prevent furcation of worlds. Causal sequences form *chains*. While choice may make it look as if there are alternatives, causation in fact blocks any such alternatives. But law-like linkages on their own do not, as modellings of choice reveal, preclude free choice. A deterministic way has to be worked from law-like linkages to causal connections to freedom-exclusion. (The standard slide from law-like linkages to causal connections

---

[221] As to the general presupposed background hierarchy, an elaboration of ancient pre-scientific ordering of things, see for example Wilber 1983.

is prettily demonstrated in Smart's industrious failure to make sense of Campbell's concept of contra-causal freedom, Smart 1989, pp.144-5.)

Often determinism is virtually conflated with universal causation (thus for example Honderich 1979, p.240; Hospers 1954, p.263; Nozick 1981, p.292), and regarded as established with it (however that is accomplished). But universal causation and determinism diverge. On the one side, determinism does not have to be causal in kind. It could be logical, it could be functional (like the noncausal equational linkage given in mechanics). On the other more interesting side, something might be caused, or caused to happen, for example through a decision taken, without being determined. For instance, Alf freely decides to hit a ball which causes another ball to move; while thereby the ball is caused to move, and Alf causes it to move, these things are, on the face of it, not determined, because of the free choice involved. While free thinkers would accept some such description, determinists and compatibilists would quarrel with it, contending that the initial moment of 'choice' had already interrupted the causal chain, or else disputing the ordinary presentation of free choices as not causes.

At this stage, terminological manipulation of various sorts may begin, to change the face of things and to remove such disturbing divergence. For example, "determine" is *redefined*, so that what is caused is determined including 'mental causation'. Thereby, what is decided is determined, and determinism entirely loses it force, and becomes a tautological doctrine, toothless again. Or else "cause" is shunted to exclude mental processing and other undetermined 'causation', and to align causation with determination. While determinism may then coincide with universal causation, far from being tautological, it appears utterly implausible—unless decision and the like are somehow reduced or otherwise disposed of or hidden. Or more subtly, choice itself may be massaged into a form of mental causation (see Appendix 1).

A central issue as to freedom concerns choice (as *free-will* directly implies). But choice characteristically includes a mental component. This would appear to cast a major spanner into the deterministic works, since mental happenings—a critical stage in the choice process—are seemingly not deterministically law governed. Two very different types of argument are rolled out in an effort to regain deterministic control: arguments from an analysis of (rational) choice, and arguments from theoretical contraction, above all from psycho-physical reductionism.

The first type of argument, sometimes regarded as no more than cute sideline argument by reductionists, looks directly to the standard choice and deliberation situation (where the main reductionist argument may look a bit thin). In essence, the rational choice argument to determinism is that a rational agent has no choice but to maximize upon expected utility (or the equivalent).[222] For a machine that has been programmed to operate in this way there is, outside malfunction, normally no other alternative.[223] For an agent, by partial contrast, any other choice than that determined by such a maximal rule would be irrational. Thus an agent, like a machine, really has no other course than that so determined. But this cute little argument is drastically flawed. Not all agents and choosers are rational always (for detail see PA). A supplementary argument, still missing, is required to deterministically curtail irrational choice. Nor are rational choosers always without genuine options. Not only are there ties to resolve occasionally, there are alternative maximizing rules which give different directives; most important, there are rational alternatives to maximization, namely satisizing alternatives (for the detailed story, see MR; a free satisizer will not be, perforce, tracking bestness). As satisizing recipes tend to leave much room for choice beyond adequacy thresholds, the rational choice argument to determinism is entirely severed. The cute argument fails. More generally, 'rational choice leaves no alternative choice' arguments, which stretch back through Spinoza to the Stoics, depend upon rational choice being regularly single-valued, as certain over-demanding recipes will ensure. Under satisizing and like recipes, however, rational choices are pluralised. Thereupon determinism is forced to retreat towards its home ground, theoretical contraction and reductionism. And that exposes better the ideological character and linkages of determinism.

An important cluster of arguments for scientific determinism take the following schematic form: Under theory t [advanced by charismatic intellectual i] all nonphysical happenings not directly covered by physical determinism are really of kind k, i.e., insofar as they count they contract to kind k. But all happenings of kind k are determined. Thus, given that t is correct [by virtue of authority i], all happenings that matter to science are determined. Spinoza's metaphysical determinism, Freud's psychological determinism, Marx's economic determinism, Hobbes' and mainstream Anglo-America psycho-physical reductionism, and Wilson's sociobiological determinism, all belong to this cluster. Examples, merely adumbrated, include: conditioned response behaviourism, as promoted

[222] This sort of argument, often assumed dead, is resurrected and elaborated in Nozick 1981, chapter 4. While Nozick contends, correctly, that satisficing comes down to a type of maximizing, he fails to notice satisizing at all.

[223] Compare McCall's 1994 reflections on machines which (might) exhibit free-will.

by Skinner, according to which all action is controlled, responses to external stimuli; personality structure determinism, as advanced by Freud, under which a person's personality package, childhood history, unconscious motivation and so on, determine that person's actions; economic determinism, as advocated by Marx, according to which a person's life and culture is fully governed by material conditions and economic factors (all these examples are developed in some nice persuasive detail, if rather uncritically, in White 1991, pp.113-126, pp.468-475).

Missing so far are impressive reasons for being rationally compelled, still less deterministically overwhelmed, by competing arguments of this cluster: libertarians can remain substantially unperturbed. Firstly, the theories involved are not merely highly speculative, but, as scientific theories, mostly highly disconfirmed, arguments from authority notwithstanding. The theories concerned do not pertain to our actual world. While no doubt there are fully deterministic worlds (though perhaps not those of the theories), these appear significantly distanced from actual ones. Secondly, these theories tend to omit relevant normal happenings and action; often they are based upon exceptional cases or pathological behaviour, whence disconfirmation by normal free actions. Thirdly, the theories do not strictly show full determination of normal actions so much as manifold influences upon and delimiting actions. Granted there are often influences, some of them hidden from actors. But arguments therefrom to determinism turn upon familiar slides: influences become determinants, as elsewhere certain controls had become total controls, and constraints get expanded and become complete controls. While it may be a popular error to underestimate constraints upon choices, which are many and various and sometimes unnoticed, it is a serious error to extend them to eliminate choice alternatives, to (try to) gain uniqueness. In these respects, the more scientific arguments resemble popular arguments for determinism, which usually involves slides, such as that from some action operating under external influences to all actions being completely governed by such influences (for several examples of circles of expanding influences allegedly forcing the leap to the deresponsibilizing embrace of determinism, see White 1991, pp.103-108). Finally, even if, what seems remote, some arguments of the cluster could be put into classical order, they would lack relevant validity for the same reasons as before.

A major subsidiary argument for empirical determinism depends upon reductionism of some sort, specifically upon reduction of full choice complexes to neurophysical happenings. Each reducing complex, it is then imagined, succumbs to standard arguments for physical determinism. So there are two grand faults in the subsidiary argument: not just the relevantly flawed argument to physical determinism, but further the reduction leap. At present, it appears that the reduction proposal is no more than that: a proposal. Less kindly, it is mere speculation, and accordingly it cannot be used to support more than a speculative determinism, or more than a speculative form of its soft subcase, compatibilism. And speculative determinism is no serious threat to free-will, any more than speculative theism is to atheism.

While a strong reductionism (conferring identity or the like) is speculative, a weaker linkage such as mere correlation of mental with physical is inadequate to transmit determination, N or whatever. Suppose, as is widely assume, A, representing mental processes including mental decisions, is correlated with B, brain processes. What is then supposed (for example by Honderich 1979) is that where B is determined so is A; that is, that determinism distributes over the correlation. The trouble with such a supposition is that the required inference is out of the same box as principles already discredited. For if the correlation is, for instance, of material implication strength, then the underlying implication comes to $A \equiv B \,\&\, NB \rightarrow NA$, which rests upon $NB \,\&\, B \supset A \rightarrow NA$. But these princples are not even classically valid, reflecting a critical weakness in main cases for determinism: *that plausible psycho-physical correlations are of quite inadequate logical strength to carry determinist arguments.* Suppose then, what is venturing back into very speculative territory, the correlation is of lawlike or deterministic strength. If the connection reaches deterministic strength, then the implication reverts to relevantly invalid forms, $N(A \equiv B) \,\&\, NA \rightarrow NB$ and $NB \,\&\, N(B \supset A) \rightarrow NA$.

Not only is the higher psychological side of determinism seriously flawed, the lower physical side, often taken for granted, is not as sound as presumed; in fact it is also fraught with difficulties. Some of these difficulties can be exposed by looking at several diverse supplementary considerations that enter to reinforce determinism and extend it beyond its small classical heartland. It is not much use that a sort of determinism (such as unique predictability) should succeed for a universe of classical particle dynamics, when our actual universe is not like that. To be sure the immediate successor of classical physics, relativity theory, looked highly amenable to determinism, by contrast with the later quantum theory. But appearances are proving deceptive, since more and more solutions to the equations of general relativity have begun to appear. One of the many misleading images to be duly removed is a "static" Minkowskian representation drawn from relativity theory. There time is taken to be a further noninvariant dimension along with space, and so future objects are taken to exist (timelessly) in all their determinate features. Future indeterminacy, naturally associated with an open future, fertile for choice, disappears. This Minkowskian picture is not only problematic, it is not compulsory, and there are good reasons for not accepting it. A rival many worlds model structure restores future indeterminacy, nonsingularity, and opportunities for choice (see McCall 1994, and CS).

One of the supplementary arguments Smart advances in his case (a typically sketchy case) for determinism, perhaps with 'free will', turns (once again) on a mad scientist example, sketched in a remote typically reductionist setting (mad scientists are popular in philosophy of science as they offer a medium for remote hypothetical examples). The mad scientist captures a female poet and fully rewires her, altering brain connections and memory, as an engineer. Smart 1989 concludes (p.149) that 'we can go two ways, either using "free will" in the compatibilist sense, ... or else going the other way and denying we have free will at all'. Either 'way' it is determinism at bottom (with or without a 'free will' gloss).

But in fact the mad scientist example is supposed to enforce determinism and cast doubt, at least, upon compatibilism. A main part of the argument, that 'the one-time poet but present engineer' is not 'perfectly possessed of free will', despite her extended deliberation and choices of her own, is this: 'The engineer's actions were determined in part by her character, but her character was determined for her by the mad scientist' (pp.147-8). But that, and indeed the whole development of the example, *assumes* that reductionism succeeds. Accordingly it assumes a substantial part of what is really at issue: *whether* what determines such items as brain connections and brain processes determines deliberation and choices. Against main intended targets, then, the argument begs the question.

There is also an important lacuna in the argument, concerning the critical quoted phrase, "in part". Even as the argument is given, the engineer's actions were not determined, but only determined *in* part, which could mean: *not* determined. (If Smart were a little more realistic about what a mad scientist might do, then he would concede that the said scientist only helped to determine—only *in part* determined—the engineer's character, since much went unchanged, such as the abilities to speak and read, and more generally to act in such and such ways.)

But it is well known that partial determination need not tell against free choice at all, but is in certain respects essential to it. If a chooser could not rely, for example, on any law-like connections, then effective choice would sometimes at least be threatened. In any case, it is generally acknowledged that agents make choices *within the setting* (physical, environmental, etc.) they are in, as *constrained* (but not determined) by it. If this constraint background is changed, as with the engineer, then different choices may of course be made. Thus what is common is choice *of* so-and-so *against* such-and-such a background. Constraints, background, define but do not thereby determine choice. There may still be—there are—partially constrained free choices. What does not follow is that background becomes foreground and fixes choices.

Helpful here is once again a constrained choice model. Choice looks to the directive formula:

P (factors) subject to N (constraints).

Determinism has N swallow up P (whereas utilitarianism has P absorb N).

But now this schema corresponds neatly to a modelling for freedom, which can conveniently be run out for groups or communities as well as individuals. Consider first, after T.H. Green, *making the most of ourselves*, which Green (in Nettleship 1906) construes in established maximizing fashion:

- Making the most of our group freedomwise
  = maximizing upon empowering factors P
  subject to constraints N.[224]

But freedom is not *that* demanding. A more leisurely approach looks to the following formulation:-

- Satisfying group freedom
  = satisizing P
  subject to N.

P will include factors of (self-)integrity, interests, etc. N will include physical restrictions. Roughly P gives positive freedom, while N formulates negative.

---

224 Approaching the issue of modelling freedom through this recipe was suggest by I. Hunt's stimulating conference paper, "Isiah Berlin's attempt to distinguish between positive and negative freedom", Australasian Association of Philosophy Conference, Sydney University, 1990. But contrary to Hunt, freedom does not correspond to features of interests, for example a capacity of individuals to pursue them. For interests may fluctuate day to day (for example, getting across an impeding range of mountains), whereas freedom does not.

14.3 Appendix 1: Discerning a reasonable libertarianism

The broad outflow of the essay is a modest libertarianism. Free choice occurs, indeed is often exhibited, in free actions. So, to express it a bit awkwardly:

(+) there are free actions in the natural world.

Accordingly (as already seen through branched-universe modellings) the natural world cannot be a (fully) deterministic system. For

(-) free actions cannot belong within a deterministic system.

The difficult questions, for the thesis (-) of incompatibilism so presented—of *what* counts as a deterministic system and *which* natural systems are so deterministic—need not delay us. For already inconsistency is supposed to loom, locking out a reasonable libertarianism.

One entertaining argument (heavily advanced in Bishop 1989, who will serve as prime antagonist) signals onto the field an important further player: the *causal theory of action* (CTA). It is asserted:

1. Validation of (+) implies some version of CTA.
2. Validation of (-) implies no version of CTA, i.e., is inconsistent with any version of CTA.
3. A reasonable libertarianism implies validation of (+) and of (-).

Therefore, a reasonable libertarianism is impossible. (Bishop also drags naturalism into the argument, in his 1991, provocatively equating *naturalism* with *reasonableness*, but it is inessential, unlike CTA. And all reasonableness here requires is a certain coherence.) What reasonableness involves proves immaterial because 3 is not contested here.

To expose the troubles with assertions 1 and 2, let us introduce ETA, the *expanded causal* theory of action, or the *explanational* theory of action. According to CTA, for one 'to act is to be caused to behave by mental states of one's own—[glossed] mental states that make the behavior reasonable in the circumstances' (Bishop 1989, p.10), and with causation normally construed. By contrast, according to ETA, for one to act intentionally is for there to be an explanation in terms of mental states of one's own. The stock validation of (+), the usual route taken to establish it, is through ETA. A free action, under an agent's own control, is one that can be explained through the agent's own choice, that modelled in an information-values (or wrongly knocked down, belief-desires) way.

That sort of explanatory modelling hardly supports CTA. For there is no commitment to the counter-intuitive core of the CTA, concerning mental states *causing* action. *Explanation through* is one thing, *causation* another. It is one thing to allow that an agent's intentional actions can be explained through mental states of the agent, as ETA does, but quite another to insist that the agent's actions are (even are thereby) *caused* by the agent's mental states. That appears to involve a quite remarkable, and repulsive, transformation. At the very limit it appears to be playing fast and loose with causation. That is one of several things happening.

To begin, there is a familiar slippage on the elastic notion of *cause* between a wider usage comprehending explanation and a narrower usage, usually of materialist cast, implying determination. The wider usage, which is nowadays a *stretched* usage (by contrast with Aristotelian usage), would not normally guarantee deterministic features. Let us signify stretched causes as *scauses*, roughly reasons. Much of what philosophers who invoke causal theories of action or insist upon universal causation have really fastened onto are scauses, not what they suppose they are talking about, causes in the narrow sense. This can be brought out through a *dilemma* between causes in the narrow usage, appropriate to deterministic systems, *ncauses* (pronounced *nauseas*) and causes in the stretched usages, scauses. Then ncauses are a very restricted subset of scauses, as exhibited sets are of properties-in-extension or individuals are of things. Accordingly, universal scausation, a rational postulate par excellence, has much better prospects than universal ncausation (it also enjoys prospects of trivial truth if explanations are not at all circumscribed). Now for the dilemma: With ncauses, though determinism holds, much action escapes the net, while with scauses, though action is scausally covered, determinism breaks down.

It is a fundamental mistake, promulgated by contemporary empiricism, that all nonlogical explanation is ultimately causal. There is a classification and contraction of all kinds of explanation to too few kinds: logical, namely deductive, and perhaps probabilistic (or statistical), and nonlogical, namely causal, and perhaps functional, and random (then there are further attempted reductions within these few kinds). What is called *intentional explanation*, which appeals to an agent's intentions, motives, plans and so on, does not fit easily into these kinds (no doubt it can be rendered deductive in form, as even induction can, by infiltration of fat enough assumptions). Consider, for instance, a politician who allows himself to be compromised or blackmailed, *because* he is not prepared to risk serious political embarrassment. This is an intensional explanation; but though evidently nonlogical it is not of the narrow kinds (as McCall 1994 in effect explains, chapter 9). In particular, the politician's action is not causally confined, because the alternative of not giving in to blackmail may remain open and choosable up to the moment of final decision. Therein rendered accessible through the realisation that intentional explanation is not any sort of narrowly causal explanation is, according to McCall, 'the "central mystery" ... of the problem of free will'.

There are those who imagine there is no such mystery (under naturalism there are really no philosophical mysteries); a compatibilist overlay upon causal determinism 'a thoroughly believable solution to the problem' (Bishop 1991, p.1). Intentional explanation is nothing over and above causal, as judicious internal transformation will show. Bishop's stunt, learnt from that master stuntsman Davidson, consists in transforming "agent *a*'s intentional actions are explained through mental states and processes of the agent" to "*a*'s actions are caused by *a*'s own mental states". Thus, for instance, the politician's action in allowing blackmail is caused by his desire to avoid risk of embarrassment (coupled with appropriate beliefs). Then it is conveniently assumed that such scausation is nausation, so rendering, by a philosophical sleight of hand, abetted and just tolerated by some ordinary usage of causal talk, what was an analytic connection into a substantial, and substantially false, thesis. It is the same stunt that transforms ETA to CTA.[225]

Now is ETA (or a version of ETA) a version of CTA? If it is, what Bishop effectively assumes, then 1 holds, but 2 however does not. If it is not, what contemporary commonsense confirms, 1 does not hold, but 2 does. Either way the entertaining argument breaks down, and a reasonable libertarianism is on the loose again.

As for Bishop's assumption (the same sleight of hand as before), it is revealcd in thc way he manages to pin a theme intimately tied to 1 onto Slote. 'In order to establish ... the falsity of [unavoidability transmission, CND], Slote has to assume the truth of a causal theory of action' (Bishop 1991, p.14). Why? '... Slote is maintaining that [something] ... is unavoidable for agent M [if and only if it] ... has been brought about *other than through* an

225 Of course having arrived at CTA, *much* effort is expended erecting defences around it and repelling objections. Here Bishop ventures beyond Davidson, who appears pessimistic about naturalizing action through CTA. Davidson's pessimism appears to me warranted (though my reasons differ from his). Fortunately issues surrounding CTA can ultimately, it seems, be avoided here. For Bishop's thesis about the role of CTA looks wrong, and investigation will no doubt show it wrong. Bishop advances the provocative, and rather paradoxical, thesis that 'the problem of freedom and determinism ... is misdescribed. The core of the problem is neither about freedom nor about determinism. Rather, it is *the problem of natural agency*—the problem of showing how it is possible for an agent's actions to belong to the natural causal order' (Bishop 1991, p.2; similarly his 1989, p.10).

explanatory (that is to say, surely, causal) chain which passes through M's own "desires, etc." ...' (p.14). The "surely" is the sure give-away: *explanatory* is just equated with *causal*. Whatever may be thought of what Slote is said to maintain, it did not invest in CTA, but at most in ETA. The same stunt is pulled in Bishop's book under the heading "Slote's Implicit Appeal to CTA", in the section where the most detailed case for principle 1 is developed.

> Slote's view would therefore seem to be that behavior counts as action only when it terminates in a causal chain (for that surely is what his talk of explanatory chains must come to) that passes through certain of the behaver's recent mental states ("desires, etc") (Bishop 1989, p.55).

The situation has resulted then that 1 holds because Slote's validation uses a version of ETA which given Bishop's assumption is a version of CTA. But thereby 2 is blown away, because validation of (–) may very well use a version of ETA. By expanding *version* to catch 1, Bishop loses 2, because some of the further versions are not duly deterministic, unlike CTA with ncauses. On the other hand, narrowing versions to recover 2 sacrifices 1.

14.4 Appendix 2: Advancing formal logical directions

For all that a relevant reformulation raises questions that should have been asked long ago about the detail of arguments to deterministic conclusions and the adequacy of these arguments as detailed—for all that, still some of the argumentative issues can be approximated in a classically-based multiple modal framework. Once the issues have been aired in this setting (what Slote can be seen to be about, albeit in an informal way), then classical simplifications and falsifications can be left behind. But there is much detailed work to be done.

The multiply modal setting adds to a classical basis—with connectives, &, $\vee$, $\sim$, $\supset$, and quantifiers so far as need be, along with standard modal operators $\Box$ and $\Diamond$ for logical modalities—as many further functors as are required to represent the deterministic arguments against free-will: N, N_1, N_2 .... These functors will include at least "N", "it is unavoidable that", and whatever other functors that differ from "N" that are deployed in setting down deterministic arguments, including "it is inevitable that" (if it differs materially), "that ... would not have eventuated had the agent chosen otherwise", and variants thereupon ("done" for "chosen", future tensing, etc.), "that ... was causally produced", etc. Some rather neglected points, of the first importance, deserve to be recorded at this stage:

- The standard problem of free-will depends for its adequate statement on the availability of a range of intensional functors (some in fact highly intensional and exceeding standard modal bounds).

Thus, interestingly, for many hard-line determinists who wish to keep their discourse sparklingly anti-ecologically extensional, there are real difficulties in stating the problem *at all*. Accordingly, the business of getting to grips decently with the detailed problems is made impossibly difficult—exactly for those whose problems they primarily are.

- Without doubt *several* functors are involved. Free-choice-excluding determinism, for example, depends upon several discreet sideways shifts from functor to functor.

Working out the logics and interrelations of these functors is a matter of the first importance for philosophical issues in the freedom-determinism arena (for a small relevant start, see PA; for older informal investigations see Stevenson 1944, chapter XIV, and references to work therein, especially, that of Broad). I commend it to anyone with logical skills with free time.

14.5 Appendix 3: Another intuitive counter-example to disjunctive syllogism

Scene setting: The conundrum is set up like this:

> The past is determined, not all the present is determined. But the present is connected to the past by law-like connections.

Symbolisation:

Let P be a complete description of the past, say at 1900.

Let L be the set of laws describing the evolution of the universe.

Let A abbreviate $P \,\&\, L$.

Let B be a description of some present free act, i.e., of something like an agent's choice that is not determined.

Background data (assumptions):

The past-present connection is material (at least). That is, P and L together guarantee present distribution of things, and that is determined. So where N symbolises determination, for example "It is determined that", "It is unavoidable that", we have $N(A \supset B)$. Also from the scene-setting, NA and $\sim NB$. Further N of course distributes over entailment, $\rightarrow$, and conjunction, $\&$. That is:

$$E \to F \,/\, NE \to NF;\ NE \,\&\, NF \to N(E \,\&\, F).$$

(Thus N is systematic in the sense of PLI.)

Argument:		Hypothesize DSyll is intuitively correct.
So	1.	$A \,\&\, (A \supset B) \to B$
Distributing (etc.) then	2.	$NA \,\&\, N(A \supset B) \to NB$
But	3.	$NA \,\&\, N(A \supset B) \,\&\, {\sim}NB$

So 2 is countered, and 1 has to be rejected.
It is, if you like, an *indirect* counterexample, but a direct refutation.

15 Knowledge as Justified True Belief—Relevantly Justified

This essay reproduces a manuscript completed in February 1995. [Editors]

> According to the most widely accepted definition, knowledge is justified true belief. (Quinton 1967, p.345.)

That long accepted definition is widely assumed to have blown apart by Gettier 1963. For his part Gettier does give the impression that he has shown that the accepted definition is inadequate, that justified true belief 'does not state a *sufficient* condition for someone's knowing a given proposition' (p.123). Yet Gettier makes no connection whatsoever between the question entitling his article, "Is justified true belief knowledge?", and the contents of his article.[226] The body of the article comprises two examples, directed against one unattributed *explication* of "*s* knows that *P*", and alleged to similarly counter two other attributed explications. There is accordingly *prima facie* ground for suspicion that the accepted definition has been taken apart only under defective explication. The present exercise aims to confirm that suspicion. All that Gettier has upset, disestablished, are certain decoupled explications, not the widely accepted definition.

For what Gettier supplies, at most, are examples where explications of *s* knows that *P* such as the unattributed one are satisfied, but *s* does not know that *P*. The general discrepancy can be distinguished as the *broad Gettier* problem; *Gettier-type* examples more generally are *certain* examples of this sort upsetting specific explications. What Gettier does not attempt to show, and does not show, is that there are examples where *s* has a justified true belief that *P* but *s* does not know that *P*.

---

[226] What we are witnessing at subtle work here is a favoured, and highly successful, philosophical stratagem, stretching back to Socratic and pre-Socratic practice. Namely, prominently pose a question which is never in fact addressed, but give the impression that it is answered through discussion of a more amenable, less testing question. Naturally the procedure can be iterated, as splendidly illustrated in Evans 1978, "Can there be vague objects?", which stands to vagueness as Gettier's does to knowledge. The dialectical method of sideways-question-shuffle has been transformed into a modern art-form by Wittgenstein, Nietzsche and many recent acolytes.

The *prime explication*, the only one Gettier 1963 argues against explicitly, takes the following form (p.121):

PE. s knows that P iff (i) P is true,
(ii) s believes that P, and
(iii) s is justified in believing that P.

Something *like* such a biconditional, explicating *knows that*, used to be one of the platitudes of American textbook philosophy; it remains, though unattributed, a prime exhibit, the 'standard analysis' of knowledge. (The very familiarity may help explain why American philosophers have had difficulty in attributing what is uncontroversially the standard analysis. The useful expression "standard analysis" is itself attributed to Gettier by Shope 1983, at the very outset of a helpful book-length presentation of 'a decade of research' on 'Gettier-type' problems for 'the analysis of knowing', though Gettier himself does not use the phrase. As for some of the difficulties in attribution, Gettier writes awkwardly that 'attempts' at stating necessary and sufficient conditions for knowledge 'have often been such that they can be stated in a form similar to the following ...'—at which point there is then a footnote to Plato. Shope 1983 exposes the flimsiness of even an oblique attribution to Plato, and proposes instead on a similarly tenuous basis, to attribute the standard analysis to another honorary American, Kant (p.3, p.19). The evident upshot is that there is no satisfactory attribution of what purports to be the standard analysis to *anyone.*[227] What there are are, firstly, similar analyses with, however, both belief and justification perhaps displaced by alternative epistemic notions such as acceptability, adequate evidence and rights to be sure, and, secondly, different analyses from competing epistemological paradigms, for instance from pragmatism—for example Dewey's dreadful characterisation of knowledge as successful practice—and from representational realism—for example Rorty's entanglement with knowledge as accurate representation. Among similar analyses, PE itself is toyed with in Griffin and Harton 1981, despite Gettier. Otherwise the closest I managed to uncover occurred in the first edition of Hospers' 1954 textbook, where clauses (i) and (ii) of PE are essentially retained, but (iii) is displaced by "(3) There must be complete evidence that P is true"; presumably, in subject-relativized form (iii′): s has complete evidence that P is true. That is an extraordinarily tough requirement, which, construed literally, would appear to fail all Gettier-counter-examples—see below—and much else.)

227 The fiasco is based on a straw man?

By way of comparison with PE, the widely accepted definition expands "knowledge that P is justified true belief that P", by straightforward steps through the intermediary, to the recipe:

JTB. s knows that P iff s has (a) justified true belief that P.

The steps involve, for instance, subject instantiation, and biconditional equation of s has knowledge that P with s knows that P. Or to enter into detail differently: transform "s knows that P", analytically, to a substitution receptive nominalized form "s has knowledge that P", with the intended sense. Then substitute in the identity: "s has knowledge that P iff s has knowledge that P". Then reassemble, thus giving JTB.

There is no such straightforward logical expansion of the JTB recipe into the prime explication PE. To reach such an explication would require a fallacious, and now largely discredited, extensional or conjunctive theory of modifiers. That theory takes the basic form: $\phi f(x)$ iff $f(x) \,\&\, \phi(x)$, whence under first iteration: $\varphi\phi f(s)$ iff $f(s) \,\&\, \phi(s) \,\&\, \varphi(s)$. Application of this to the RHS of JTB yields: s has a belief that P & s has (it) true that P & s has justification that P, i.e., reshuffling and using yet more orthodox fudging: P is true & s believes that P & s has justification that P, i.e., roughly PE.

The conjunctive theory of modifiers is entirely mistaken, as two familiar consideration will show:

- The theory induces contradiction. Consider the true claim "Some small elephants are not small animals" for example. Under the theory this becomes, in obvious symbols, $\exists x(s(x) \,\&\, e(x) \,\&\, {\sim}(s(x) \,\&\, a(x)))$. Thence, expanding and distributing, $\exists x(s(x) \,\&\, e(x) \,\&\, {\sim}s(x)) \vee \exists x(s(x) \,\&\, e(x) \,\&\, {\sim}a(x))$, both clauses of which are contradictory given that elephants are animals.
- The conjunctive theory allows free commutation of order of adjective, altering truth values or yielding nonsense. While there are many spectacular examples ("large cold pretentious Dutch country house", "very tall young man", "little old Irish lady"), consider what is helpful for present purposes: "s challenged an alleged defamatory proposition", which goes haywire when the order of modifiers is reversed.

More generally, what extensional analyses of modifiers do is to sever connections that often matter for concatenated modifiers. There is an alternative and superior combinatory theory of modifiers, an intensionally compatible logical theory, which analyses $\phi f(x)$ in terms of combinatorial *application* of modifier ϕ to $f(x)$, a theory worked out in detail in Rennie

1974 and in others. Under that theory, presumed here, application does not collapse into conjunction.[228]

Lessons recently learnt about order of modifiers, and from logics of application and modification, do not appear to have been sufficiently appreciated. Philosophers continue to rush headlong into extensional analyses. Manifestly these lessons have not been properly applied to the concatenation "justified true belief". The accepted definition for *knowledge* was *justified true belief*, indeed the definition has become known under this, illuminatingly bracketed *justified (true belief)*. What was offered was accordingly not *true justified belief* or *believed and justified and true*, what Gettier-type examples were directed against; the order and connection of modifiers matters. The correct order and relevant connections get lost in erroneous expansions of the slogan such as the prime explication.

Though the recipe has now been distanced from the prime explication against which Gettier-type examples have been primarily directed, they may still succeed against the JTB recipe. The next main stage in the argument aims to show that they do not, that the recipe is immune to them, from which it then follows that PE *does not explicate* JTB.

Now Gettier-type counter-examples, heralded as putative counter-examples to the accepted JTB definition, endeavour to show that justified true belief is not sufficient for knowledge in the following way, the *same* in form in virtually every case:– An agent with justified belief in some statement which is also true, does not in fact know that statement, because the agent's 'justification' for the belief in question is *irrelevant* to its truth; it is only accidentally, coincidentally, or fortuitously related to the truth of the belief. Such a *general feature* will recur in all Gettier-type examples; it is a pervasive irrelevance of requisite justification to the true belief concerned.[229]

It is worth illustrating that feature immediately:

> Suppose that Smith has just signed the papers to buy the Ford in the dealership showroom. Though Smith doesn't know it, the dealership

---

[228] The difference corresponds to a distinction, appreciated in relevant logic but not recognised classically, between two sorts of couplings of premisses: merely concatenated and extensionally replicated, through conjunction, and, by contrast, bunched premisses, explained in terms of application, and intensionally represented through fusion (i.e., 'intensional conjunction'). For requisite details, and some philosophical development, see Slaney 1990.

[229] The general point is virtually attained, but not applied, by Schreiber 1987 on p.50 of a suggestive article. Unfortunately Schreiber himself shoots off wildly in several unhelpful directions, claiming falsely that Gettier examples are informally incoherent, that a merely sufficient condition for not-knowing has been elevated into a necessary condition, even (what is politics, not philosophy) that such hypothetical example-making is potentially treacherous!

> does not have clear title to the car. However, moments before and far away, Granny Smith died, and title to her old Ford passes to Smith. So Smith believes he owns a Ford, and his belief is both justified and true. Does Smith *know* he owns a Ford?[230]

No, Smith does not know, because his belief, which is as it happens true, lacks relevant justification. What renders Smith's belief true is disconnected information he does not have, concerning an inheritance. What justification *is* relevant to Smith's belief, namely the dealership situation, does not ensure its truth.

With the irrelevance feature identified, a main theme can be stated. The traditional recipe for knowledge, expressed through the slogan "justified true belief", is rendered relatively unproblematic by relevant relocation and restatement, that is presentation in a generous relevant logic setting, where relevant connections are not lost. In broad approach the procedure is substantially uniform with applications elsewhere made of relevant logical theory to remove paradoxes and puzzles from a wide range of notions rendered problematic under stock (classical) treatment.

Observe next that counter-examples do not stand up just on their own. They depend for success on various assumptions and upon certain arguments, often little arguments that exponents run past the astonished, and even past true disbelievers, very swiftly (even, in fast philosophy, as if they were not there at all). These arguments and assumptions are, however, of the first importance: exposing them will often reveal that the counter-examples do not succeed generally. Such is the case with at least some Gettier-type examples; they do not work in an appropriate relevant setting. And *all* depend upon certain underexposed or unexposed epistemic inferential principles (concerning transmission of justification and belief, reordering or insertion of modifiers, and so on).

Now Gettier 1963 does expose two assumptions but as if they were obvious points that we are simply obliged to accept.[231] One is this:

---

[230] Stitch 1992 is in fact using this 'well known example', and others for rather different, and ultimately grander, purposes. Grander purposes include these speculations: Not just knowledge but '*no* commonsense concept that has been studied has turned out to be analysable into a set of necessary and sufficient conditions' and, more sweepingly, that 'the traditional philosophical method of proposing definitions and hunting for intuitive counterexamples will have to be abandoned', or might as well (p.250), theses Stitch reiterates elsewhere. Reinstatement of the widely accepted definition of knowledge would stop such damaging speculation.

[231] Observe that the distinction between the recipe JTB and its prime explication impinges on Gettier's 1963 first 'obvious' point: '1°. It is possible for a person to be justified in believing

2°. If s is justified in believing P, and P entails Q and s deduces Q from P and accepts Q as a result of this deduction, then s is justified in believing Q (p.121).

Unfortunately for obviousness, it is not difficult to discover that this principle is problematic, especially classically. Suppose s, a typical naive male university graduate, has been persuaded by a course on standard logic that any statement entails any necessary truth and has learnt a simple proof of this. Let P be some falsehood which however s is justified in believing, and Q some abstruse necessary truth, the concepts involved in which s does not even understand (however s is assured of the necessity of Q by mathematicians whom he trusts entirely in such matters). Then the conditions are satisfied to apply 2° to conclude that s is justified in believing that Q. But surely he is not. How can he be justified in believing something he does not even properly understand? The last point also indicates that 2° cannot be repaired by making the 'entailment' in 2° into a good one, i.e., a relevant entailment. For such an entailment can still insert concepts beyond the grasp of a given agent, through entailments as elementary as $A \rightarrow. A \vee B$. What is required, other analogous exercises show, is not entailment but relevant containment. Accordingly let us substitute relevant containment for entailment in 2° to give a new linkage 2^c.

The precise form of the linkage is a material issue that will recur. It is worth trying to break 2^c down into components, and to reformulate it in a less controversial determinable way. Investigation indicates that principles like 2^c and 2° can be expressed in a basic rule form:

2R. $$\frac{P\ s\!>\ Q}{\text{if } s \text{ is justified in believing } P \text{ then } s \text{ is justified in believing } Q}$$

Then the problem is how to explicate "$P\ s\!>\ Q$", to be read "P *s-guarantees* Q". It presumably proceeds something like this in the case of 2^c, with relevant containment:

a proposition that is in fact false, *in* that sense of 'justified' in which s's being justified in believing P is a necessary condition for s 's knowing that P' (p.121, order reversed).

While it is certainly possible for a person to be justified in believing a proposition that is in fact false (contrary to a certain Cartesianism and rationalism), to concede as much is *not* to grant the disconnected necessary condition of the prime explication, as all there is to justification.

> P relevantly s-guarantees Q iff P relevantly contains Q and s makes the connection (and so, accepts the connection and is prepared to apply it).

Then the determinable linkage takes the form:

> P $s>$ Q iff P determinably implies Q and s makes the connection.

Separating out the elements is important because other epistemic functors are transmitted under $s>$, for instance, belief and justification. (A creature may have justification for something without believing it.) Thus, for instance, P $s>$ Q / if s has justification for P then s has justification for Q, and so on. Whence, P $s>$ Q, s has justification for P / s has justification for Q. And so on.

The second example Gettier 1963 constructs (it is his case II), which we shall call an *outer type*, depends upon a truncated version of assumption 2°. In outline the example, which affords a nice warm-up, runs as follows:– Agent s has strong evidence for:

P1. Jones owns a Ford.

Although s has no information on the whereabouts of Brown, he is entitled to float propositions that Brown is in this or that city, Boston, Barcelona, Brisbane, Bangkok. By virtue of assumption 2° he is then justified in believing, according to Gettier, many disjunctions, among them:

Q1. Jones owns a Ford or Brown is in Barcelona.

Now by virtue of two infelicitous accidents, Jones does not own a Ford, and Brown is 'by the sheerest coincidence' in Barcelona. So certainly Q1 is true. Furthermore, Gettier claims, *all* the conditions for a PE explication of knowledge are satisfied. But nonetheless s does *not* know Q1. Why not?

Remarkably, Gettier does not say. It is not that s does not know either disjunct, for there are cases where an agent may not know either disjunct in a disjunction yet may know the disjunction, as with instances of excluded middle, R or not-R. Rather it is to do with the business of irrelevance; because of the 'sheerest coincidence' of Brown's being in Barcelona utterly unknown to s, the justification required for knowledge is lacking. Only that is not how Gettier can put it. But we can. For s does not appear justified in believing that Q1 is true. And while 2° may imply that he is, *that* principle has been rejected in favour of 2^c, which gives no such assurance.

There are also grounds for suspecting that an epistemically covered version of the relevantly rejected disjunctive syllogism, $\sim A \,\&\, (A \vee B) \to B$, is playing a role here in the rejection of knowledge claims as to Q1. The implicit argument goes like this. Agent s does not know P1, i.e., in symbols $\sim K_s P1$. Now suppose s did know Q1, $P1 \vee B1$ in obvious symbols, that is suppose $K_s(P1 \vee B1)$. Then, by an epistemic analogue of disjunctive syllogism (namely $\sim KA \,\&\, K(A \vee B) \to KB$), it follows, $K_s B1$. But, as has been emphasized, s does not know B1. Accordingly the supposition is false; $\sim K_s(P1 \vee B1)$, i.e., $\sim K_s Q1$. However the principle is incorrect, *even* modally. (For take B again as $\sim A$, with $\sim A$ not known.)

In short, the general irrelevance aspect emerges plainly in little discussed, outer examples. What the "entailment" principle 2° enables is *justifications which are not pertinent for what is proposed as known*, which is why knowledge claims (though they *could* be enforced, for example as *lax* knowledge) are unwarranted.

While Gettier's other example—case I, supplying an *inner type* example—also applies a version of assumption 2°, *it* is not stopped by upgrading to 2^c, to closure under relevant containment. The case does however exhibit, in a conspicuous way, that *disconnection* apparently characteristic of Gettier-type examples, and it exposes the suppressed condition for *not knowing* that is regularly assumed. Let us see, by sketching the case. Smith is presented as having 'strong evidence for the following conjunctive proposition:

Q2. Jones is the man who will get the job, and Jones has ten coins in his pocket' (p.122).

Not only does Q2 entail:

P2. The man who will get the job has ten coins in his pocket

(through the identity principle $x = y \,\&\, Ax \to .\ Ay$, good in relevance logic), but the further conditions for 2° are presumed satisfied. So 'Smith is clearly justified in believing' P2. Now, however, the following accidental circumstances obtain in the imagined case: unknown to Smith, not Jones but Smith himself will get the job, and, also unknown to Smith, Smith has ten coins in his pocket. The upshot is that while P2 is true, in the circumstances, Q2, which provided Smith's evidential basis, is false. Whereupon Gettier claims:

- *Standard* necessary and sufficient *conditions for* Smith's *knowing* P2 are met.

For P2 is true, Smith believes P2, and Smith is justified in believing that P2 (by virtue of 2°).

• Smith does *not know* P2. Why? Because Smith's erroneous informational basis is sharply at variance with what makes P2 true; justification is disengaged from truth. For P2 'is true in virtue of the number of coins in Smith's pocket' which Smith does not know, basing his belief on false information concerning Jones. Does a *further* necessary condition for knowledge lurk hereabouts, one not anticipated in the accepted definition? No, it is only a further condition as regards the prime explication. In such Gettier-type examples the believer has only a partial justification, a *subjustification* of the belief without the requisite truth guaranteeing features, which is why both knowledge and justified true belief are lacking.

Only a naive philosopher would imagine that Gettier had supplied a decisive counter-example to accounts of knowledge as justified true belief, a knockdown argument. There are three areas of vulnerability (and just three given the way the Gettier problem was initially set up). Firstly, true believers in the standard conditions for knowledge may dig in, asserting that, not least because the PE conditions are met, Smith *does* know that P2. Justificatory informational bases do not, it is claimed, have to be relevantly linked to truth-making features. While not accepting this approach (suggested in some authors, for example Schreiber 1987), something should be conceded to it, perhaps a piece of new terminology, *slack knowledge*. While Smith may not know that P2, Smith *does* slackly know that P2. Slack knowledge, of one sort or another, is not only commonplace where ordinary claims to knowledge are concerned; it is also widespread in claims made in the practice and theory of science (witness especially usage of many influenced by Popper's notion of scientific 'knowledge', and the fashionable account in cognitive sociology of 'knowledge' as any 'collectively accepted system of belief').

Secondly, it may be denied that the standard explicative conditions for Smith's knowing that P2 are met (an approach taken by Smith 1989, and in effect by Armstrong 1973). What is most likely to be contested is that Smith is justified in believing P2. That in turn is open to challenge on two scores: the justificatory argument (what was contested in outer cases), and the initial 'strong evidence', as it happens the president of the company's assurance that Jones would get the job, so Gettier tells us. In the light of the emerging circumstances that assurance does not look particularly reliable. However, as Gettier in effect notes (p.121), the usual sense of "justified" and routine justificatory practices let through false and unreliable information as justificatory. And while the argument could be challenged (strengthening 2°

to a tighter connexive linkage), that does not look so plausible with inner type cases.

More promising is a third approach: that the explications of knowledge as justified true belief that Gettier lifts out of the literature are defective. They are defective because they leave out relevant connections intended in the original JTB recipe. A critical connection in justified *true* belief, omitted in justified belief (which is all that the standard explication which Gettier aims to satisfy requires), is the linkage between what justifies the given agent in believing that P and what makes P true. Put differently, justification and truth conditions should track in tandem; but such standard explications as PE allow that linkage to be snapped. A relevant explication *maintains* the linkage. Thereby it reflects precisely the reasons given as to why Smith does not know P2, because his justificatory basis does not proceed through what makes P2 true, but has become detached from it and irrelevant to truth.

The relevant approach should be categorically distinguished from a *strengthened definition* tack, quickly advanced by members of the Brown school, Chisholm and colleagues, according to which "justified true belief" has to be strengthened by yet another—apparently so far undiscovered—modifier. For example, Chisholm proposes *nondefective* JTB (a patently theory-saving addition), Lehrer *defensible* JTB, Lehrer and Paxson *undefeated* JTB.[232] That approach effectively concedes to Gettier, accepts Gettier-type examples as *countering* a JTB recipe, and offers a repair. The relevant approach makes no such concession to Gettier, who suggests in his article (from the title on, remember) that knowledge is not justified true belief, and who is widely supposed to have established as much. He has not. The recipe is satisfactory, and stands. No further modifiers are required. All that is needed is a satisfactory explication of the accepted recipe that does not, through stock logical treatment, sacrifice relevant connection.

Now circulating in an extensive North American literature are many many variations on Gettier's two examples, and a concomitant epicycling array of proposed rectified analyses of knowledge. Most of the examples vary case I, which involves a mistaken identity (namely, Jones = the man who will get the job). All these examples can be treated, a survey appears to disclose, in

---

232 For these and other examples, and alleged shortcomings of one and all but one, see again Shope 1983. Yet another example, but with justification displaced, Nozick's 1981 interesting (but unsatisfactory) proposal of TB *with tracking*, may appear to more closely approach what is suggested below, engagement. However while tracking may sound superficially like engagement, it is very different—belief tracks truth, whereas justification engages with true-belief—and (though supposedly covering for due justification conditions) tracking offers overly strengthened justification.

essentially similar fashion to the examples already considered. Systematic illustrative purposes, though not completeness, will be served by one further inner case. In this example an agent mistakes a dog in a field for a sheep, when there is a sheep in the field unseen by the agent. In this situation, proponents of such examples claim: The agent *s* believes:

P3. There exists a sheep in the field,

and *s* is justified in believing this, on the basis of inference from perceptual and other evidence. The inferences involved can be set down in first approximation thus:

1. (a dog = a sheep) & a dog is in the field →. a sheep is in the field, and

2. a sheep is in the field →. there exists a sheep in the field.

Further, 'on the strength of' separate information, irrelevant to what supports the agent's belief, P3 is true. So, according to proponents, *s* has justified belief in P3 and independently P3 is true, so the standard conditions of PE are ensured. But the agent does not know that P3. Why not? The trouble is, once again, that the tendered evidence does not link up relevantly with what is presumed to be JTB known. So JTB is not ensured, after all. Observe that in this sort of example, the offering of a relevant explanation (the final step to be taken) and the questioning of the agent's justification come together. For on the basis of a mistaken perception (who versed in country ways would mistake a dog for a sheep!) an agent is not duly justified in believing the claim made.

A mere survey, and disposal of exponentiating examples, would not satisfy purpose of generality. In order to show that the proposed relevant dissolution succeeds generally, it is important to uncover the general form of Gettier-type examples. So far as I have been able to ascertain (admittedly without an exhaustive search) this has not been satisfactorily done, despite the extensive research effort on these sorts of examples. However the exercise has been attempted, with less than adequate results, and these attempts, which aim to cover the array of examples developed, can be built upon. A promising conjecture is that the immense American effort to repair PE, through further clauses (which would reinstate relevance), failed because the general form has not been satisfactorily grasped.[233]

---

[233] For examples of unsatisfactory presentations, see for example Shope 1983, p.4 and p.222; Sosa 1979, p.81. Naturally, there is now no difficulty in presenting the sweeping Gettier

In every case there is a springboard statement O, which is false but which 'catapults' the subject s to the statement P which satisfies the conditions for PE (i.e., P is true, believed by P and s is justified in this belief) but which s does not know, at least if the *counter*-example is sustained, *because* s's (only) justification for P is disengaged from the truth of P. It is this last *critical* clause in particular that tends to get left out of attempts to state the general form of PE counter-examples. But it is just this clause that encapsulates the irrelevance feature: that the justification s has is disconnected, dissociated from, irrelated to the truth of P, including, since it is a *verification* procedure that is at issue, what makes P true.

The general form of purported counter-examples to PE explication of s knows that P can be summarized:

- s believes P which is true; so at least two PE conditions are met.
- there is an O such that
  - $O\ s> P$, i.e., O s-guarantees P,
  - s is (or would be) justified in believing that O, and
  - O is not known, typically because it is false.
- s's justification for P (which normally derives from the preceding clause) is disengaged from the truth of P.

These then are the requirements for Gettier-type examples. They are approaching a minimal set, which is one reason why counter-examples may not eventuate. Note that it has to be required that s believes P, as this does not follow from features of O, *should* s believe it. The justification condition among PE conditions may not follow either, inducing a divide. There are two types of putative counter-examples (between which there is a certain interplay), according as "s-guarantee" does transmit justified belief to P, or not. In *outer* cases it does not, normally because relevance demands on the implicational connection are vitiated. In *inner* cases it does transmit such justification, so PE conditions are met, giving a bona fide Gettier problem.

Wherever the general form is satisfied there is a purported Gettier problem. For the conditions for PE are purportedly met but s does not know that P. Let us now concentrate upon inner cases where PE conditions are met, so PE is countered. Even so, the resulting counter-examples to PE do *not* tell against JTB. That is for two reasons. Firstly, because of disengagement, the examples do not exhibit *justified true belief*. That fails along with knowledge,

---

problem: that conditions like those of PE are satisfied when the agent involved does not know the proposition concerned. But that is only a problem while there are associated counter-examples, and it is *this* class that requires circumscription.

naturally, and for the same sorts of reasons, namely irrelevance of one sort or another, and ensuing but partial justification. Secondly, by virtue of modifier logic, there is no arriving back at PE from JTB. The critical issue devolves on the following inference scheme:

$$\text{NA.}\qquad \frac{s \text{ has a justified belief that } P \qquad P \text{ is true}}{s \text{ has a justified true belief that } P}$$

which is rejected along with:

$$\text{NAK.}\qquad \frac{s \text{ has a justified belief that } P \qquad P \text{ is true}}{s \text{ knows that } P}$$

Rejection of NA, as Not Acceptable, some may find *hard* to accept, especially people inclined to extensionalized reasoning. However some familiar epistemic substitutions may make it easier to appreciate, for instance of "well founded" or "adequately grounded" for "justified". Likewise substitutions for "true" help both in such appreciation and in making it evident that the claim is not some sort of special pleading to exonerate JTB. Consider for instance:

$$\text{NAIr.}\qquad \frac{s \text{ has a well-grounded belief that } P \qquad \text{(but) } P \text{ is irrational}}{s \text{ has a well-grounded irrational belief that } P}$$

Similarly for such substitutions as "inconsistent", "stupid", "prejudiced", "ideological", "unorthodox", "defamatory", and corresponding positive appraisal terms.

The case for rejection of NA includes the following points:– Firstly, the general pattern of the inference, essentially:

$$\frac{\phi f(x) \qquad \varphi(x)}{\phi\varphi f(x)}$$

is *in*valid. Consider, for instance: Tina looks like an elephant, Tina is large / Tina looks like a large elephant; or similarly with a first premiss: s is justified in reporting a sighting of a black rhinoceros. Or consider inferences like: the Class Abstraction Axiom is an intuitively obvious proposition, the Class Abstraction Axiom is directly inconsistent / the Class Abstraction Axiom is an intuitively obvious directly inconsistent proposition. Secondly, scheme NA cannot be validated through expansion of the ordered modifiers in the conclusion statement (as was previously seen). It might be claimed that extensional expansion only fails with relational features like "large", where criteria for the feature depend upon a background class, which can vary according to application. That is not so. Moreover, while "justified" is not so conspicuously relational as "large", nonetheless criteria and standards *do* vary considerably. What is appropriate in certain law courts differs from what will do in a mathematics seminar (and will vary there should the mathematics be intuitionistic or relevant), and that from what is acceptable in a seminary. Thirdly, having come this far, it is not difficult to explain *why* the inference in question fails. In the conclusion the justification must be relevant to truth as well as to belief; it has to apply to *both*, to *make both right*, i.e., justi-fyed together. That too is why Gettier-type justifications, which circuit through irrelevant disengaged falsehoods, though they may justify a belief, do not justify a duly connected true belief, and do not deliver knowledge. Because of disengagement Gettier-type justification is but partial.

No doubt there is a countercase, developing from the often highly contingent way truth and falsity attach to beliefs, and often but superficially colour them. While, admittedly, it commonly makes no difference to a justification of a belief whether it is false or true, it does make a difference to a justification that justification is of a true belief, a rational belief, or some other modified type of belief. No doubt too an inference resembling NA can be instated by invoking an extensionally flattened notion of justification: *subjustification*. Then the following inference is:

$$\text{OK.}\qquad \frac{s \text{ has a (sub)justified belief that } P \qquad P \text{ is true}}{s \text{ has a subjustified true belief that } P}$$

because subjustified true beliefs are just beliefs that are justified and also true. But subjustification is not justification, as we know from the logic of modifiers.

Furthermore, regular justification is not subjustification but, rather, *relevant* justification. Because that is what is meant by justification, relevant justification, there is no need to accede to demands to *add* to the accepted definition of knowledge, to advance a strengthened slogan. Some modifications, though they may have a point, add nothing, as Kant observed: real cream is just cream. Justification is relevant justification.

Surely the accepted definition sustains some fuller explication, the modifiers are not for ever locked away (as in unsatisfactory adverbial theories) in an unanalysable claim. Of course it admits expansion, provided connection is maintained, for which some strengthening may be apposite. The trouble with such expansions of the JTB recipe as the prime explication is, as we have seen, that they snap the connection between *what justifies* s in believing that P and *what makes* P true, that they permit *justified true belief* to degenerate extensionally to *justified belief which is true*. A way to retain requisite connection lies through a different cumulative expansion of the recipe, for instance as follows:

s knows that P iff	s believes that P
where	what s believes, namely P, is true
and where	s is duly justified in believing the combination, namely the true belief that P.

Here, furthermore, *duly justified* implies that s's justification does *engage* with what makes P true. Due justification is relevant justification of the combination, of truth applied to belief. The term *combination* is intended to suggest combinatory methods, production of a single element of application. Of course that combinatory result yields a single uncombined item. Now that cumulative definition, operating within a relevant setting, appears capable of despatching all Gettier-type examples. For where in any inner case justification is disengaged from truth, the requirement of due justification fails, only partial justification is achieved.[234]

The problems posed by Gettier-type examples have been greatly magnified, as epicycling research attempting to solve the problems has proceeded. The problem (for inevitably it has been assumed to contract to one grand problem, normally finding the right patch for punctured knowledge definition) has even been inflated to Kantian dimensions. Thus for instance Lehrer's 1979 eloquent statement of 'just what the problem is', in which initial problems are just recognisable:

> It is ... the problem of showing that a fallibilistic theory of epistemic justification is possible. For, the problem arises in certain cases in which a person is justified, whether he knows it or not, in believing or accepting some false proposition which transmits justification to some true proposition. Thus, for a Cartesian who held that epistemic justification must proceed from what is certain by certain steps to arrive

[234] In a way this only repeats what Russell 1948 (and Meinong) had already emphasized (but Nozick 1981 neglected): that there is a difficulty of accidents and irrelevance as regards a 'true belief' recipe, which is removed by requiring due justification of the combination. Russell has regularly been seen as anticipating Gettier-type examples, with his particular counter-examples to definitions of knowledge as true belief. Further, it seems clear that he would not have seen true beliefs in such examples as (duly) justified. Certainly 'some good first-year students with no previous philosophy are also inclined to this very plausible view' (communication from M. Pendlebury; similar sentiments from others).

> what is known, the problem would not arise. For a philosopher who avers, as I do, that epistemic justification is fallible, the problem is to articulate a theory of fallible epistemic justification which allows us to distinguish between those cases in which justification, though fallible, yields knowledge, and those in which some false proposition deprives one of obtaining knowledge from justification (p.65).

A resolution of the main problem is already supplied from the JTB slogan (*without* need of further patching conditions): namely, ensure engagement, look to due justifications of true beliefs. That is no big deal. Accordingly such a grand problem only remains *given* 'the problem ... Gettier formulated is ... still unsolved' despite the research effort (p.65), what remarkably Lehrer 1979 does assume.

While this grand problem presumably disappears, others do not. Even if problems generated by Gettier-type examples can be dissolved through *careful* application of JTB, there remain many other problems concerning knowledge that cannot be avoided, without much additional work (for instance, some of those problems repeatedly raised by Russell). In a way then, even if this sort of dissolution of Gettier difficulties succeeds, we are substantially back where we would have been before Gettier derailed the North American epistemological enterprise.[235]

[235] Innovative application of new technology from deep relevant logic does however enable significant advances to be made on other, sometimes neglected, epistemic problems. But that is business for other occasions.

See Sylvan 200+ for further details. [Editors]

16 Ubiquitous Vagueness Without Embarrassment: Logic liberated and fuzziness defuzzed (i.e., respectabilized)

Co-authored with Dominic Hyde, this essay originally appeared as Sylvan and Hyde 1993. [Editors]

Although puzzles surrounding vagueness are ancient, they have been much intensified since the rise of standard 'classical' logic. A main trouble is that standard logic proceeds to *validate* both ancient defective arguments (Sorites paradoxes) and modern defective arguments (modal paradox analogues). But much evidence has recently accumulated as to the severe philosophical inadequacy of standard logic. Plausibly then ways out of such logic-induced puzzles which work elsewhere, relevant ways which work well, will succeed with puzzles of vagueness also. So it proves. By letting the puzzles constrain the logic, so producing their own demise, we reach elegant logical resolutions with the *very* logics adopted elsewhere for removing other logic-induced problems.

This is not so remarkable. Puzzles surrounding vagueness are not isolated from other puzzles. Even if new functors like *vaguely* and *definitely* enter, the puzzles use the same logical particles—*if, same, object*, etc.—which presumably conform to the same logics. By good fortune it turns out that features of deep relevant logic, already unearthed independently, enable requisite work of puzzle dissolution to be accomplished.

Better than banning much ordinary language involving use of vague predicates, language that is ubiquitous, as logically incoherent; better than confining logic to regimented coherent discourse where the puzzles do not arise; better than these is adjusting the theory and logic to the data; still better is the course of adopting and elaborating a logical theory already substantially adequate to the puzzling data. That logical theory is relevant logic.

What already falls naturally out of that logical theory are two forms of *modus ponens*; a fully reliable form for entailment, $\rightarrow$, but only a locally reliable form for conditionality, $>$. However the validity of $>$ *modus ponens* need not be contested, only its truth-preserving powers. But there is an array of independent arguments for this, it is not (just) *ad hoc*. This way too leads to a desirable uniformity: not a (relativistic) mishmash, with a different sort of logic for different sorts of applications (nor, like Slaney 1989 and others, dress out in an irrelevant system if vagueness looks like being present). Certainly, there *is* a plurality of logics, and a plurality of ways of accommodating vagueness, but no stable person wants to be changing dresses or hats too often or for bad reasons. Put differently, virtually every philosophical problem has a range of potential resolutions. A *coherent* philosophy selects and develops its resolutions uniformly, with connected solutions for related problems, not a different sort of solution of every (sort of) problem.

16.1 A focal puzzle: the Sorites paradox

As accounts of this paradox abound, we can be exceedingly brief in setting up the puzzle, and briefer still in assessing the stock range of 'solutions'.[236] A formally amenable variant of this well-rehearsed paradox is couched in mathematical induction form:

base premiss:	$f(1)$	
inductive premiss:	*if* $f(n)$ *then* $f(n+1)$	n a positive integer
particular conclusion:	$f(z)$	z suitably large.

Some examples of f: "grains of millet fall noiselessly to the ground when dropped" (Zeno of Elea), "minutes afternoon is noonish" (Sorenson 1991, p.55), "is a small number" (Wang).

The detailed argument to the conclusion proceeds standardly through the $(z-1)$ intermediate conclusions $f(2), f(3) \ldots f(k), f(k+1) \ldots f(z)$ with each obtained from its predecessor by an elementary 'modus ponens' inference of the form, for step k to $k+1$:

$$f(k), \text{if } f(k) \text{ then } f(k+1) \quad / \quad f(k+1) \qquad \textit{CD step},$$

the first premiss having been already obtained and the second being an application, by variable change, of the inductive premiss. The CD step itself

[236] A fine introduction to, and critique of, this stock range is given in Slaney 1989.

makes application of one of the (few and) crucial bits of logic involved in the paradox, conditional detachment:

CD. $A, A > B \,/\, B$

where the conditional $>$ symbolises *if ... then.*

The paradox arises because though the premisses appear perfectly acceptable, the conclusion, to which seemingly impeccable logic forces us, is not. For we seem logically obliged to conclude that six hundred minutes after noon is noonish? 10^{100} is a small number! and similar.[237]

But *is* the logic impeccable, or only in possession of impeccable classical references? Consider some of the derived inferences to which the 'impeccable' procedure commits us: cases of chain transitivity like:

$$f(1) > f(2), f(2) > f(3), \ldots, f(z-1) > f(z) \,/\, f(1) > f(z),$$

where although each premiss may be acceptable, its output certainly is not. For its output may just be such results as that *if* 1 *is a small number then* 10^{100} is *a small number.* In any case, it is now fairly well appreciated that transitivity, as a truth-preserving rule for the conditional, does not obtain generally; and there are in circulation several well canvassed and quite widely accepted counter-instances to it (see for example RC).

While that puts a serious dent in certain versions of the Sorites, it hardly finishes off the main problematic version which applies CD. For the failure of transitivity does not undermine CD. For instance, while transitivity lapses in modal conditional logics, CD does not (a situation that is readily altered, as Chellas' 1975 most basic conditional logic, which also flunks CD, makes plain).

---

[237] While all these examples can be said to involve cumulative features, holistic elements are explicitly present in Zeno's puzzle: 'If one gain of millet is dropped it falls noiselessly to the ground; and so, too, with any numbers of gains taken *singly*; but if a basket of millet is poured out it falls with a great noise. Yet it is *paradoxical* that 10,000 instances of noiselessness should produce any noise at all' (Reese 1980, p.641; italics added). Inferences from *each* to *all* are not generally valid; for example each ticket can win / all tickets can win; each may be first / all may come first. Inferences from *any singly* to *all together* are notoriously invalid, classically or not: for example, each stamping singly had no impact on the building, all stamping together shook it down; with each consumer operating separately the bank (business, market) flourished, with all operating in concert the bank collapsed; each bug separately was inaudible from here, all in unison produced a great sound. (Zen reverses Zeno. Several hands clapping make a fine sound; but?). Standard quantification theory fails to accomodate such holistic quantifiers.

The issue of holistic summation arises in more sensitive technical form in other paradoxes of magnitude, can sum to a fixed finite magnitude. Here non-standard paralogical analysis can supply a technique; see also below.

Relevant conditional logic flunks CD in certain circumstances. A conditional for which CD holds in called *ponible*. Ponibility, though the normal state, is not universal. In the context of vagueness the situation can be described as follows: While CD is for the most part *locally* reliable, while it preserves truth in nearby applications, it is not reliable for long or distant purposes. Compare a car that is reliable for running about town, but becomes unreliable for long journeys on the open road.

That is, while CD is locally reliable, its nonlocal derivatives are not, notably an *iterated* form becomes unreliable for n large:

ICD. $A_1, A_1 > A_2, \ldots, A_{n-1} > A_n \,/\, A_n$

But ICD is derivable from CD, by repetition, as the following diagram reveals:

D.
$$\begin{array}{llllll} A_1 & & & & & \\ A_1 > & A_2 & & & & \\ & A_2 & & & & \\ & A_2 > & A_3 & & & \\ & & \ldots & A_{n-1} & & \\ & & & A_{n-1} > & A_n & \\ & & & & A_n. & \end{array}$$

ICD of course gives the effect of chain transitivity. So it yields in a similar way, 10^{100} is a small number. Since ICD is not truth preserving, neither is what yields it, CD. But naturally that does not prevent CD from being locally satisfactory. In interesting respects the situation resembles that for the rule MD, of material detachment (i.e., $A, A \supset B \,/\, B$) in entailment theory; that rule is admissible only in certain consistent contexts.[238]

[238] The newcomers' reaction to the Sorites puzzle of rejecting the induction premiss gets pounced upon by Sorenson, who savages it as asserting, contrary to vagueness, 'that there is a precise division point between times that are noonish and times that are not'. Such a result he reaches, however, by equating the negation of the induction step, formulated universally for noonnish, with the statement: '2′ There is an n such that n minutes after noon is noonish and $n+1$ minutes after noon is not noonish' (Sorensen 1991, p.56). That simply *assumes* that the conditional involved is material, an outrageous assumption. Newcomers, who are onto something, cannot be rubbished so readily.

In setting up a parallel 'fiction sorites', Sorenson does make his reliance upon classical logic explicit. That forces the induction premiss—'2. If Sherlock Holmes had at least n hairs on his head, then he had at least $n+1$ hairs on his head' (p.56)—into material form: either he did not have at least n hairs or he had at least $n+1$. The only way that could be false would be through his having precisely n hairs up top, but such 'precision is impossible here'. So 2, as a material conditional, is presumably true. As a satisfactory conditional it is not true; for

There is no problem in explaining why reliability gives out. Connection weakens and finally dissolves over long chaining. A comparison relevance illuminates this. Relevant conditionality is something like:

$A \gg B$ and B is relevant to A,

where $\gg$ is a modal conditional. Under iteration relevance becomes more and more tenuous and finally disappears. Where it finally vanishes is itself a vague feature. So it would be a mistake to try, as is done with certain many-valued logics, for exactness or numerical definiteness in these matters.

A feature like this relevance emerges directly from the semantics for the conditional, which is given by the evaluation rule:

$$(A > B, d) = 1 \quad \text{iff} \quad s(d, [A]) \subseteq [B],$$

explained in background work (for example PLI, p.294 ff). Here s is a selection function, from worlds and sets of worlds, which is often interpreted in terms of similarity. Thus, tracking diagram D above, there is a sequence $s_1, s_2, \ldots, s_{n-1}$ (with labelling corresponding to the antecedent of the relevant conditional in D). It is well attested that similarity does not transmit over distance connections; whereas close relatives, for instance, may be similar, distant ones are not. In fact the semantics appropriately reflects the very sort of vagueness that the theory is concerned to resolve problems concerning (a point later important as regards avoiding 'higher order' and other derivative vagueness).

The modelling condition for CD, *as* truth preserving, can likewise reflect semantically the vagueness lurking in the syntax. That condition, $V(x)$ for short, is:

for $x \in O$, (i.e., x a regular world), when $x \in \alpha$ then $x \in s(x, \alpha)$.

Now iteration of $>$ in ICD corresponds to chaining of V. So vagueness reemerges in V. In a chaining of V, it passes through a penumbra of cases where it holds, to end somewhere in failure.

Because CD eventually gives out in some concatenations, that only exposes it as *not always truth preserving* under extensions. By contrast, the

---

instance set $n = 10^{100}$. So such 'fiction sorites' lapse; they have no force, no puzzling logical compulsion.

What of the remaining degenerate material form? Then the argument is invalid, since it requires a form of Material Detachment (MD) drawing upon Disjunctive Syllogism, $A \,\&\, ({\sim}A \vee B) \rightarrow B$. But why in consistent circumstances does MD fail as truth preserving? Because in fictional settings primeness cannot be universally expected.

validity of CD, widely regarded as sacrosanct, need not be contested. That is, where A and $A > B$ are provable (mirroring logical truth) so also is B. That may well obtain (it will until genuinely vague logical systematizations are introduced). What goes down is CD in iterated truth preserving form: that is, where A_1 is true and all of $A_1 > A_2, \ldots, A_{n-1} > A_n$ are likewise true, A_n may not be.

To sum up so far, and to look ahead. The Sorites argument has been defeated, not by disputing its premisses, not by embracing its conclusion, but by repudiating its logic. The operative principle CD (*modus ponens* for *if*) is not reliable over the sequences of inference the argument involves. Now to remove possible misunderstanding:

- *modus ponens* is valid, in its main role, for entailment, $\rightarrow$;
- the validity of *modus ponens* for conditionality, $>$, is *not* contested.
- the truth-preserving reliability of $>$ (not $\rightarrow$) under testing iteration is what is challenged. But the logic already adopted for other purposes, for conditionality (in an entailment setting), adequately allows for just this.

What the logical theory does not so far offer—though it does not in any way exclude it—is a nice simulation of the appealing features of Łukasiewicz fuzzy logics: that unreliability accumulates so that after sufficient iteration reliability slips into unreliability. However the semantical theory can no doubt be *enriched* to accomplish such features. One approach produces a *further* semantics with indefinitely many values, which can reflect the features of reliability. At least for entailment logics such semantics are available in the shape of De Morgan-valued models; and these can presumably be extended to encompass conditionality, and a reliability measure upon its application. Another more direct approach simply adjoins a reliability measure, a measure upon statements and arguments. For statements this could be characterised like a relevant probability measure, i.e., a function r from statements to the open interval $(0, a)$ satisfying requisite postulates (on which see JB, p.941 ff). For arguments, if $A_1, \ldots, A_n \,/\, B$ is universally truth-preserving, then $r(A_1, \ldots, A_n \,/\, B) = 1$ and derivatively if $r(A_1) = \ldots r(A_n) = 1$ then $r(B) = 1$. But if $A_1, \ldots, A_n \,/\, B$ is only locally reliable, $r(A_1, \ldots, A_n \,/\, B) = 1 - \delta$, with δ a small distance increment. Under iteration δ's will accumulate, giving a merely locally reliable rule less and less reliability.

Measure function r does not introduce degrees of truth, but it does insert quantitative elements, concerning reliability, where only rough qualitative ones seem intuitively to figure. However there are apparently persuasive arguments that truth does come in degrees, for example what Slaney 1989 presents as 'the Simple Argument' (another *prima facie* argument for abandoning classical logic in favour of multiple-value logic). According to this Simple Argument,

> "Grass is green" is true (in English) if and only if grass is green. That is, the truth of "grass is green" marches together with the greenness of grass. But greenness is a matter of degree; so truth is a matter of degree (Slaney 1989, p.5).

Upon reflection, there is little reason to be fully persuaded by this argument, which is less compelling than the Sorites argument. Consider the rich and the poor march together in this life (or better, and historically fuller, masters and servants); the rich are well off, therefore, so are the poor. Correlation is not in general a sufficient basis for property transfer. But of course the degree of greenness does enter, *inside* the truth attribution.

How then do we accommodate the qualitative and comparative (*not* quantitative) judgements we do make about *truth*, such as "more or less", "far distant from", "roughly", "approximately", "mostly"? *Through* the functors involved, which should figure in the syntax anyway, not simply spoil semantics. This too is the way to try to deal with what said about the indeterminacy of truth, about approximating and receding from truth and similar, in the Sorites and elsewhere. The first of these sorts of functors, indeterminacy and the like, we shall need for later major tests for a logic adequate to vagueness, identity puzzles. But first, the tireless and sometimes tiresome Sorites has not been exhausted.

16.2 Further issues devolving from Sorites puzzles: indiscriminability and the character of border regions

Seemingly unfortunately for the preceding nice resolution, there are forms of the Sorites argument which involve *no* conditional or implication, and *directly* neither transitivity or ponibility. Consider the following:

Situation I. Let the sequence $a_0, \ldots, a_n$ be described as follows:
a_0 is red; a_n is not red;
but successive pairs of the sequence (a_{i-1}, a_i) are not (colour property) discriminable for each $i, 1 \leq i \leq n$.

Now apply the rule:

$$x \text{ is not discriminable from } y,\ x \text{ is red} \,/\, y \text{ is red}$$

Then situation I is *inconsistent*, i.e., the Sorites so presented supplies an 'impossible state-of-affairs'.

Proof. a_n is not red (given)
a_n is red (by induction over sequence).

Given a_{i-1} is red, show a_i is red. The result follows since the pair is not discriminable, by the rule. Plainly too this inferential form avoids implication or conditionality.

What is the *upshot?* The situation is in order, if inconsistent. It is all alright, but perhaps a *little* surprising, since most ancients thought they were describing a consistent (hardly a quite everyday) situation. But once the proof has been grasped, it is not even surprising.

One question emerging runs: How do we describe consistent situations like I *may* have been intended to be? The single discrimination rule has to be modified. An obvious way to achieve this is to let the rule supply a *conditional*, which breaks the induction argument. In fact, the recent relevant theory of conditionals will do exactly this; so once again no new logic is required. The conditional rule is, in evident symbols:

$$x \sim_d y \,/\, fx > fy \quad \text{(for suitable } f\text{)}.$$

As it happens, however, $\sim_d$ will supply an appropriate conditional (given an elementary switch from subjects to statements):

$$A \triangleright B =_{df} A \,\&\, B \sim_d A.$$

Since, for each adjacent pair (a_{i-1}, a_i), $a_{i-1} \sim_d a_i$ but $\sim(a_0 \sim_d a_n)$, $\sim_d$ is not transitive. *Ergo*, neither is $\triangleright$. So, therefore, it is not globally ponible. A further corollary is that indiscernibility $\sim_d$ does not yield normal identity, which is transitive. While (an) identity could be taken as the transitive closure of indiscernibility, the proper point is that such indiscernibility is intensional (because subjective, appearances that matter may vary from subject to subject) and does not exclude *non* identity. Put differently, psychological indiscriminability differs from trait indiscriminability, which is what conserts with identity.

Suppose now a_n is not red because yellow. That is, at some stage, right around border case a_b, the sequence is changing from red (through orange) to yellow. What colour is a_b? Red, or yellow, or neither (but perhaps some other colour, perhaps not), or both. Small considerations rule out the first two answers, for instance symmetry in nice cases suggests it cannot be one (for example, just red) without the other, and the sequence cannot just jump from one to the other. The logical theory invoked and developed (at bottom four valued) does not force an answer. Most important, it does not exclude the formerly appalling, but increasingly an appealing, answer: both. In some cases at least there are good reasons for this shocking answer. For example, we make ourselves a vague two-dimensional object with our new crayon colouring set by colouring a rough circular area properly within a square red, and the remainder of the square yellow at the same time overlapping the circular area. Then surely the border regions of our colour patch which is orange, is both red *and* yellow. For that is how we just coloured it.[239]

Of course "both red and yellow" is not yet inconsistent, without further linkage such as "red entails not yellow". But it does open the way to a wealth of discourse that standard logics preclude, beginning with familiar low-level descriptions of indeterminate situations in inconsistent terms (for example, "it is and it isn't", "it's raining and not raining"). For more high-powered issues hereabouts, we should look to other of Zeno's paradoxes, puzzles concerning 'the continuum', and the like. Consider briefly the paradox of the flying arrow, which cannot be flying,

> for if the arrow is moving it must either be moving in a place where it is, or in a place where it is not. It cannot be moving in a place where it is, or it would not be there; and it cannot be moving in a place where it is not, for it is not there (Reese 1980, p.641).

Let us say the obvious, that *it is moving where it is* (the point of the arrow, say, is in vague micro-interval i). Then, as it is moving, it is not there, so the appealing argument has it. But that is alright. It is both there (in i) because that is where it is moving through, and not there because it is on the move (and out of i). The arrow paradox is paralogically neutralised.[240]

Ordinary ways of reporting vagueness do appear to take a paralogical turn, and to lead into inconsistency. People report that these individuals are neither tall nor not tall, that those arid lands are neither desert nor not desert. In symbols, for certain objects and appropriate predicates, neither sP nor not sP. That is, not (either sP or $\sim sP$), $\sim(sP \vee \sim sP)$, whence, by a De Morgan

---

[239] It is easy to produce vague exhibits of both *over*- and *under*-determined types, as well as mixed types; not merely those where we overcolour a neutral background, but those where we do not colour borders but simply merge into the background, and so on.

[240] What is said represents but a *beginning* on such neutralisation. For *one* further development, see Priest 1987. A *paralogic* is a logic that is both paraconsistent and paracomplete; roughly, it admits both truth-value gluts and truth-value gaps. Deep relevant logics are paralogics.

principle, $sP \,\&\, {\sim}sP$. Alternatively, by direct representation of "neither ... nor", and using again the familiar (but not always admissible) transformations $s{\sim}P$ iff ${\sim}sP$, $sP \,\&\, {\sim}sP$, inconsistency. Moreover, if we do not allow for such gluts, then we are precluded from saying some of what we want to say and ordinarily do say; 'we are stranded on the shores of ineffability' (Peña 90, whom this whole paragraph follows).

Taking a paralogical turn does not resolve all difficulties in the everyday transition from red, through red and yellow, to yellow. There remain awkward questions about the first element that is yellow, or also not red, and about the last that is red? If we *were* to respond positively to such requests with specific elements, then we should be introducing exactness into indeterminate situations where it appears inappropriate (see the next section). To remain appropriately vague all the way across the border, the response should be that there are no such first and last elements (our sequence may thus appear in danger of defying Zorn's Lemma).

Where they are even recognised, there is heavy prejudice against paralogical ways of doing things (no doubt because they destabilise so much that includes or is based upon standard logics). But often they are not recognised. In particular, vagueness is defined so as to allow under-determinacy only, not over-determinacy ever. A natural logical symmetry gets entirely hidden. New definitions and accompanying theory are accordingly needed.

16.3 On characterising vagueness, and disposing of another philosophical pseudo-problem: higher-order vagueness

Unfortunately, all definitions of vagueness hitherto (not a dense supply) appear inadequate. For none allow appropriately for both overdetermined and underdetermined varieties. Locating a satisfactory characterisation is not however an easy task. It helps to divide a search into two stages: first spatial, and second generalisation thereupon. As it happens, vagueness has mostly been explained in broadly spatial terms, from its very inception in wandering (Latin *vagus*). More recently it has been explained in terms of lacking sharp borders, border-lines or limit-lines, as boundarylessness, through lack of distinct outlines, indistinctness, and so on: in short, in spatial terms.

Conveniently, most items which are candidates for vagueness either have spatial dimensions, or else (like concepts) have a spread of instances which can be viewed quasi-geometrically: they have an *extension*. In both cases vagueness tends to arise around the border regions (more generally over marginal elements in the extension), issuing in questions as to whether something is or is not a part of or an instance of the item concerned. It issues further in embarrassment sometimes as to answers, not being able to give a definite Yea or Nay. Sometimes it both is and is not, or neither is nor is not. So gluts and gaps are *part* of the story; part only. It is unsatisfactory to leap to an account of vagueness in terms of truth-value gluts or gaps. For there are many other items which may not be conspicuously or at all vague, incomplete items and overcomplete items of various sorts (for example, exact antinomic items of dialethic set theory). Incompleteness and overcompleteness of *sharp* sorts are, so to say, orthogonal to vagueness.

What *more* is required? Let us try an informal approach from relevant topology. After all it is geometry, perhaps without dimensions, that sets the main scene for vague items. Consider then items that can be represented topologically, through their extensions. Such an item is *vague* if its interior and exterior are not exclusive and exhaustive. Plainly the account depends upon preliminary characterisations of *interior* and *exterior* (independent of gapping and glutting). Topology early introduces an axiomatic characterisation of an interior operator (or of closure). No doubt we could attempt something similar with both interior and exterior, in a relevant setting.[241] For the present nothing so grand is required; even less than an appropriately vague circular characterisation will serve. We shall operate with preanalytic notions of *interior* and *exterior* (taking these to include where appropriate limit points). Now a vague object, such as a shadow, a rainbow or a forest, has both an interior and an exterior and *more*. The interior of a forest comprises the region within, inside the forest; its exterior comprises the neighbouring region without, outside the forest. But these regions do not exhaust the encompassing region. There will be places where a watcher is neither within the forest nor outside, or both within and outside. In simple (incompleteness) cases we can suppose that a topological-like object has a

---

[241] In the propositional setting of relevant intensional logic, I is a systematic interior functor with $S4$-like properties. As explained in PLI, systematic functor I satisfies these conditions: $A \rightarrow B \,/\, IA \rightarrow IB$ and $IA \,\&\, IB \leftrightarrow. \; I(A \,\&\, B)$. What further distinguishes an interior functor is the S4-ish feature $IIA \leftrightarrow IA$. Set operator I applied to sets, in relevant set-theoretical topology, will have analogous properties. An exterior functor E will behave logically like the interior of an opposite; $EA \leftrightarrow I\overline{A}$. So whereas I is reminiscent of a necessity functor, E resembles rather an impossibility functor.

Standard topological theory, where I is characterised, in effect defines topologies set-theoretically, *given* a Boolean algebra of sets. Relevant topology can define topologies analogously, in terms of closure under positive set operations, *but given* a De Morgan lattice of sets (therewith too misplaced objections to sets as sharp objects can be circumvented).

margin, its *terior*. Then the *interior* of an object is everything with*in* its terior, and its *exterior* is everything *out* side its terior.

In more complex cases, the terior is displaced by *neither* (N) and *both* (B) regions. In general there is no sharp cut-off from interior and exterior to other regions; like old soldiers, they just fade away. With a predicate, such as red or child, the interior comprises the items that have it, the exterior the items that do not, the neither-space the items that neither have it nor do not, and the both-space the items that both have it and do not. The whole class of items involved, those of the super-region, can be comprehended through the *super-predicate* of the predicate, which supplies its significance range.

While vague items are, naturally, not characterisable through sharp necessary and sufficient conditions, there are other ways in which they can be explained, ways which match fading away, and consequent boundarylessness, of interiors and exteriors. For example, some of these items can be characterised, as Wittgenstein more or less explained, through family resemblances, or by way of clusters, others through paradigms and similarity thereto, or the like (for a little more on such accounts, see Sainsbury 1991, pp.178 ff).

An object which is not vague, whose interior and exterior are exclusive and exhaustive, might well be called *sharp*, or *classical*. For the properties involved resemble those of classical consistency and completeness (and they match those of classical negation). There are accordingly two primary types of vagueness (which may co-occur in the one item):

- *non-exhaustiveness*, or underdetermination: there are elements which belong to neither interior nor exterior;
- *non-exclusiveness*, or overdetermination: there are elements which belong to both interior and exterior.

In generalizing away from topology, interior and exterior are replaced, respectively by *inclusion* and *exclusion* classes; and so on. But for graphicness we shall stick with a topological schematization.

Sharp items have *boundaries*. An *inside* boundary is a set of elements within the object which are not within other elements within the given object. Correspondingly for an *outer* boundary. Inner and outer boundaries need not in general coincide. But for standard sharp objects they will. Vague objects, by contrast, turn out to lack complete boundaries. They are thus, at least in part boundaryless, as a consequential feature. Thus a vague predicate 'classifies without drawing boundaries' (Sainsbury 1991, which differently insists that this boundaryless feature, only consequential and but partial for us, is 'the right way to characterize the vagueness of a predicate', p.179).

It is such a boundaryless feature that enables the so-called problem of 'higher-order vagueness' to be dissolved. For the genesis of this puzzle invariably proceeds through an explanation of vagueness by way of (yet further) bounded regions. Vagueness around a boundary, or borderline, is in each case supposed to be resolved, and accommodated, by introducing *two* boundaries enclosing an intermediate region, as diagrammed:

$L \mid R$		$L \mid I \mid R$
vagueness around boundary	$\rightarrow$ resolved $\rightarrow$	by further bounded intermediate regions

But each such new boundary produces a higher level of vagueness around it. Thus, through iteration, is launched a sequence, tending to infinity, of ever higher levels of vagueness.[242] This too is the route to infinitary values of fuzzy logic.

In the absence of boundaries this generative process cannot operate.

> [I]f an idea is bad, as is the idea of thinking of vagueness in terms of boundaries [or of divisions], you make it no better by iterating it. Even an infinite iteration of the bad falls short of the good. ...A boundaryless predicate draws no boundary between its positive and negative cases [initial L, R cases], between its positive cases and its borderline cases [for example, L and I cases], between its positive cases and those which are borderline cases of borderline cases [and so on]. The phenomena which, from a classical viewpoint, lead to notions of 'higher-order vagueness' are accounted for by boundarylessness ...there is no real hierarchy here. (Sainsbury 1991, pp.178-9).

Philosophy quite regularly generates internal problems, problems supervenient on their own way of approaching problems. These supervenient problems, which typically arise on the basis of a particular way of approaching problems, break loose from the particular way, and become

---

[242] For details and illustrative examples see Sainsbury 1991, pp.168-9. There is a similar route to an infinite hierarchy of vague metalanguanges, ascending in an effort to resolve the paradoxes engendered by classical principles forever seeming to be in operation and making trouble at the next level up; like a self-replicating virus, the solution at any level always gives birth to new paradoxes at another and so the infinitary task is tendered—a solution always tantalizingly close but semmingly one step away, of methodological necessity. For descriptions of this infinite hierarchy in operation in supervaluation logics of vagueness, see Fine 1975, p.297 ff.

installed as 'problems' in their own right, irrespective of background approach. While an occasional genuine problem, previously overlooked, may be turned up in this way, often these 'problems' are *pseudo* -problems, which do not really arise for other different approaches. Then part of the problem, for outsiders, consists in trying to find out what the problem is supposed to be and why it is troublesome. A striking recent example of such a pseudo-problem is that of transworld identity.[243]

Higher-order vagueness, of which much is made in British discussions of vagueness, belongs to this class of philosophical pseudo-problems. In this case the problems are generated by approaching vagueness already equipped with classificatory preconceptions, of classifications as involving boundaries and sharp divisions (no interiors fading out into nothing). Such preconceptions are of course built into applications of classical and many-valued logics, especially of their superstructures such as classical and fuzzy set theory. The broad source of these problems of vagueness lies in assuming such logical systematizations apply; they are generated by defective logical theory. Unfortunately it is not widely realised, even by those philosophers who have seen what the theory is doing, that such theorizing is neither compulsory nor compatible with implicit theory incorporated in ordinary often vague discourse.[244] There are logics beyond classical and many-valued logics, which answer better to discourse, and set theories beyond classical and many-valued set theories. Vagueness does not force abandonment of logical or set theoretic practice, only of entrenched forms.

Philosophers still have a taste for propounding outrageous propositions as if, for all their repugnance, there were no avoiding them. Thus for instance, it is confidently announced,

> ...there is no set of red things. This seems to me as certain as anything in philosophy yet it can be a bitter pill to swallow, by non-philosophers and by philosophers alike. In some debates about abortion, one can face a real sense of shock at the realisation that there is no set of persons: the concept of *person* is vague at just the relevant point (Sainsbury 1996, pp.252-3).

---

243 For details see the dismissal in JB, pp.593–4.

244 Thus Wittgenstein who thought that there was nothing for it but to jettison all logic. Logic, with its crystalline clarity, just was restricted to sharp objects; and apart from logic, mainstream logic, there was no alternative or implicit theory, no theory. Similarly, in a less extravagant form, Sainsbury 1996 imagines that features of vagueness scupper all set theoretic approaches.

Similar claims are advanced for all other vague concepts, and indeed for all notions that do not conform to narrow classical logical constraints. The set of all truths is a favourite candidate for this outrageous treatment: 'there is no set of truths' (Sainsbury 1996, p.253; Grim 1991, p.91[245]). *If* this sort of treatment were correct, much more radical logical changes than those we have proposed (already criticised as too radical) for theories of non-classical items, such as vague objects, intentional items and inconsistent items, would be mandatory. For instance, set-theoretical semantics for the logics involved would be vitiated. Fortunately it is not: philosophers have no monopoly on such common notions as *set, truth, person* and so on. Ordinarily, more remarkably, there appear to be no insurmountable problems with such objects as the set of persons, the set of all red things, and similar. These problems too are theory dependent, generated by defective classically inspired logical theories. They do not even extend as far as fuzzy logic, which does not debar us from a set of red things or generally from sets abstracted from vague concepts.

The arguments for these shocking propositions are likewise only impressive given classical assumptions, and are removed with them. For example, arguments to no sets answering to vague concepts are premissed on the assumption that sets are sharp, thereby excluding immediately imprecise and fuzzy sets. Thus the following style of argument:

> Sets have sharp boundaries, or, if you prefer, are sharp objects: for any set, and any object, either the object quite definitely belongs to the set or else it quite definitely does not. Suppose there were a set of things of which "red" is true: it would be the set of red things. However, "red" is vague: there are objects of which it is neither the case that "red" is (definitely) true nor the case that "red" is (definitely) not true. Such an object would neither definitely belong to the set of red things nor definitely fail to belong to this set. But this is impossible, by the very nature of sets. Hence there is no set of red things (Sainsbury 1996, p.252).

We reconstrue such arguments as *reductio* arguments against the initial assumption that sets are sharp objects. Relevant sets, like ordinary sets, are often not sharp; set membership statements can, like other statements,

[245] While Sainsbury considers only aspects of the incompleteness side, Grim concentrates upon the inconsistency side of the business. While all Grim's arguments can be stopped in their tracks in dialectic relevant logic (or indeed in other logical theories with adequate procedures for containing problematic self-reference), showing how would divert us far from our present topic.

be indeterminate (taking value $\{\emptyset\}$ or *neither*). As it is indeterminate whether that magenta object is red, so (by the set abstraction axiom) it is indeterminate whether it belongs to the set of red objects. It is also not definite whether the statement, that the object is red, belongs to the set of truths. Therewith we collapse too Sainsbury's similar argument, again from excessive sharpness constraints, that there is no set of truths; and therewith 'no adequate set-theoretic description of vague predicates... [because] a set-theoretic description of a language typically ends up identifying a set of truths [by virtue of the Tarski T-schema]' (Sainsbury 1996, p.253). Natural set-theoretic approaches stand, and need not infiltrate sharp boundaries.[246]

16.4 Identity, another indeterminacy paradox, and intensional functors

So far the emphasis has fallen on vague statements, and the statemental logic of vagueness (i.e., just ordinary relevant logic). But vague statements are commonly about vague objects, such as our colour patch, and otherwise they typically involve vague attributions, implying vague attributes (such as redness, raininess), more vague objects. Either way, by virtue of their ultimate subject-predicate structure, vague statements normally mean vague objects, which are reckoned major troublemakers, in more ways than those we have already observed. The troubles naturally include identity.

'No entities without identity!' So the challenge comes: what are the identity criteria for vague entities? The challenge is easily turned: the same as for other objects, coincidence in all extensional features. Surely vague objects imply indeterminate identity between objects, which implies lack of (appropriate) identity criteria? Both implications fail. The second, bad failure is already evident: indeterminate identity simply means indeterminate coincidence in traits, not the breakdown of criteria. The first failure is more interesting, because it has been widely assumed that vague objects in the (accordingly vague) world yield indeterminate identity—which is pronounced inconsistent or incoherent (thus Evans 1978). But it is not difficult to model simple possible worlds containing but few isolated vague

[246] As before, there are eventually repercussions. In particular, it would be a mistake to expect a neatly classical epi-theory (or meta-theory) to accommodate fully whatever decent logic of vagueness finally emerges. That theory, and the eventual semantics, will also manifest vagueness. For example, a satisfactory 'semantic theory can quite legitimately be *homophonic*, that is, can re-use in the metalanguage the very expressions whose object-language behaviour it is attempting to characterize' (Sainsbury 1996, p.260). Relevant semantical theory can already pull off that simple trick.

objects where identity can accordingly be fully determinate (consider a one-object world, worlds with two remote objects, three isolated items, ...). Thus logical implication to indeterminate identity is countermodelled.

While it looks, furthermore, as if indeterminate identity can be kept out of a vague actual world (for example for proof, model Tye's 1990 hard-line stance), doing so looks like artifice. For one thing, where there are abundant vague objects, where vagueness is ubiquitous, some identity statements are vague also, because vagueness of similar objects spills over onto questions of their identity. For another, the arguments to vague objects from paradigmatic examples appear to extend to cases of identity, where there are numerous examples of indeterminacy (the ship of Theseus, personal identity puzzles, and so on, many examples being dismissed as merely epistemic of course).

What is wrong with conceding the vaguely obvious, concerning vague objects in that actual world: namely, instances of indeterminate identity? A compact argument from Evans is considered a very formidable obstacle. Without wishing to be too disparaging, this much cited argument is but a transformed modal paradox (even the symbolism $\triangle$ and ∇ appears but relocated from noncontingency and contingency in modal logic). Moreover it succumbs to exactly the same objections as the well-known modal paradoxes (on which see JB p.96 ff). So once again *no* new logic is required; only persistence with the same ('ultralogic as universal') logic already used for a range of other purposes (see Chapter Three, section 2, and JB). To sketch relevant details:-

Suppose, for a *reductio* argument, there is some true but indeterminate identity, $a = b$ say. Although the argument is usually intended to apply to individuals, a and b can be any sorts of items at all, for instance properties or propositions to take interesting cases. As $a = b$ is true, $a = b$. Let the functor ∇ represent indeterminacy (i.e., it reads "it is indeterminate that", "indefinitely", or similar). Then the statement that $a = b$ is indeterminate can be symbolised $\nabla(a = b)$. Whereas $\nabla(a = b)$ is assumed true, $\sim\nabla(a = a)$ is not; self-identity remains determinate. Now according to a classical identity principle—utter indiscernibility of identicals:

(†)	$a = b \rightarrow . A(b) \rightarrow A(a)$	for suitable schema $A(x)$,
whence	$a = b \rightarrow \sim(A(b) \,\&\, \sim A(a))$,	so contraposing,
	$A(b) \,\&\, \sim A(a) \rightarrow . \sim(a = b)$.	

But just take $A(x)$ as $\nabla(a = x)$. Then as $\nabla(a = b)$ and $\sim\nabla(a = b)$, by detachment $\sim(a = b)$, contradicting $a = b$. (Alternatively, positively, all indeterminate identities are false.[247])

As with modal paradoxes, the basic critical trouble lies with (†). There (†) would falsely ensure, given a true contingent identity $a = b$, that when $a = b$ is contingent so is $a = a$. Here analogously (†) falsely assures given a true indeterminate identity $a = b$ that where $a = b$ is indeterminate so is $a = a$. In all those circumstances the resolution is simple: not abandoning (†), but not admitting it in the first place. (As argued in JB, all resolutions lead in effect to qualification of (†), perhaps a well-hidden qualification. But once again we consider only our preferred treatment of the puzzle, but the range of initially feasible escape routes.). All this is depressingly familiar. Perhaps the only minor surprise is that ∇ should yield nonextensional predicates. In view of the intensionality of ∇ (like most statement functors it distinguishes materially equivalent statements) and its straightforward combination features, that should be a very minor surprise (and no surprise at all to any who regard as plausible supervaluational representations of indeterminacy).

Intensional functors may be general sources of 'intensional paradoxes', but *only* in combination with excessive identity assumptions, which are sensibly shed. Certainly intensional functors are crucial for the fuller development of vagueness theory. For it is through such functors that we express many central features of vagueness: those of indefiniteness, approximation, more-or-less-ness, and so on. In this way, too, we accomplish—but properly, right there in the surface discourse, not hidden away in a semantics—much of what a main rival, fuzzy logic, was advertised as accomplishing. For all its marvellous technological sophistication, and occasional real-life applications to transport systems and similar, fuzzy logic should not be regarded as a troublesome rival (or even, in some way, as a rival at all, as relevant logics yield many-valued matrices, enabling similar applications). Fuzzy logic has been around long enough for its obvious weaknesses to be appreciated and documented in the AI and philosophical literature (thus for example Ramsay 1991, Smuts 1988, Hyde 1994, Sainsbury 1991, and Slaney 1989). Few, apart from some engineers, are satisfied with fuzzy logic as more than a very approximate way of treating vagueness.

---

[247] The argument treated diverges immaterially for present purposes from Evans' original, the precise thrust of which is left open (in unanswered interrogatives: Can there be vague objects? Might the world itself be vague? Is the idea of a world containing objects with fuzzy boundaries coherent?). As well, Evans' argument included a further stage, appealing to an unlikely $S5$ principle, designed to indict remaining (false) indeterminate identities. While the $S5$ principle is dubious for individuals, it does enjoy some plausibility for properties and for propositions, whose frequent vagueness is accordingly more of a classical embarrassment than that of individuals.

The problems discussed do not appear to extend to apply against a relevant approach.

All that is lacking relevantly that fuzzy logic appears to accomplish through (infinitely) many values can be alternatively accomplished by way of functors just as in natural languages, with such functors as "It is approximately so that", "It is definitely right that", and so forth. Consider, for instance, the matter of grades of truth, or falsity, as illustrated with approximate truth. It is substantially similar with forms like "It is true to such-and-such an extent that". Now,

that A is approximately true iff "approximately A" is true;

so approximate truth is accommodated through the functor "approximately". Features of approximate truth can then be read off functorial behaviour. For example, approximate truth does not conform to the normal semantical principle for conjunction, as approximately B & approximately C does not coentail approximately $(B \,\&\, C)$. In fact both halves of the coentailment break down as easy examples show.

16.5 Vague objects in a vague actual world?

Naturally that is exactly our situation. How better to argue the case than by paradigm case argument á la Moore? *There* are some vague objects: high cumulus clouds, a rainbow that breaks in the middle, ... Or *here* are a few designs of vague objects we have run off for you: a red sun with a Magenta corolla on a blue background, a yellow square with an orange penumbra merging into red, ... These are vague objects. They exist in this world, and that is actual. Therefore ... .

It is insinuated that anyone who says this *is* stargazing, and is bound to fall over logically into a deep ditch. But we already defused the arguments with which the entrapping ditch was blown; and eroding sand has since gradually filled the now only vaguely outlined ditch. We can continue stargazing.

Vagueness is alright as it is. A side corollary is that the pressure, from a vocal minority, for exact philosophy is misconceived. Inasmuch as this includes demands for displacement of vague items and notions, or else replacement of them by precise (i.e., nonvague) ones, it will remove much of philosophy, which like normal discourse is essentially vague, still in the ditch up its neck in vagueness. In fact 'almost all concepts lack boundaries, so that the classical picture is of very little use to us' (Sainsbury 1996, p.252); exact philosophy deploying classical logic is similarly rather useless.

While vagueness has a full, proper and logically unembarrassing place in our rich world, we do not try to insist that others are logically compelled to accept this splendidly vague world. There are stratagems, which do not admit of decisive refutations, for banning vagueness from an actual world. No doubt these are awkward stratagems, which locate vagueness in language, or in relations of language to the world, which isolate logic (which may give the true structure of the world) from the vagaries and vagueness of language or mind, which construe 'vagueness in the world' as at best projection or conceptual imposition on a clean tidy exact world, or, more likely, as confusion. If philosophers are hooked to some or other of these alternatives, there may be no changing them. Live and let live.

17 Part IV Fragments

The draft of a typescript on the 'Enlightenment Project' was found amongst fragments for Sylvan's proposed monograph. Whilst it is unclear exactly where this was to be included, we have reproduced the fragment below. It was last worked on only a few days before his death. [Editors]

17.1 The impact of alternative systems on the Enlightenment Project

There is a recent debate, arising out of older debates, concerning logic, rationality, truth, science and, more sweepingly, the Enlightenment Project. It is a debate where the concentrated intellectual superpower formed from those substantially committed to the Project is matched against dispersed anti- or counter-Enlightenment forces, and so confronted with protracted guerilla activity. Historically furthermore, the scattered and diverse counter-movement has taken various different forms at different times. In the main respects the recent debate has crystallised into one between definite realism and an egalitarian relativism (as revealed in a useful working example, the debate in Albert 1992). Let's call these sides *realism* and *relativism* for short, though the divisions are more complex than that, as will appear (for one reason, because 'realism' comprehends an idealist 'anti-realism', and 'relativism' includes 'constructive', 'deconstructive' and diffuse 'post-modern' forms). Let us consider first a compact condensed form of the debate, before venturing on to fuller, more diffense issues.

17.1.1 The conveniently simplified debate

The contrast is invariably presented (as in the debate in Albert 1992) as between *the one* correct theory, that ultimately delivered by Science, on the one side, and *many* stories or narratives, many points of view perhaps equally good, on the other.[248] But, presented thus, it is almost obvious that what is offered affords a *false* contrast. For omitted are alternatives with

[248] See Albert's Editorial, Albert 1992, p.2, and the clearly stated position of 'the three defendents', namely Albert, Chomsky and Ehrenreich. Essentially the same false contrast pervades ancient Greek philosophy, in the debate between dogmatism and (Pyrrhonian) scepticism. It is both curious and revealing that the debate in Albert 1992 should have taken this polarized course, given the question addressed to 'the six critics' (of the Enlightenment Project). Namely, 'Do we need to reject/transcend/escape an old way of thinking variously

more than one correct account (some of these accounts perhaps better than others), but with many other stories faulted. For instance, neglected is an attractive *middle* way—intermediary betwen definite realism and egalitarian relativism—which recognises many correct accounts, none absolute, and many defective stories.[249]

There is nothing intrinsically difficult about plural correctness. For example, there are many correct proofs of a theorem, several right solutions to a puzzle along with many faulty proposals. Correct/incorrect, right/wrong contrasts remain intact, but both sides are plural. Such further pluralistic alternatives are not glimpsed for several reasons: one is a false underlying absolute (one) *versus* relative (many) dichotomy;[250] another is a false (also absolute) character accredited to received logic and scientific methodology. Thus a longer subtitle for the present exercise might well run: the *conflict* concerning logic, rationality, science, truth and all that: the Enlightenment Project and its Enemies, a pluralistic resolution

Realists take their stand on the basis of received logic and scientific methodology, typically as if that were an indisputable given, in great intellectual shape (which it isn't). A couple of statements will confirm this:

> ... we do not have to transcend science or rationality as a prerequisite to developing worthy vision and strategy... logic and the roles of evidence are here to stay. Moreover, they're on our side. (Albert 1992, p.51.)

Even some small criticism of science is permitted and justified; for instance 'science is sometimes narrow, mechanical, colonizing or hypocritical'.[251] What lie beyond such critical reach are rationality, reason and logic; '... to critique reason or logic as being at the root of sciences' many ills is wrong' 'and has no role in making the world better'. Despite

called Western Rationality/Science/Logic if we are to attain a better world' (quoted, p.17). Put differently, what is the role of received 'enlightenment' in an Amelioration Project?

Because of these different issues there is much (loose) arguing at cross-purposes in Albert 1992.

249 Such a way—splitting into two such ways, radical and deep, and more—is elaborated in TM. The image of such a middle way borrows, of course, from ancient axial wisdom concerning virtuous practice, especially from Taoism and Aristotelianism.

250 Conspicuous in Albert's own efforts. Albert attempts to recast the multi-faceted debate in these elementary white and black terms: rationality and science ('real science') which is one and unified) versus anti-rationalism, which is utterly relativist. Anti-rationalism he can then hammer as abandoning truth as well as science, as giving away rational and intelligent (and effective) methods, and as accordingly self-defeating in arguing for ameliorating social change (see Albert 1992, p.51).

251 One sample from half a page of 'balancing criticism' in Albert 1992, p.50.

extensive left political activity, 'I have not once criticized rationality or logic, and anti-intellectualism has always been anathema to me'.[252] Such criticism of logic, reason and so on can have a significant role in making with world better. Others appear fractionally more cautious about reason and logic (not Science!):

> ... in a discussion ... there are certain familiar ground rules: those of rational inquiry, ... what seems to be under discussion here is whether we should abide by these ground rules at all (trying to improve them as we proceed[253]). If the answer is that we are to abide by them, then the discussion is over: we've implicitly accepted the legitimacy of rational inquiry. If they are to be abandoned, then we cannot proceed until we learn what replaces the commitment to consistency, responsibility to fact, and other out-dated notions (Chomsky, in Albert 1992).

Before a final quoted sentence rounding off this passage is presented, notice the (defective) equations made:

procedures in a discussion (ground rules) = methods of rational inquiry
= commitment to consistency, responsibility to fact, etc.

The last are variants upon (Albert's) methods of standard logic (consistency-based) and rules of evidence. Discussion or debate has accordingly been twice narrowed (in broad form):

1. by a rationality constraint
2. by explication of rationality through a standard system (by consistency, maximum expected utility, etc.).

Pictorially:

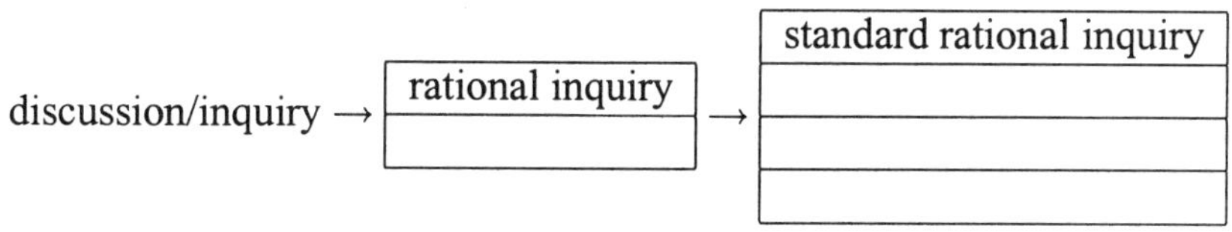

Now to continue:

---

[252] Albert 92, p.56. Such criticism of logic, reason and so on can have a significant role in making the world better.

[253] Like Neurath in his boat, a favourite image of the American intellectual establishment. But at least room for improvement is here conceded.

> Short of some instruction on this matter ... new procedures or ideas to replace the old ... we are reduced to primal screams. (Chomsky in Albert 1992, p.52.)

That is hardly so. Almost everyone is familiar with inquiries, debates, arguments and generally procedures that do not proceed rationally, by any means or standards, but which do not reduce to screams (though some no doubt do). Regrettably few, by contrast, are familiar with *alternative* systems of rationality and logic by which inquiry and so forth can proceed, quite rationally. These alternatives confound and upset realism and the Enlightenment Project.[254]

That impact of alternatives on the Enlightenment Project—an impact not recognised or acknowledged by realists—is a main topic to be briefly pursued. But first the uniform, definite realist Project.

17.1.2 The Project itself which persists

The Enlightenment Project can be drawn out of standard accounts of the Enlightenment and the world discerned under it. The multi-stranded Project involved proceeding in accord with Enlightenment practices and in bringing its vision to fruition. First, then, a revealing picture of the world discerned by the Enlightenment and the associated ideology (both presented in conveniently over-simplifying black-and-white caricature by Berlin), and secondly the sort of accompanying Project which that yields.

In the main, the world discerned was a rational scientific world tending to conform to 'materialism, utilitarianism, ethical naturalism and atheism':

> Whatever the [many significant] differences that divided the French *philosophes* and their disciples in other countries ... there existed nevertheless a wide *consensus* : it rested on an acceptance of what was in effect, a secular version of the old natural law doctrine according to which the nature of things possessed a permanent unalterable structure, differences and changes in the world being subject to universal and immutable laws. These laws were discoverable in principle by the

[254] The useful (if problematic) term Enlightenment Project coalesces in the course of the debate. In the course of trying to shame and chastise left intellectuals and left counterparts for their neglect of human concerns, of emancipation, removal of oppression and similar, Chomsky assigns a central place to the 'project of the Enlightenment' in serving emancipation, in displacing oppressors, removing monopolies of the powerful, and supplying liberating information and technology. That expression which he has drawn from essays he is commenting upon, quickly contracts to 'the' Enlightenment project (same page, p.57).

use of reason and controlled observation, of which the methods of the natural sciences constituted the most successful application. The most powerful instrument in the acquisition of knowledge was held to be mathematics ... *the true path to knowledge was that of the natural sciences*; that is to say, all statements with claims to truth must be public, communicable, testable—capable of verification or falsification by methods open to and accepted by any rational investigator. From this it followed that *all other types of authority were to be rejected*, and in particular such foundations of faith as sacred texts, divine revelation and the dogmatic pronouncements of its authorised interpreters, tradition, prescription, immemorial wisdom, private intuition and all other forms of non-rational or transcendent sources of putative knowledge. This principle was held to apply to both the human and the non-human world: to abstract disciplines, such as logic or mathematics, to the applied sciences which established the laws of the behaviour of inanimate bodies, plants, animals and human beings, and to the normative disciplines which revealed the true nature of ultimate human goals, and the correct rules of conduct, public and private, social and political, moral and aesthetic.

According to this doctrine, *all genuine questions were in principle answerable: truth was one*, error multiple; the true answers must of necessity be universal and immutable, that is, true everywhere, at all times, for all men, and discoverable by the appropriate use of reason, by relevant experience, observation and the methods of experiment, logic, calculation. A logically connected structure of rules, laws, generalisations, susceptible of demonstration or, at least in practice, of a high degree of confirmation (and, where required, of application appropriate to differing circumstances) could, at least in principle, be constructed, and could replace the chaotic amalgam of ignorance, laziness, guesswork, superstition, prejudice, dogma, fantasy, and, above all, what Helvetius called 'interested error', which enabled the cunning and the strong to dominate and exploit the stupid, ignorant and weak, and had throughout human history been largely responsible for the vices, follies and miseries of mankind. Only knowledge, that is, the growth of the sciences, could rescue mankind from these largely self-induced evils. Some believed that certainty in empirical matters was attainable, others that no more than high probability could be achieved; some were pessimistic about progress towards virtue or happiness, others were more sanguine. But the majority of the *philosophes* were

> agreed that if irrational passions could be controlled, and ignorance, prejudice, fear and greed diminished, an end could be made to the worst confusions in human thought and feelings, which led to blind fanaticism in thought and savage barbarism in practice. (Berlin 1979, pp.162-4.)

Though we should contest very many of those claims, it would be foolish to deny that there was *much* that was enlightened, by comparison with what had gone before, in Enlightenment proposals.[255] In many respects, the proposals were socially and otherwise enlightened—though definitely not environmentally so, ideas and (limited) ideals of social justice not being matched similar imputs regarding environmental justice. My own opinion is that Chomsky has grossly overestimated the positive role of the Enlightenment Project in achieving emancipation and social amelioration. Patently too, he similarly underrates environmental negativities and externalities of implementations of the Project.

The Enlightenment Project affords a brief convenient way of alluding to the prime aims and objectives of the Enlightenment, the core of which largely coincide with those of modern secular science broadly construed. (No doubt there are lots of different themes intermixed under the Enlightenment, themes that come apart but that span more than science.) Central among emphasized objectives were: pursuit of the truth; exact deployment of rational methods, including logical argumentation; reliance on science and especially rigorous empirical procedures. A suitable mnemonic would be TRASH, (short for truth and reason and science and (atheistic) humanism).

The Project also included these components:

- as regards *practice*:
  acceptance of rational and scientific practice and methods, and *rejection* of other methods, such as those depending upon closed authority, private or organisational revelation, magic and so on.
- as to bringing to *fruition*:
  rendering the intelligent inhabitants of the Earth (humans) *enlightened*, in particular in their practice. Corollaries are extensive, and include *educational* and *social* programs (for example humans have to be in a position to use thier intelligence, and not ill, disabled, enslaved, etc.); *obtaining* that accessible correct information.

---

[255] See TM, final chapter. A detailed critique of both the Enlightenment project and its Anti-Enlightenment alternatives is offered in this work.

For all its apparent merits, the project had, and has, a large downside. For example, the Project underwrites monopolies and elites, namely those who held control of the correct methods, the right procedures. It brings in its train all those strait-jacketing features now so familiar in scientific and intellectual life: experts, technocracy, the peer group process, procedural correctness, scientific accountability, and so on. (Plurallistic criticism will pull the blocks from under all this familiar rigmarole.)

What looks even worse (from outside), main oppositional groups assume a similar set of assumptions. Present monopolies and elites are simply mistaken about the truth, which they have no monopoly upon and do not even hold. Nonetheless truth will out (for example, with enough publicity or disguised propaganda, when the public wake up or become informed, or whatever). Then that opposition, in possession of truth and rationality (or its relevant variants, facts and consistency, etc.) will triumph. These oppositional groups have premissed their opposition upon similar monistic illusions (but nor is *the* pluralistic truth hidden away in the background).[256]

17.1.3 Criticism of the Project: trashing TRASH

As a sidelight, the Project was *unachievable* in principle. It implied that there was a set of methods by which any sought piece of knowledge could be obtained. But given the logical systematization the Project presumed (with its consistency requirements, etc.), there are no such methods; there are inevitable limitations to knowledge.[257] The Project is not merely incomplete, but incompletable.

The problems with the Project are not restricted to knowledge and information. The idea that all genuine questions are in principle answerable is similarly flawed. As there are (classically) unknowable truths, so there are unanswerable questions.

---

[256] Most of my own experience confirms the contention (defended at the beginning of JB) that truth, even in fairly degenerate forms, will not out. Certainly not on its own; but I grant that we should be working and agitating to force truth (plural) out. Even then truth may not emerge and be recognised in time, before people are hanged, or species or cultures lost. The traditional tag *magna est veritus et prevalebit* is itself infected with falsity.

The erroneous assumption that truth (similarly reason) will win out on its own, without agitation or political activity is also a prime reason, or excuse, for much academic and scientific inactivity and complacency. An associated illusion in that they have grasped the truth or, if not, have a better grasp upon it (now singular) than any others. It is an illusion shared with—what enjoys significantly less respectable evidence—monistic religion.

[257] Details are given in NL.

As a result, the Project has to withdraw, to retract some of its sweeping claims—else it fragments, admitting different logical theories which skirt the limitations, in different ways. Much as with naive set theory, where consistencizing repairs yield many competing alternatives, none intellectually decisive, so it is elsewhere. Paradoxes and puzzles produce a plurality of 'solutions'. Likewise, and in any case, the Project fragments, or rather plurallizes,[258] so it can be argued. But, most important for present purposes the Project fragmented, and was *bound* to fragment.

The Project breaks up into a plurality—of projects, frameworks, ideologies, and so forth—because of *alternative systems*, of one sort or another. Fundamental among these are alternative systems of logic, alternative correct systems, and therewith of alternative frameworks of reason and rationality. As von Wright, formerly a major proponent of the classical consistency dogma, has now conceded, 'there is not one logic which is the *true one*'.[259] But from such a plurality of logics derives a plurality of reasoning and rational procedures, and therewith of all theoretical domains deploying logical methods. From these flow, directly or by analogous arguments, alternatives elsewhere, indeed everywhere in sciences and arts.[260] Therewith repealed also are all absolutes, both logical, including obstructing laws of thought and limiting commitments to consistency, and others, including the latest revelations of objective science. They are repealed without inducing a debilitating relativism.

Alternative *systems* are intended to take in a wide sweep of alternatives: alternative logics, alternative arithmetics, alternative mathematical systems, alternative theories or theoretical systems more generally, alternative cultural arrangements, and shifting definitely down to a semantical level, alternative actual worlds, alternative realities, and so on.

There is an important corollary for value theory: ethics is not in such an invidious position as compared with logic and science as has been broadcast. Science and ethics resemble each other more than has been recently disseminated; even superior systems involve in their adoption many (largely concealed) evaluative choices among alternatives. Moreover, as, and because, there are no absolutes concerning truth and correctness, so there

---

[258] *Talk of 'plura<u>ll</u>izing' is almost certainly here intended. It derives from a neologism of Sylvan's, "plurallism", describing a preferred species of pluralism argued for in detail in* TM*—a book completed by Sylvan just before his death, originally with the working title of "Deep Plurallism". [Editors]*

[259] Havas 1986, agreeing with von Wright's thesis, p.149. Requisite, considerable details again get assembled in TM.

[260] Detailed arguments are assembled in TM.

are none concerning right and what is morally good. Even if we are right, others may be right also. We have no automatic mandate to change them, to stamp out what they hold, or similar.[261] Likewise if they are not right, as with adherents of mainstream logic.

An obvious reaction—had all the alternatives now (beginning to be) appreciated been systematically assembled at once, in an appropriate show of alternative strength—should have been pluralization of the Enlightenment Project. Each viable system of alternatives would have its own project or set of projects. There would be enlightenments.[262]

But, for one thing, the alternatives did not appear historically all at once, but in dribs and drabs; (for example, of a Kantian triad of absolute *a priori* sciences, geometry a century ago, arithmetic recently this century, logic next century?) For another, some have not yet been widely appreciated at all (for example, alternative logics, alternative rules of evidence, alternative rational methods). So alternatives and diversity could be fought off or assimilated.[263] For yet another, some unwarranted reactions occurred, which both neglected and further information did little to justify. These included for instance, outright rejection of the Project, and debilitating relativisation.[264]

Most philosophers, indeed most intellectuals of scientific disposition, are still caught up in the Enlightenment Project, in what, as a unified or singular scientific enterprise, should be a defunct Project.

---

[261] The need to be right is coupled with the drive to power. Rightness is taken, under a widespread assumption, to give a mandate to eradicate what is wrong.

[262] It seems a similar plural development should have occurred with oriental religions, such as Buddhism, advocating enlightenment. But religions tend to fossilize.

[263] Consider too the search for uniformity and commonality, which is often implausible as at more than a superficial level, in ethics or aesthetics.

[264] Rival forms have evolved and developed. Thus what was Anti-Enlightenment became Post-modernism, including both Nihilism and Egalitarian Relativism (see Gare 1994).

Bibliography

BG Sylvan, R. (1989), *Bystanders' Guide to Sociative Logics*, Research Series in Logic and Metaphysics #9, Australian National University.

CL *Compendium Logicae Porretnam, Cahiers de l'Institut du Moyen-Age Grec et Latin*, 46 (1983), Copenhagen.

CS Sylvan, R. (1986), "Towards an improved cosmo-logical synthesis", *Grazer Philosophische Studien*, 25/26, pp. 135-178.

DRL Norman, J. and Sylvan, R. (eds) (1989), *Directions in Relevant Logic*, Dordrecht, Kluwer.

EA Martin C. (1987), "Embarrassing arguments and surprising conclusions in the development of theories of the conditional in the twelfth century", in J. Jolivet and A. De Libera (eds), *Gilbert de Poitiers et ses Contemporains*, Bibliopolis, Naples, pp. 377-400.

ENT Anderson, A.R. and Belnap, N.D. Jr. (1975), *Entailment*, Vol. I, Princeton, Princeton University Press.

FD Routley, R. and Routley, V. (1972), "The semantics of first degree entailment", *Noûs*, 6, pp. 335-359.

IC Priest, G. (1987), *In Contradiction*, The Hague, Nijhoff.

JB Routley, R. (1980), *Exploring Meinong's Jungle and Beyond*, Departmental Monograph #3, Australian National University.

LM I De Rijk, L.M. (1962), *Logica Modernorum Vol 1.: On the twelfth century theories of fallacies*, Assen, Van Gorcum.

LM II De Rijk, L.M. (1967), *Logica Modernorum Vol 2.: The origin and development of the theory of supposition*, Assen, Van Gorcum.

MD Routley, R. and Plumwood, V. (1984), "Moral dilemmas and the logic of deontic notions", Discussion Papers in Environmental Philosophy #6, Australian National University. Reprinted in PL.

MR Sylvan, R. (1989), "Modern myths concerning rationality", in S. Biderman and B. Scharfstein (eds), *Rationality in Question*, Leiden, Brill.

NC Routley, R. and Plumwood, V. (1984), "Negation and contradiction", Research Series in Logic and Metaphysics #2, Australian National University. Reprinted in *Proceedings of the Fifth Latin American Symposium on Mathematical Logic, Revista Colombiana de Mathemáticas*, 19 (1986), pp. 201-231.

NL Routley, R. (1981), "Necessary limits to knowledge: unknowable truths", in E. Morscher *et al.* (ed.), *Essays in Scientific Philosophy*, Comes-Verlag, pp. 97-118.

OED *Oxford English Dictionary*, Oxford, Clarendon Press (1971).

OP Priest, G. and Routley, R. (1984), *On Paraconsistency*, Research Series in Logic and Metaphysics #1, Australian National University. Reprinted as the introductory chapters of PL.

PA Sylvan, R. (1992), "Process and action: relevant theory and logics", *Studia Logica*, 51, pp. 379-438.

PL Priest, G., Routley, R. and Norman, J. (eds) (1989), *Paraconsistent Logic*, Munich, Philosophia Verlag.

PLI Sylvan, R. (1989), "Philosophical and linguistics inroads: multiply intensional relevance logics", in DRL, pp. 269-304.

RC Sylvan, R. (1996), "Relevant conditionals, and relevant applications thereof", in S. Akama (ed.) (1996), *Logic, Language and Computation*, Dordrecht, Kluwer, pp. 193-246.

RCR Sylvan, R., Goddard, L., and da Costa, N.C.A. (1989), *Reason, Cause and Relevant Containment with an application to Frame Problems*, Research Series in Logic and Metaphysics #3, Australian National University.

RD Pylyshyn, Z.W. (ed.) (1987), *The Robot's Dilemma: The frame problem in artificial intelligence*, New Jersey, Abel, Norwood.

RLR I Routley, R., Meyer, K., Plumwood, V. and Brady, R. (1982), *Relevant Logics and Their Rivals: Vol. I*, California, Ridgeview.

RLR II Sylvan, R., Brady, R., *et al.* (forthcoming), *Relevant Logics and Their Rivals: Vol. II*, Aldershot, Ashgate.

SM Routley, R. (1976), "The semantical metamorphosis of metaphysics", *Australasian Journal of Philosophy*, 54, pp. 187-205.

TE *Tractatus Emmeranus de Impossibili Positione*, in L.M. De Rijk (ed.), *Vivarium*, 12 (1974), pp. 117-123 (transl. by C. Martin, 1982).

TM Sylvan, R. (1997), *Transcendental Metaphysics: From radical to deep plurallism*, Cambridge, White Horse Press.

TT Sylvan, R. (1993), "Other withered stumps of time", in J. Copeland (ed.), *Logic and Reality*, Oxford, Oxford University Press.

UC Routley, R. (1979), "The choice of logical foundations: non-classical choices and the ultra logical choice", *Studia Logica*, 39, pp. 77-98.

UU Routley, R. (1977), "Ultralogic as universal?", *Relevance Logic Newsletter*, 2, pp. 50-90, 138-175. Reprinted as an Appendix in JB.

Abaelard, P. (1956), *Dialectica*, L.M. De Rijk (ed.), Assen, Van Gorcum.

Achinstein, P. (1983), "Evidence: additional topics", in P. Achinstein, *The Nature of Explanation*, New York, Oxford University Press.

Ackermann, W. (1956), "Begründung einer strengen Implikation", *Journal of Symbolic Logic*, 21, pp. 113-128.

Adams, E.W. (1975), *The Logic of Conditionals*, Dordrecht, Reidel.

Albert, M. (ed.) (1992), *Z Papers*, 1(4).

Anderson, A.R. and Belnap, N.D. Jr. (1962), "The pure calculus of entailment", *Journal of Symbolic Logic*, 21, pp. 19-52.

Anderson, J. (1962), *Studies in Empirical Philosophy*, Sydney, Angus and Robertson.

Angell, R.B. (1962), "A propositional logic with subjunctive conditionals", *Journal of Symbolic Logic*, 27, pp. 327-343.

Aristotle (1924), *Rhetorica*, in W.D. Ross (ed.), *Works Vol. II*, Oxford, Oxford University Press.

Aristotle (1928a), *Analytica Priora*, in W.D. Ross (ed.), *Works Vol. I*, Oxford, Oxford University Press.

Aristotle (1928b), *De Interpretatione*, in W.D. Ross (ed.), *Works Vol. I*, Oxford, Oxford University Press.

Aristotle (1928c), *Topica and De Sophisticis Elenchis*, in W.D. Ross (ed.), *Works Vol. I*, Oxford, Oxford University Press.

Armstrong, D.M. (1973), *Knowledge, Truth and Belief*, Cambridge, Cambridge University Press.

Armstrong, D.M. (1983), *What is a Law of Nature?*, Cambridge, Cambridge University Press.

Arnauld, A. (1964), *The Art of Thinking: Port Royal Logic*, translated by J. Dickoff and P. James, New York, Bobbs-Merrill.

Arruda, A.I. and da Costa, N.C.A. (1970), "Sur le schéma de la séparation", *Nagoya Mathematical Journal*, 38, pp. 71-84.

Ashworth, E.J. (1974), *Language and Logic in the Post-Medieval Period*, Dordrecht, Reidel.

Ashworth, E.J. (1983), "English *Obligations* Texts after Roger Swyneshed", in P.O. Lewry (ed.), *The Rise of British Logic*, Toronto, Pontifical Institute of Mediaeval Studies, pp. 309-334.

Attfield, R. (1979), *God and the Secular*, Cardiff, University College Cardiff Press.

Ayer, A.J. (1946), *Language, Truth and Logic*, second edition, London, Gollancz.

Bacon, J.E. (1985), "The completeness of predicate-functor logic", *Journal of Symbolic Logic*, 50, pp. 903-926.

Bass, W.A. (1987), "Relevant to Grue?", *International Logical Review*, 35, pp. 32-36.

Batens, D. (1980), "Paraconsistent extensional propositional logics", *Logique et Analyse*, 23, pp. 195-234.

Batens, D. (1987), "Relevant implication and the weak deduction theorem", *Studia Logica*, 46, pp. 239-245.

Belnap, N.D. Jr. (1960), "Entailment and relevance", *Journal of Symbolic Logic*, 25, pp. 144-146.

Belnap, N.D. Jr. (1977), "A useful four-valued logic", in J.M. Dunn and G. Epstein (eds), *Modern Uses of Multiple-Valued Logic*, Dordrecht, Reidel.

Bennett, J.F. (1954), "Meaning and implication", *Mind*, 63, pp. 451-463.

Berlin, I. (1979), *Against the Current: Essays, in the history of ideas*, H. Ward (ed.), London, Hogarth Press.

Bhaskar, R. (1978), *A Realist Theory of Science*, Sussex, Harvester Press.

Bishop, J. (1989), *Natural Agency*, Cambridge, Cambridge University Press.

Bishop, J. (1991), "Why there can't bc a reasonable libertarianism", typescript, Australian National University.

Blum, A. (1989), "An anomaly in the D-N model of explanation", *British Journal for Philosophy of Science*, 40, pp. 365-367.

Bochenski, I.M. (1961), *A History of Formal Logic*, translated by I. Thomas (ed.), Notre Dame, University of Notre Dame Press.

Boehner, P. (1952), *Medieval Logic*, Manchester, Manchester University Press.

Boethius, A.M.S. (1880), *Commentarii in Librum Aristotelis Peri Hermenias Pars Posterior*, C. Meiser (ed.), Leipzig, Teubner.

Boethius, A.M.S. (1969), *De Hypotheticis Syllogismis*, L. Obertello (ed.), Brescia, Paideia.

Boethius, A.M.S. (1988), *In Ciceronis Topica*, translated by E. Stump (ed.), Ithaca, Cornell University Press.

Boh, I. (1982), "Consequences", in Kretzmann 1982, pp. 300-314.

Bollen, A. (1985), "A relevant reasoner", Logic Research Paper #18, Australian National University.

Boole, G. (1951), *The Laws of Thought*, New York, Dover.

Brady, R. (1984), "Depth relevance of some paraconsistent logics", *Studia Logica*, 43, pp. 63-73.

Bridges, D.S. (1994), *Computability: A mathematical sketch book*, New York, Springer Verlag.

Broadie, A. (1983), *George Lokert: Late-scholastic logician*, Edinburgh, Edinburgh University Press.

Bronstein, D.J. (1933), *Necessity, Implication, and Definition*, Ph.D Dissertation, Harvard University.

Bunder, M. (1983), "On Arruda and da Costa's logics J_1 to J_5", *Journal of NonClassical Logic*, 2, pp. 43-48.

Buridan, J. (1976), *Tractatus de Consequentiis*, H. Hubien (ed.), Louvain, Pulications Universitaires de Louvain.

Burks, A.W. (1951), "The logics of causal propositions" *Mind*, 60, pp. 363-382.

Carnap, R. (1956), *Meaning and Necessity*, second edition, Chicago, University of Chicago Press.

Carnap, R. (1962), *Logical Foundations of Probability*, second edition, Chicago, University of Chicago Press.

Charniak, E. and McDermott, D. (1985), *Introduction to Artificial Intelligence*, Massachusetts, Addison-Wesley.

Chellas, B. (1975), "Basic conditional logic", *Journal of Philosophical Logic*, 4, pp. 133-153.

Chihara, C. (1981), "Quine and the confirmational paradoxes", in P.A. French *et al.* (eds), *The Foundations of Analytic Philosophy*, Minneapolis, University of Minnesota Press.

Cleave, J.P. (1974), "An account of entailment based on classical semantics", *Analysis*, 34, pp. 118-122.

Copeland, B.J. (1997), "The broad conception of computation", *American Behavioral Scientist*, 40, pp. 690-716.

Copeland, B.J. (1998a), "Turing's O-machines, Searle, Penrose, and the brain", *Analysis*, 58, pp. 128-138.

Copeland, B.J. (1998b), "Even Turing machines can compute uncomputable functions", in C. Calude *et al.* (eds) *Unconventional Models of Computation*, Singapore, Springer-Verlag.

Copeland, B.J. and Sylvan, R. (1999), "Beyond the universal Turing machine", *Australasian Journal of Philosophy*, 77, pp. 46-66.

Copi, I.M. (1954), *Symbolic Logic*, New York, Macmillan.

Copi, I.M. (1978), *Introduction to Logic*, fourth edition (first edition 1953), New York, Macmillan.

Corcoran, J. (1973), "A mathematical model of Aristotle's syllogistic", *Archiv für Geschichte der Philosophie*, 55, pp. 191-219.

Corcoran, J. (ed.) (1974), *Ancient Logic and its Modern Interpretation*, Dordrecht, Reidel.

Curry, H.B. (1963), *Foundations of Mathematical Logic*, New York, McGraw Hill.

Daniels, C.B. (1986), "A story semantics for implication", *Notre Dame Journal of Formal Logic*, 27, pp. 221-246.

Davidson, D. (1980), "Causal relations" reprinted in his *Essays on Action and Events*, Oxford, Oxford University Press.

De Morgan, A. (1847), *Formal Logic: Or the calculus of inference, necessary and probably*, London, Taylor and Walton.

Dennett, D. (1984), "Cognitive wheels: the frame problem of AI", in C. Hookway (ed.), *Minds, Machines and Evolution*, Cambridge, Cambridge University Press.

DeRijk, L.M. (ed.) (1986), *Translations of the Sanlet Emmersan Treatise on Positio Impossibilis*, typescript, Stony Brook, State University of New York.

De Soto, D. (1529), *Summule (or Introductiones Dialecticae)*, Burgis.

Deutsch, H. (1979), "The completeness of S", *Studia Logica*, 38, pp. 137-147.

Deutsch, D. (1985), "Quantum theory, the Church-Turing principle and the universal quantum computer", *Proceedings of the Royal Society*, Series A, 400, pp. 97-117.

Diogenes Laertius (1958), *Lives of Eminent Philosophers*, translated by R.D. Hicks, London, Heinemann.

Donagan, A. (1973), "Determinism in history", *Dictionary of the History of Ideas*, Vol. II , New York, Charles Scribner & Sons.

Dummett, M. (1975), "Wang's paradox", *Synthese*, 30, pp. 301–324. Reprinted in Keefe and Smith 1996.

Dummett, M. (1981), *Frege: Philosophy of language*, second edition, Cambridge, Harvard University Press.

Duncan-Jones, A.E. (1935), "Is strict implication the same as entailment?", *Analysis*, 2, pp. 70-78.

Dunn, J.M. (1980), "A sieve for entailments", *Journal of Philosophal Logic*, 9, pp. 41-57.

Dunn, J.M. (1987), "Relevant predication 1: the formal theory", *Journal of Philosophical Logic*, 16, pp. 347-381.

Dunn, J.M. (1990), "Relevant predication 2: intrinsic properties and internal relations", *Philosophical Studies,* 60, pp. 177-206.

Eaton, R.M. (1931), *General Logic*, New York, Scribner.

Ebbeson, S. (1981), *Commentators and Commentaries on Aristotle's Sophistici Elenchi*, Vol. 1, Leiden, Brill.

Ellis, B. (1979), *Rational Belief Systems*, Totowa, Rowan and Littlefield.

Emch, A.F. (1936), "Implication and deducability", *Journal of Symbolic Logic*, 1, pp. 26-35.

Evans, G. (1978), "Can there be vague objects?", *Analysis*, 38, p. 208. Reprinted in Keefe and Smith 1996.

Fine, K. (1975), "Vagueness, truth and logic", *Synthese*, 30, pp. 265-300. Reprinted (in part) in Keefe and Smith 1996.

Fine, K. (1980), "Analytic implication", in J. Dancy (ed.), *Papers on Language and Logic*, Keele, University of Keele Press, pp. 54-70.

Fitch, F.B. (1952), *Symbolic Logic: An introduction*, New York, Ronald Press.

Frede, M. (1974), "Stoic vs. Aristotelian Syllogistic", *Archiv für Geschichte der Philosophie*, 56, pp. 1-32.

Fuhrmann, A. (1988), *Relevant Logics, Modal Logics and Theory Change*, Ph.D. Thesis, Australian National University.

Galtung, J. (1984), *There are Alternatives! Four roads to peace and security*, London, Spokesman.

Gandy, R. (1980), "Church's thesis and principles for mechanisms", in J. Barwise, H.J. Keisler and K. Kunen (eds), *The Kleene Symposium*, Amsterdam, North-Holland, pp. 123-48.

Gärdenfors, P. (1978), "On the logic of relevance", *Synthese*, 37, pp. 351-367.

Gardner, M. (1983), *The Whys of a Philosophical Scrivener*, Brighton, Harvester.

Gare, A. (1994), *Nihilism Incorporated*, Bungendore, Eco-Logical Press.

Geiss, M. and Zwicky, A.M. (1971), "On invited inferences", *Linguistic Inquiry*, 2, pp. 561-566.

Gettier, E.L. (1963), "Is justified true belief knowledge", *Analysis*, 23, pp. 121-123.

Glymour, C. (1975), "Relevant evidence", *Journal of Philosophy*, 72, pp. 903-926.

Glymour, C. (1980), "Hypothetico-deductivism is hopeless", *Philosophy of Science*, 47, pp. 322-5.

Goddard, L. (1977), "The paradoxes of confirmation and the nature of natural laws", *Philosophical Quarterly*, 27, pp. 97-113.

Goddard, L. and Sylvan, R. (1989a), "Relevance and reasoning, part I" (typescript). Subsequently published in *Dialogue and Universalism*, V (1995), pp. 37-63.

Goddard, L. and Sylvan, R. (1989b), "Relevance and reasoning, part II" (typescript).

Gödel, K. (1946), "Remarks before the Princeton Bicentennial Conference on problems in mathematics", in M. Davis (ed.), (1965), *The Undecidable*, New York, Raven.

Goodman, N. (1965), *Fact, Fiction and Forecast*, second edition, New York, Bobs-Merrill.

Goodman, N. (1972), *Problems and Projects*, Indianapolis, Bobbs-Merrill.

Gould, J.B. Jr., (1967), "Chrysippus: on the criteria for the truth of a conditional proposition", *Phronesis*, 12, pp. 152-161.

Green-Pedersen, N.J. (1983), "Early British treatises on consequences", in P.O. Lewry (ed.), *The Rise of British Logic*, Toronto, Pontifical Institute of Mediaeval Studies, pp. 285-308.

Grice, H.P. (1967), "Logic and conversation", William James Lectures, Harvard University, in D. Davidson and G. Harman (eds), *The Logic of Grammar*, Encino, Dickenson, pp. 64-75.

Griffin, N. and Harton, M. (1981), "Sceptical arguments", *Philosophical Quarterly*, 31, pp. 17-30.

Grim, P. (1984), "There is no set of all truths", *Analysis*, 44, pp. 206-208.

Grim, P. (1991), *The Incomplete Universe: Totality, knowledge, and truth*, Cambridge, MIT Press.

Gumanski, L. (1988), "Remarks on Cantor's diagonal method and some related topics", in *Types of Logical Systems and the Problem of Truth*, Sofia, Bulgarian Academy of Sciences, pp. 45-56.

Hacking, I. (1991), "How should we do the history of statistics?", in G. Burchell, C. Gordon and P.M. Miller (eds), *The Foucault Effect*, London, Harvester Wheatsheaf.

Halldèn, S. (1951), "A question concerning a logical calculus related to Lewis's system of strict implication, which is of special interest for the study of entailment", *Theoria*, 14, pp. 265-269.

Hamblin, C.L. (1970), *Fallacies*, London, Methuen.

Hare, R.M. (1978), "Moral relevance", in A.I. Goldman and J. Kim (eds), *Values and Morals*, Philosophical Studies, Series in Philosophy, Vol. 13, Dordrecht, Reidel.

Harper, W.L. and Hooker, C.A. (eds) (1976), *Foundations of Probability Theory, Statistical Inference and Statistical Theories of Science*, Vol. 1, Dordrecht, Reidel.

Havas, K. (1986), "Differences in the unity", *Logique et Analyse*, 14, pp. 149-150.

Hay, W.H. (1969), "Stoic use of logic", *Archiv für Geschichte der Philosophie*, 51, pp. 145-157.

Hempel, C.G. (1965), "Studies in the logic of confirmation", reprinted in *Aspects of Scientific Explanation and Other Essays in the Philosophy of Science*, New York, Free Press, pp. 1-51.

Hesse, M. (1974), *The Structure of Scientific Inference*, London, Macmillan.

Heyting, A. (ed.) (1959), *Constructivity in Mathematics*, Amsterdam, North-Holland.

Hiż, H. (1959), "Extendible sentential calculi", *Journal of Symbolic Logic*, 24, pp. 193-202.

Hogarth, M.L. (1992), "Does general relativity allow an observer to view an eternity in a finite time?", *Foundations of Physics Letters*, 5, pp. 173-181.

Hogarth, M.L. (1994), "Non-Turing computers and non-Turing computability", *PSA*, Vol. 1, pp. 126-138.

Honderich, T. (1979), "On determinism", in T. Honderich and M. Burnyeat (eds), *Philosophy As It Is*, Harmondsworth, Penguin, pp. 239-275.

Hosiasson-Lindenbaum, J. (1940), "On confirmation", *Journal of Symbolic Logic*, 5, pp. 133-148.

Hospers, J. (1954), *Introduction to Philosophical Analysis*, New York, Prentice-Hall.

Hunter, G. (1983), "Conditionals", *Proceedings of the Aristotelian Society*, Supplementary, 57, pp. 1-15.

Iseminger, G. (1980), "Is relevance necessary for validity?", *Mind*, 89, pp. 196-213.

Hyde, D. (1994), *Being Coherently Vague*, Ph.D Thesis, Australian National University.

Jeffreys, H. (1942), "Does a contradiction entail every proposition?", *Mind*, 51, pp. 90-91.

Jevons, W.S. (1874), *The Principles of Science: A treatise on logic and scientific method*, 2 Vols, London, Macmillan.

Joseph, H.W.B. (1946), *An Introduction to Logic*, second edition, Oxford, Clarendon Press.

Keefe, R. and Smith, P. (eds) (1996), *Vagueness: A reader*, Massachusetts, MIT Press.

Kielkopf, C.F. (1977), *Formal Sentential Entailment*, Washington, University Press of America.

Kleene, S.C. (1952), *Introduction to Metamathematics*, Amsterdam, North-Holland.

Kneale, W. and Kneale, M. (1962), *The Development of Logic*, Oxford, Clarendon Press.

Kretzmann, N., Kenny, A. and Pinborg, J. (1982), *The Cambridge History of Later Medieval Philosophy*, Cambridge, Cambridge University Press.

Kritzman, L.D. (ed.) (1988), *Michael Foucault: Politics, philosophy, culture*, New York, Routledge.

Lakatos, I., and Musgrave, A. (eds) (1968), *Problems in the Philosophy of Science: Proceedings of the International Colloqium in the Philosophy of Science, London, 1963*, Amsterdam, North-Holland.

Lakoff, G. (1975), "Pragmatics in natural logic", in E.L. Keenan (ed.), *Formal Semantics of Natural Language*, Cambridge, Cambridge University Press, pp. 283-286.

Leblanc, H. (1972), "Popper's 1935 axiomatisation of absolute probability", *Pacific Philosophical Quarterly*, 63, pp. 133-145.

Lehrer, K. (1979), "The Gettier problem and the analysis of knowledge", in G.S. Pappas (ed.), *Justification and Knowledge*, Dordrecht, Reidel, pp. 65-78.

Leman, M. (1984), "On the logic and criteriology of causation", *Logique et Analyse*, 107, pp. 245-266.

Lenzen, W. (1988), "Liebnitz's calculus of implication", in J. Srzednicki (ed.), *Initiatives in Logic*, Dordrecht, Martinus Nijhoff, pp. 1-35.

Lewis, C.I. (1960), *A Survey of Symbolic Logic*, second edition, abridged, New York, Dover.

Lewis, D.K. (1973), *Counterfactuals*, Oxford, Blackwell.

Lewis, D.K. (1976), "Probabilities of conditionals and conditional probabilities", *Philosophical Review*, 85, pp. 297-315.

Lewis, D.K. (1981), "Are we free to break the laws?", *Theoria*, 47, pp. 113-121.

Long, A.A. (1974), *Hellenistic Philosophy: Stoics, Epicureans, Sceptics*, London, Duckworth.

Łukasiewicz, J. (1951), *Aristotle's Syllogistic from the Standpoint of Modern Formal Logic*, Oxford, Clarendon Press.

Łukasiewicz, J. (1970), *Selected Works*, L. Borkowski (ed.), Amsterdam, North-Holland.

MacColl, H. (1906), *Symbolic Logic and Its Applications*, London, Longmans, Green, and Co.

Mackie, J.L. (1969), "The relevance criterion of confirmation", *British Journal for Philosophy of Science*, 20, pp. 27-40.

Mackie, J. (1980), *The Cement of the Universe*, Oxford, Clarendon Press.

Mannoia, V.J. (1980), *What is Science?: An introduction to the structure and methodology of science*, Lanham, MD, Univeristy of America Press.

Martin, C. (1986a), "William's machine", *Journal of Philosophy*, 83, pp. 564-572.

Martin, C. (1986b), *Translations of the Sanlet Emmersan Treatise on Positio Impossibilis*, typescript, StonyBrook, State University of New York.

Martin, C. (1987), "Something amazing about the Peripatetic of Pallet: Abaelard's development of Boethius' account of conditional propositions", *Argumentation*, 1, pp. 419-436.

Martin, C. (1991), "Boethius and the logic of negation", *Phronesis*, XXXVI, pp. 277-304.

Martin, C. (1999), "Non-reductive arguments from impossible hypotheses in Boethius and Philoponus", *Oxford Studies in Ancient Philosophy*, XVII, pp. 279-302.

Martin, E.P. (1983), "Noncircular logic", preprint 7/83, University of Melbourne Philosophy Department, Preprint Series.

Martin, E.P. and Meyer, R.K. (1983), "*S* for Syllogism", typescript, Canberra.

Mates, B. (1961), *Stoic Logic*, second edition, Berkeley, University of California Press.

McCall, S. (1966), "Connexive implication", *Journal of Symbolic Logic*, 31, pp. 415-433.

McCall, S. (1967), "Connexive implication and the syllogism", *Mind*, 76, pp. 346-356.

McCall, S. (1994), *A Model of the Universe: Space-time, probability, and decision*, Oxford, Clarendon Press.

McGee, V. (1985), "A counterexample to *modus ponens*", *Journal of Philosophy*, 82, pp. 462-471.

Mendelson, E. (1964), *Introduction to Mathematical Logic*, New York, Van Nostrand.

Merrill, G.H. (1979), "Confirmation and prediction", *Philosophy of Science*, 46, pp. 98-117.

Meyer, R.K. (1978), "Why I am not a relevantist", Research Papers of the Logic Group # 1, Australian National University.

Mill, J.S. (1884), *A System of Logic: Ratiocinative and inductive*, London, Longmans Green.

Montgomery, H. and Routley, R. (1968), "On systems containing Aristotle's thesis", *Journal of Symbolic Logic*, 33, pp. 82-96.

Montgomery, H. and Routley, R. (1971), "Conventionalist and contingency modal-oriented logics", *Notre Dame Journal of Formal Logic*, 12, pp. 131-152.

Moody, E.A. (1953), *Truth and Consequence in Medieval Logic*, Amsterdam, North-Holland.

Moody, E.A. (1965), *The Logic of William of Ockham*, New York, Russell and Russell.

Morgan, C., (1982), "Simple probabilistic semantics for propositional $K, T, B, S4$ and $S5$", *Journal of Philosophical Logic*, 11, pp. 443-458.

Mueller, I. (1969-70), "Stoic and peripatetic logic", *Archiv für Geschichte der Philosophie*, 51, pp. 173-187.

Nelson, E.J. (1929), *An Intensional Logic of Propositions*, Ph.D Thesis, Harvard University.

Nettleship, R.L. (ed.) (1906-1908), *Works of Thomas Hill Green*, London, Longmans Green.

Nozick, R. (1981), *Philosophical Explanations*, Oxford, Clarendon Press.

Nute, D. (1984), "Conditional logic", in D. Gabbay and F. Guenthner (eds), *Handbook of Philosophical Logic*, Vol. II, Dordrecht, Reidel, pp. 387-439.

Nute, D. (1980), *Topics in Conditional Logic*, Dordrecht, Reidel.

Ockham, W. (1957), *Philosophical Writings*, translated by P. Boehner (ed.), Edinburgh, Nelson.

Pappas, G.S. (ed.) (1979), *Justification and Knowledge*, Dordrecht, Reidel.

Parry, W.T. (1932), *Implication*, Ph.D Thesis, Harvard University.

Parry, W.T. (1933), "Ein axiomensystem für eine neue art von implikation (Analytische Implikation)", *Ergebnisse eines mathematischen Kolloquiums*, 4, pp. 5-6.

Paul of Venice (1984), *Logica Parva*, A.R. Perreiah (ed.), München, Philosophia Verlag.

Paul of Venice (1988), *Logica Magna* Part IV, fascicule 4, G.E. Hughes (ed.), Oxford, Oxford University Press.

Peña, L. (1990), "Fuzziness entails contradictions", typescript, Australian National University.

Popper, K. (1972), *The Logic of Scientific Discovery*, London, Hutchinson.

Popper, K. (1974), "Replies to my critics", in P.A. Schilpp (ed.), *The Philosophy of Karl Popper*, Illinios, LaSalle Open Court.

Post, E.L. (1965), "Absolutely unsolvable problems and relatively undecidable propositions: account of an anticipation", in M. Davis (ed.), *The Undecidable*, New York, Raven.

Priest, G. (1979), "The logic of paradox", *Journal of Philosophical Logic*, 8, pp. 219-241.

Priest, G. (1980), "Sense, entailment and modus ponens", *Journal of Philosophical Logic*, 9, pp. 415-436.

Priest, G. (1994), "Is arithmetic consistent?", *Mind*, 103, pp. 338-349.

Priest, G. and Crosthwaite, J. (1988), "Truth, meaning and relevance", in DRL, pp. 377-397.

Prior, A.N. (1962), *Formal Logic*, Oxford, Clarendon Press.

Quine, W.V. (1981), "Empirical content", in his *Theories and Things*, Massachusetts, Harvard University Press.

Quinton, A. (1967), "Knowledge and belief", in P. Edwards (ed.), *Encyclopaedia of Philosophy*, Vol. 4, New York, Macmillan, pp. 345-352.

Ramsay, A. (1991), *Formal Methods in AI*, Cambridge, Cambridge University Press.

Reese, W.L. (1980), *Dictionary of Philosophy and Religion*, New Jersey, Humanities Press.

Reid, T. (1895), *Works*, Vol. 2, Edinburgh, J. Thin.

Rennie, M.K. (1974), *Some Uses of Type Theory in the Analysis of Language*, Departmental Monograph #1, Australian National University.

Rescher, N. (1976), *Plausible Reasoning*, Assen, Van Gorcum.

Rescher, N. (1964), *Introduction to Logic*, New York, St Martins Press.

Restall, G. and Slaney, J. (1995), "Realistic belief revision", Technical Report TR-ARP-2-95, Australian National University.

Robinson, D.S. (1929), *The Principles of Reasoning*, New York, Appleton & Company.

Rogers, H. (1967), *Theory of Recursive Functions and Effective Computability*, New York, McGraw-Hill.

Rorty, R. (1980), *Philosophy and the Mirror of Nature*, Oxford, Blackwell.

Rosenkrantz, R.D. (1977), *Inference, Method and Decision*, Dordrecht, Reidel.

Rosenkrantz, R.D. (1982), "Does the philosophy of induction rest on a mistake?", *Journal of Philosophy*, 79, pp. 78-97.

Routley, R. (1972), "A semantical analysis of implicational system *I* and of the first degree of entailment", *Mathematische Annalen*, 196, pp. 58-84.

Routley, R. (1982), "How disjunctive syllogism can be seen as a fallacy of relevance", *Bulletin of the Section of Logic, Polish Academy of Sciences*, 10, pp. 144-151.

Routley, R. (1983), "Nihilisms and nihilist logics", Discussion Papers in Environmental Philosophy #4, Australian National University.

Routley, R. (1984a), "Maximizing, satisfying, satisizing: the difference in real and rational behaviour under rival paradigms", Discussion Papers in Environmental Philosophy #10, Australian National University.

Routley, R. (1984b), "Relevantism, material detachment, and the disjunctive syllogism argument", *Canadian Journal of Philosophy*, 14, pp. 167-188.

Routley, R. and Urbas, I. (1989), *Factorisation Logics 1968-88*, Research Series in Logic and Metaphysics #5, Australian National University.

Routley, R. and Routley, V. (1975), "The role of inconsistent and incomplete theories in the logic of belief", *Communication and Cognition*, 8, pp. 185-235.

Russell, B. (1948), *Human Knowledge: Its scope and limits*, New York, Simon & Schuster.

Russell, B. (1954), *Mysticism and Logic, and Other Essays*, Harmondsworth, Penguin Books.

Ryle, G. (1950), "If, so and because", in M. Black (ed.), *Philosophical Analysis*, Cornell, Cornell University Press, pp. 323-340.

Sainsbury, M. (1991), "Is there higher-order vagueness?", *Philosophical Quarterly*, 41, pp. 167–182.

Sainsbury, M. (1996), "Concepts without boundaries", in Keefe and Smith 1996, pp. 251-264. Originally available as a typescript of an Inaugural Lecture, given at King's College London, November 1990.

Salmon, W.C. (1975), "Confirmation and relevance", in G. Maxwell and R. Anderson Jr. (eds), *Induction, Probability and Confirmation*, Minneapolis, University of Minnesota Press, pp. 3-36.

Scheffler, I. (1964), *The Anatomy of Inquiry*, London, Routledge & Kegan Paul.

Schreiber, D.S.G. (1987), "The illegitimacy of Gettier examples", *Metaphilosophy*, 18, 49-54.

Schurz, G. (1991a), "Relevant deduction: from solving paradoxes towards a general theory", *Ekenntnis*, 35, pp. 391-437.

Schurz, G. (1991b), "Relevant deductive inference: criteria and logics", in Schurz and Dorn 1991, pp. 57-86.

Schurz, G. and Dorn, G. (eds) (1991), *Advances in Scientific Philosophy*, Amsterdam, Rodopi.

Schutz, A. (1970), *Reflections on the Problem of Relevance*, New Haven, Yale University Press.

Scott, D. (1973), "Background to formalization", in H. Leblanc (ed.), *Truth, Syntax and Modality*, Amsterdam, North-Holland.

Segerberg, K. (1982), "The logic of deliberate action", *Journal of Philosophical Logic*, 11, pp. 233-254.

Sextus Empiricus (1960), *Outlines of Pyrrhonism, Complete Works Vol. 1*, translated by R.G. Bury, London, Heinemann.

Sextus Empiricus (1967), *Against the Logicians*, translated by R.G. Bury, London, Heinemann.

Sheperdson, J.C. (1956), "On the interpretation of the Aristotelian syllogistic", *Journal of Symbolic Logic*, 21, pp. 137-147.

Shope, R.K. (1983), *The Analysis of Knowing: A decade of research*, Princeton, Princeton University Press.

Sidgwick, A. (1883), *Fallacies: A view of logic from the practical side*, London, Kegan Paul.

Skyrms, B. (1974), "Contraposition of the conditional", *Philosophical Studies*, 24, pp. 1-21.

Slaney, J.K. (1989), "Vagueness revisited", Technical Report TR-ARP-15/88 ARP, Australian National University.

Slaney, J.K. (1990), "A general logic", *Australasian Journal of Philosophy*, 68, pp. 74-88.

Slater, B.H. (1979), "Aristotle's propositional logic", *Philosophical Studies*, 36, pp. 35-49.

Slote, M. (1982), "Selective necessity and the free-will problem", *Journal of Philosophy*, 79, pp. 5-24.

Slote, M. (1990), "Response to van Inwagen", *Social Theory and Practice*, 16, pp. 291-295.

Smart, J.J.C. (1989), *Our Place in the Universe*, Oxford, Blackwell.

Smiley, T.J. (1959), "Entailment and deducability", *Proceedings of the Aristotelian Society*, 59, pp. 233-254.

Smith, J.M. (1990), "The Emperor's New Mind ... by Roger Penrose", *The New York Review*, 37(4), pp. 21-25.

Smith, J.W. (1989), "Justified true belief is knowledge", in *The Worms at the Heart of Things*, Adelaide, Flinders University, pp. 82-86.

Smuts, P. *et al.* (1988), *Non-Standard Logics for Automated Reasoning*, Copenhagen, Academic Press.

Sommers, F. (1970), "The calculus of terms", *Mind*, 74, pp. 1-39.

Sorenson, R.A. (1991), "Fictional incompleteness as vagueness", *Erkenntnis*, 34, pp. 55-72.

Sosa, E. (1979), "Epistemic presupposition", in G.S. Pappas (ed.), *Justification and Knowledge*, Dordrecht, Reidel, pp. 79-92.

Spade, P.V. (1982a), "Three theories of obligations: Burley, Kilvington and Swyneshed on counterfactual reasoning", *History and Philosophy of Logic*, 3, pp. 1-32.

Spade, P.V. (1982b), "Insolubilia", in N. Kretzmann, A. Kenny and J. Pinborg, (1982), *The Cambridge History of Later Medieval Philosophy*, Cambridge, Cambridge University Press.

Sperber, D. and Wilson, D. (1986), *Relevance*, Oxford, Blackwell.

Steel, L. (1979), "Reasoning modelled as a society of communicating experts", M.I.T. AI Laboratory Technical Report TR 542.

Stevenson, C.L. (1944), *Ethics and Language*, New Haven, Yale University Press.

Stich, S. (1992), "What is a theory of mental representation?", *Mind*, 101, pp. 234-261.

Strawson, P.F. (1952), *Introduction to Logical Theory*, London, Methuen.

Strode, R. (1517), *Consequentie Strodi cum commeto Alexandri Sermonete*, Venice.

Stump, E. (1985), "The logic of disputation in Walter Burley's treatise on obligations", *Synthese*, 63, pp. 355-374.

Stump, E. (1987), "Boethius's *In Ciceronis Topica* and Stoic Logic", in J.F. Wippel (ed.), *Studies in Medieval Philosophy*, Washington, Catholic University of America Press, pp. 1-22.

Sullivan, M.W. (1967), *Apuleian Logic*, Amsterdam, North-Holland.

Swinburne, R. (1973), *An Introduction to Confirmation Theory*, London, Methuen.

Sylvan, R. (1985), "Introducing polylogue theory", *Philosophica*, 33, pp. 53-69.

Sylvan, R. (1987), "Language, thought and representation of 'the' world", *Revue Internationale de Philosophie*, 41, pp. 64-96.

Sylvan, R. (1988), "Relevant containment logics and certain frame problems of AI", *Logique et Analyse,* 31, pp. 11-25. Reproduced herein as Chapter 7.

Sylvan, R. (1989), "Idealism: its perennial message and its persistent misrepresentation", in *Idealism and Coherence: One rehabilitation attempt*, Research Series in Unfashionable Philosophy #3, Australian National University.

Sylvan, R. (1990), "On making a coherence theory of truth true", *Philosophica*, 46, pp. 301-330.

Sylvan, R. (1992), "Grim tales retold", *Logique et Analyse*, 139-140, pp. 349-374.

Sylvan, R. (1995), "Freedom without determinism: decent logics of relevance and necessity applied to problems of free-will", *Acta Analytica*, 13, pp. 7-32. Reproduced herein as Chapter 14.

Sylvan, R. (200+), *Ventures in Epistemology*, N. Griffin (ed.), Aldershot, Ashgate.

Sylvan, R. and Bennett, D. (1990), "Of utopias, tao and deep ecology", Discussion Papers in Environmental Philosophy #19, Australian National University.

Sylvan R. and da Costa, N. (1988), "Cause as an implication", *Studia Logica*, 47, pp. 413-248. Reproduced herein as Chapter 12.

Sylvan, R. and Günther, A. (1987), "Natural logic–another myth, or (just) a misnomer", (typescript) Canberra.

Sylvan, R. and Hyde, D. (1993), "Ubiquitous vagueness without embarrassment", *Acta Analytica*, 10, pp. 7-29. Reproduced herein as Chapter 16.

Sylvan, R. and Nola, R. (1991), "Confirmation without paradoxes", in Schurz and Dorn 1991, pp.5-44. Reproduced herein as Chapter 10.

Taylor, R. (1967), "Causation", in P Edwards (ed.), *Encyclopedia of Philosophy*, Vol. 2, New York, Macmillan, The Free Press, pp. 60-61.

Teixeira, J.F. and Sarmento, G.M. (1997), "Solving Turing's halting problem?", Universidade Federal de S. Carlos, Mimeo # 4.

Thagard, P. (1992), *Conceptual Revolutions*, Princeton, Princeton University Press.

Thom, P. (1981), *The Syllogism*, München, Philosophia Verlag.

Turner, R. (1984), *Logics for Artifical Intelligence*, London, Ellis Horwood.

Tye, M. (1990), "Vague objects", *Mind*, 99, pp. 535-557.

Twining, W. (ed.) (1983), *Facts in Law*, Wiesbaden, Franz Steiner.

Ullian, J. (1965), "A note on Scheffler on Nidditch", *Journal of Philosophy*, 62, pp. 274-275.

Urbas, I. (1988), *On Brazilian Paraconsistent Logics*, Ph.D Thesis, Australian National University.

Urbas, I. (1988a), "Paraconsistency and naive set theory", typescript, Canberra.

Urbas, I. (1990), "Paraconsistency", *Studies in Soviet Thought,* 39, pp. 343-354.

Urquhart, A. (1972), "Semantics for relevant logics", *Journal of Symbolic Logic*, 37, pp. 159-169.

Van Fraassen, B.C. (1976), "Probabilities of conditionals", in Harper and Hooker 1976, pp. 261-308.

Van Fraassen, B.C. (1983), "Theory confirmation: tension and conflict", in P. Weingartner and J. Czermak (eds), *Epistemology and Philosophy of Science*, Vienna, Hölder-Pichler-Temsky.

Van Fraassen, B.C. (1986), "The world we speak of, and the language we live in", in *Philosophy and Culture*, Proceedings of the XVIIth World Congress of Philosophy, Montreal, pp. 213-221.

Van Inwagen, P. (1983), *An Essay on Free Will*, Oxford, Clarendon Press.

Von Wright, G.H. (1951), *A Treatise on Probability and Induction*, London, Routledge and Kegan Paul.

Von Wright, G. H. (1983), *Practical Reason*, Oxford, Blackwell.

Walton, D. (1982), *Topical Relevance in Argumentation*, Philadelphia, John Benjamins.

Waters, C.K. (1987), "Relevance logic brings hope to hypothetico-deductivism", *Philosophy of Science*, 54, pp. 453-464.

Weingartner, P. (1987), "Remarks on the consequence-class of theories", in E. Scheibe (ed.), *Theory and Experiment*, Berlin, de Gruyter.

Weingartner, P. and Schurz, G. (1986), "Paradoxes solved by simple relevance criteria", *Logique et Analyse*, 29, pp. 3-40.

Whately, R. (1868), *Elements of Logic*, London, Longmans, Green, Reader & Dyer.

White, T.I. (1991), *Discovering Philosophy*, New Jersey, Prentice Hall.

Whitehead, A.N. (1941), *Process and Reality: An essay in cosmology*, New York, Social Science Book Store.

Wilber, K. (1983), *Eye to Eye: The quest for a new paradigm*, New York, Archer Press.

Winston, P.H. (1984), *Artificial Intelligence*, second edition, Massachusetts, Addison-Wesley.

Wittgenstein, L. (1953), *Philosophical Investigations*, Oxford, Blackwell.

Wittgenstein, L. (1956), *Remarks on the Foundations of Mathematics*, Oxford, Blackwell.

Woods, J. (1964), "Relevance", *Logique et Analyse*, 7, pp. 130-137.

Woods, J. and Walton, D. (1982), *Argument: The logic of the fallacies*, Toronto, McGraw-Hill Ryerson.

Wright, C. (1989), "The verification principle: another puncture–another patch", *Mind*, 98, pp. 611-622.

Zinov'ev, A.A. (1967), *Foundations of the Logical Theory of Scientific Knowledge*, Dordrecht, Reidel.

Index